SELECTED PAPERS OF A. H. STURTEVANT
Genetics and Evolution

Selected and Edited by E. B. LEWIS

CALIFORNIA INSTITUTE OF TECHNOLOGY

with a Foreword by G. W. BEADLE

UNIVERSITY OF CHICAGO

W. H. FREEMAN AND COMPANY

San Francisco and London

Other Books in Biology

Chester A. Lawson, Ralph W. Lewis, Mary Alice Burmester, and Garrett Hardin
Laboratory Studies in Biology: Observations and Their Implications

Graham DuShane and David Regnery
Experiments in General Biology

Adrian M. Srb and Ray D. Owen
General Genetics

Curt Stern
Principles of Human Genetics (Second Edition)

Evelyn L. Oginsky and Wayne W. Umbreit
An Introduction to Bacterial Physiology (Second Edition)

Henry J. Oosting
The Study of Plant Communities: An Introduction to Plant Ecology (Second Edition)

James Bonner and Arthur W. Galston
Principles of Plant Physiology

John Leonard Machlis and John G. Torrey
Plants in Action: A Laboratory Manual of Plant Physiology

Adriance S. Foster and Ernest M. Gifford, Jr.
Comparative Morphology of Vascular Plants

C. L. Porter
Taxonomy of Flowering Plants

N. J. Berrill
Growth, Development, and Pattern

Garrett Hardin
Biology: Its Principles and Implications

Printed in the United States of America.

Library of Congress Catalogue Card Number: 61–17422.

Foreword

I doubt if the excitement of scientific discovery can be appreciated in any real sense without personally experiencing it. Sturtevant experienced it abundantly. For as an original member of the Thomas Hunt Morgan group of Drosophila workers he early contributed actively and importantly to the development of the science of genetics. And he has continued to do so for more than half a century. Sex-linkage, linkage and crossing over, the linear order of the genes, chromosome "maps," interference in crossing over, and inversions were among the important genetic phenomena discovered and explained during the eventful decade beginning in 1910. Drosophila was ideal material in many respects, and it was expertly exploited by the group, which included Morgan, Sturtevant, Calvin B. Bridges, Herman J. Muller, and others. Advances were so rapid that it was difficult for anyone outside the group to keep up with developments. It was indeed a group effort, for all the "fly" geneticists worked in a single modest-sized laboratory. Sturtevant says that group discussion, argument, and speculation were so frequent that it was next to impossible to determine who had what ideas. E. B. Wilson, then the distinguished Professor of Cytology at Columbia, often dropped in to take part in the discussions and to help himself to one of the bananas that were kept in constant supply as "fly food."

As a postdoctoral National Research Council Fellow I was fortunate in being able to work in Morgan's Division of Biology at the California Institute of Technology for five years beginning in 1931. It was there I first became acquainted with Morgan, Sturtevant, Bridges, Ernest G. Anderson, Marcus Rhoades, Th. Dobzhansky, Sterling Emerson, Jack Schultz, Georgii Karpechenko, Cyril Darlington, and others. Drosophila, Oenothera, Neurospora, and maize were being investigated genetically at the time, and I recall vividly my confusion on being suddenly immersed in a sea of Oenothera and Drosophila terminology, almost entirely new to me. At the time I was working on corn (maize) genetics, and I shall never forget Sturtevant's comment on reading a manuscript of mine on "sticky" chromosomes. He brought it to my room and said, "Do you mind if I show this to 'The Boss'?" Morgan was always referred to with genuine affection as "The Boss," and to have a paper referred to him was a real compliment, or so I regarded it. Years later, after Morgan's death and after I had returned to "Caltech" I learned how much the expression meant. For when a postdoctoral fellow who had formerly worked as a graduate student in our laboratory at Stanford referred to me as "Boss," he was taken aside and politely but firmly told that the term was never to be used in that laboratory for anyone other than Morgan.

Later I worked with Sturtevant on the relation of inversions to crossing over and disjunction of the X chromosomes of *Drosophila melanogaster,* a study in which I shared with him the satisfaction of interpreting a seemingly complex situation in a relatively simple way. I believe this experience gave me some small measure of appreciation of the excitement of the earlier Drosophila days in Morgan's laboratory at Columbia University.

Sturtevant has worked with Drosophila during his entire scientific career and continues to do so. But he is by no means a narrow specialist. He is a biologist in a very broad sense and has maintained an active interest in many areas of this branch of science. His published work includes studies on the genetics of horses, snails, the domestic fowl, rabbits, moths, evening primroses, bearded irises, and man. He is an authority on the taxonomy of the genus Drosophila and related genera of Diptera and has published several papers on the biology of ants. He has followed developments in chemical genetics with deep interest and understanding. It was because of his observations on the nonautonomous behavior of vermilion eye color in Drosophila that Boris Ephrussi and I began our investigations on eye pigments in this fly using transplantation and chemical techniques. This in turn led Edward L. Tatum and me to extend our biochemical genetic studies to the bread mold Neurospora. In much the same way Sturtevant has kept in close touch with the recent exciting developments in bacterial and viral genetics and has contributed significantly to these areas in seminars and informal discussions with his colleagues.

This collection of the more important of Sturtevant's papers will serve a real need. It provides the serious student of genetics with a collection of important papers that are not otherwise readily accessible. They are based on clear thinking as well as on soundly conceived and elegantly executed research. They are models of simple, concise, and effective exposition, and they have lasting historical significance.

Those of us who have had the privilege and pleasure of knowing and working with Professor Sturtevant see this volume as a fitting tribute on the occasion of his seventieth birthday.

G. W. Beadle
The University of Chicago
July 1961

Introduction

This book is intended as a tribute to A. H. Sturtevant on the occasion of his seventieth birthday. The idea for such a book grew out of discussions among his colleagues and students at the California Institute of Technology and at other institutions.* It seemed to many of us that a fitting recognition of Sturtevant's work would be to make more accessible to students and research workers some of his scientific papers in genetics and evolution. The result is this collection of incisively written, first-hand accounts of some of the most fundamental and exciting discoveries which have yet appeared in these fields.

Sturtevant is author or co-author of some 137 scientific publications which include several books, one (with Beadle) a textbook of elementary genetics. A complete list will be found in the appendix. Thirty-three of his most important scientific papers are reprinted here in full and in chronological order. It will be immediately evident that only a fraction of his contributions to biology have been included. The following sketch may help to explain the basis on which papers in this volume were selected.

The first paper has always been of foremost importance. In it Sturtevant demonstrates that the genes are arrayed in a linear order. Each linear array or "linkage group" in turn corresponds to a particular chromosome (paper 3). Even the tiny fourth chromosome of Drosophila which for many years resisted efforts to be mapped finally yielded to analysis when Sturtevant discovered special conditions which allowed recombination to occur in that chromosome (32). The construction of linear maps by essentially the same

* Especially Dr. Dan Lindsley of the Oak Ridge National Laboratory.

methodology Sturtevant originally employed remains basic to all modern genetic investigations.

Sturtevant has contributed many short analytical notes on problems of inheritance in race horses, moths, fowl, irises, evening primroses, snails, rabbits, mice, and men. Only a sample of such notes is included here. Those that have been selected illustrate well how profitable it can be to test a theory by rigorous examination of the apparent exceptions to it. It is such a procedure that led Sturtevant to arrive at a simple explanation for the puzzling behavior of the albino and Himalayan coat color factors in rabbits (2), that led him to uncover in snails the classic case of maternal inheritance (12), and that led S. Emerson and him to clear up many mysteries in the extraordinarily complicated genetics of *Oenothera* (18).

An early paper (4) on sexual selection was the forerunner of an extensive line of research by other workers.

Sturtevant has always been as much concerned with the manner in which genes come to control the whole course of growth and development of an organism as with the manner in which they are transmitted from generation to generation. The sequence of scientific discoveries in developmental and biochemical genetics which followed from his paper on the behavior of the vermilion gene in gynandromorphs (8) is eloquently told by G. W. Beadle in the foreword to this book.

The way in which mosaic individuals can be used to study the developmental effects of genes (19) and the discovery of genetic control over the production of mosaic individuals (17) are milestones in a field of genetics which is now commanding the attention of many investigators, including those engaged in analyzing chromosomal mosaicism in man, transformations of cells in tissue culture, and the problems of cancer. Even more subtle and remarkable is Sturtevant's discovery of the phenomenon of position effect—a form of gene action so primary that it must be taking place at the site of the genes in the chromosome (11, 13, 16). This phenomenon, which Muller, Schultz, Dubinin, and others were soon to show takes diverse forms, is only beginning to come into prominence in the genetics of higher organisms as the result of the clear demonstration in *Oenothera* (Catcheside), and still more recently in mice (L. Russell), of the so-called variegated- or V-type position effect.

The study of sex determination has always excited the interest of biologists. Few have contributed more to our understanding of basic mechanisms in this field than has Sturtevant. Interspersed throughout this book will be found his analyses of specific genes causing intersexuality (7, 30), of maternal effects on sex determination in hybrids (31), and of genetic control of the production of sex mosaics (17). These discoveries take on renewed importance when viewed in relation to current studies on the genic and chromosomal bases of sex determination in mice and men.

A field of great current interest in genetics concerns itself with certain real or apparent exceptions to Mendel's first law. One of the first and still intriguing examples is the nonrandom segregation of the fourth chromosome of *Drosophila,* which Sturtevant analyzes in paper 21. Here will be found an admirable statement of the formal algebraic aspects of this particular problem for which no solution is as yet at hand.

In modern genetic analysis one of the most important sets of tools which the investigator has at his disposal are chromosomal rearrangements. No other type of rearrangement has been more extensively used to analyze the problems of genetics and evolution than has the inversion. It is characteristic that Sturtevant should have uncovered such a tool and then exploited it to the fullest. The history of this aberrant type of chromosome commences with Sturtevant's discovery of crossover suppressors (5) and continues with his presentations of genetic evidence that such suppressors are chromosomes containing an inversion in their gene sequence (10, 14, 15). In a monumental paper with Beadle (22), many of the unusual aspects of the transmission of inversions through the female germ line are analyzed and accounted for in a simple and precise way. Sturtevant and Mather deal, in a speculative paper (27), with some of the complex problems posed by the occurrence of inversions in natural populations. And by showing how inversions can be used to trace phylogenetic relationships Sturtevant and Dobzhansky develop one of the most remarkable concepts in modern evolutionary theory (20).

The study of speciation through a comparative study of specific gene mutations is a novel and exciting idea which traces to a paper published in 1918 (6). This led to extensive studies which culminated many years later in a paper with Novitski which brings together the known mutational parallels in the genus *Drosophila* and discusses their significance for the evolution of this group (29).

Sturtevant's command of genetics, his keen interest in evolution, and his mastery of taxonomy have enabled him to speculate profitably on the diverse mechanisms of evolution. In addition to the papers of this kind cited above, the reader will find included other papers dealing with the significance of the occurrence of lethal genes in wild populations (23, 33), with the effect of selection on genes affecting mutation rate (24, 28), with bizarre problems of selection in the social insects (25), and with one of the great puzzles of evolution—the origin of the sterility of hybrids (9, 26).

With a few exceptions, if a paper had already been published in book form it was not included in the present collection in order to make room for less accessible ones. This was, for example, the basis for not including the following works: Sturtevant's 1918 monograph on the effects of selection, a paper of great historical interest in the field of quantitative inheritance; his National Sigma Xi Lecture entitled "The Evolution and Function of Genes" (1948); and an address entitled "Social Implications of the Genetics of Man," (reprinted in *Classic Papers of Genetics,* edited by J. A. Peters, Prentice-Hall, New Jersey). Numerous taxonomic papers on flies and ants have been omitted.

For the most part, papers of a polemic nature, which were numerous in the early days when the successes of the Morgan school did not always gain immediate acceptance, have also been omitted.

It is a pleasure to acknowledge the permission of numerous publishers to reprint the articles in this collection. Wherever found, typographical errors in the original papers have been corrected. However, no substantive changes have been made; instead, wherever such changes are known to be needed, Professor Sturtevant has appended at the end of the paper concerned comments in the form of "Author's Notes, 1961." Typographic styles of the original journals have been followed as closely as possible in resetting type for this book. This, and the gradual evolution of genetic symbolism over the years, will help to explain some inconsistencies in style which are inevitable in a book of this kind. Two of the papers (1, 13) have been reprinted in *Classic Papers of Genetics,* edited by J. A. Peters and published by Prentice-Hall.

Grateful acknowledgment is also extended to my wife, Pamela, who has assumed a considerable share of the editorial work connected with the preparation of this book.

E. B. Lewis
Pasadena
August 1961

Contents

1

THE LINEAR ARRANGEMENT OF SIX SEX-LINKED FACTORS IN *DROSOPHILA*, AS SHOWN BY THEIR MODE OF ASSOCIATION †

Historical

The parallel between the behavior of the chromosomes in reduction and that of Mendelian factors in segregation was first pointed out by Sutton ('02) though earlier in the same year Boveri ('02) had referred to a possible connection (loc. cit., footnote 1, p. 81). In this paper and others Boveri brought forward considerable evidence from the field of experimental embryology indicating that the chromosomes play an important rôle in development and inheritance. The first attempt at connecting any given somatic character with a definite chromosome came with McClung's ('02) suggestion that the accessory chromosome is a sex-determiner. Stevens ('05) and Wilson ('05) verified this by showing that in numerous forms there is a sex chromosome, present in all the eggs and in the female-producing sperm, but absent, or represented by a smaller homologue, in the male-producing sperm. A further step was made when Morgan ('10) showed that the factor for color in the eyes of the fly Drosophila ampelophila follows the distribution of the sex-chromosome already found in the same species by Stevens ('08). Later, on the appearance of a sex-linked wing mutation in Drosophila, Morgan ('10 a, '11) was able to make clear a new point. By crossing white eyed, long winged flies to those with red eyes and rudimentary wings (the new sex-linked character) he obtained, in F_2, white eyed rudimentary winged flies. This could happen only if 'crossing over' is possible; which means, on the assumption that both of these factors are in the sex-chromosomes, that an interchange of materials between homologous chromosomes occurs (in the female only, since the male has only one sex-chromosome). A point not noticed at this time came out later in connection with other sex-linked factors in Drosophila (Morgan '11 d). It became evident that some of the sex-linked factors are associated, i.e., that crossing over does not occur freely between some factors, as shown by the fact that the combinations present in the P_1 flies are much more frequent in F_2 than are

† Reprinted from The Journal of Experimental Zoology, Vol. 14, 1913, pp. 43–59.

new combinations of the same characters. This means, on the chromosome view, that the chromosomes, or at least certain segments of them, are more likely to remain intact during reduction than they are to interchange materials.[1] On the basis of these facts Morgan ('11 c, '11 d) has made a suggestion as to the physical basis of coupling. He uses Janssens' ('09) chiasmatype hypothesis as a mechanism. As he expresses it (Morgan '11 c):

> If the materials that represent these factors are contained in the chromosomes, and if those that "couple" be near together in a linear series, then when the parental pairs (in the heterozygote) conjugate like regions will stand opposed. There is good evidence to support the view that during the strepsinema stage homologous chromosomes twist around each other, but when the chromosomes separate (split) the split is in a single plane, as maintained by Janssens. In consequence, the original materials will, for short distances, be more likely to fall on the same side of the split, while remoter regions will be as likely to fall on the same side as the last, as on the opposite side. In consequence, we find coupling in certain characters, and little or no evidence at all of coupling in other characters, the difference depending on the linear distance apart of the chromosomal materials that represent the factors. Such an explanation will account for all the many phenomena that I have observed and will explain equally, I think, the other cases so far described. The results are a simple mechanical result of the location of the materials in the chromosomes, and of the method of union of homologous chromosomes, and the proportions that result are not so much the expression of a numerical system as of the relative location of the factors in the chromosomes.

Scope of This Investigation

It would seem, if this hypothesis be correct, that the proportion of 'cross-overs' could be used as an index of the distance between any two factors. Then by determining the distances (in the above sense) between A and B and between B and C, one should be able to predict AC. For, if proportion of cross-overs really represents distance, AC must be, approximately, either AB plus BC, or AB minus BC, and not any intermediate value. From purely mathematical considerations, however, the sum and the difference of the proportion of cross-overs between A and B and those between B and C are only *limiting* values for the proportion of cross-overs between A and C. By using several pairs of factors one should be able to apply this test in several cases. Furthermore, experiments involving three or more sex-linked allelomorphic pairs together should furnish another and perhaps more crucial test of the view. The present paper is a preliminary report of the investigation of these matters.

I wish to thank Dr. Morgan for his kindness in furnishing me with material for this investigation, and for his encouragement and the suggestions he has offered during the progress of the work. I have also been greatly helped by numerous discussions of the theoretical side of the matter with Messrs. H. J.

[1] It is interesting to read, in this connection, Lock's ('06, p. 248–253) discussion of the matter

Muller, E. Altenburg, C. B. Bridges, and others. Mr. Muller's suggestions have been especially helpful during the actual preparation of the paper.

The Six Factors Concerned

In this paper I shall treat of six sex-linked factors and their inter-relationships. These factors I shall discuss in the order in which they seem to be arranged.

B stands for the black factor. Flies recessive with respect to it (b) have yellow body color. The factor was first described and its inheritance given by Morgan ('11 a).

C is a factor which allows color to appear in the eyes. The white eyed fly (first described by Morgan '10) is now known to be always recessive with respect both to C and to the next factor.

O. Flies recessive with respect to O(o) have eosin eyes. The relation between C and O has been explained by Morgan in a paper now in print and about to appear in the Proceedings of the Academy of Natural Sciences in Philadelphia.

P. Flies with p have vermilion eyes instead of the ordinary red (Morgan '11 d).

R. This and the next factor both affect the wings. The normal wing is RM. The rM wing is known as miniature, the Rm as rudimentary, and the rm as rudimentary-miniature. This factor R is the one designated L by Morgan ('11 d) and Morgan and Cattell ('12). The L of Morgan's earlier paper ('11) was the next factor.

M. This has been discussed above, under R. The miniature and rudimentary wings are described by Morgan ('11 a).

The relative position of these factors is B, $\frac{C}{O}$, P, R, M. C and O are placed at the same point because they are completely linked. Thousands of flies had been raised from the cross CO (red) by co (white) before it was known that there were two factors concerned. The discovery was finally made because of a mutation and not through any crossing over. It is obvious, then, that unless coupling strength be variable, the same gametic ratio must be obtained whether, in connection with other allelomorphic pairs, one uses CO (red) as against co (white), Co (eosin) against co (white), or CO (red) against Co (eosin) (the cO combination is not known).

Method of Calculating Strength of Association

In order to illustrate the method used for calculating the gametic ratio I shall use the factors P and M. The cross used in this case was, long winged, vermilion-eyed female by rudimentary winged, red-eyed male. The analysis and results are seen in table 1.

Table 1

	Long vermilion ♀—MpX MpX Rudimentary red ♂—mPX
F_1	MpX mPX—long red ♀ MpX —long vermilion ♂
Gametes F_1	Eggs —MPX mPX MpX mpX Sperm—MpX
F_2	MPX MpX, mPX MpX —long red ♀—451 MpX MpX, mpX MpX —long vermilion ♀—417 MPX —long red ♂—105 mPX —rudimentary red ♂—33 MpX —long vermilion ♂—316 mpX —rudimentary vermilion ♂—4

It is of course obvious from the figures that there is something peculiar about the rudimentary winged flies, since they appear in far too small numbers. This point need not detain us here, as it always comes up in connection with rudimentary crosses, and is being investigated by Morgan. The point of interest at present is the linkage. In the F_2 generation the original combinations, red rudimentary and vermilion long, are much more frequent in the males (allowing for the low viability of rudimentary) than are the two new or cross-over combinations, red long and vermilion rudimentary. It is obvious from the analysis that no evidence of association can be found in the females, since the M present in all female-producing sperm masks m when it occurs. But the ratio of cross-overs in the gametes is given without complication by the F_2 males, since the male-producing sperm of the F_1 male bore no sex-linked genes. There are in this case 349 males in the non-cross-over classes and 109 in the cross-overs. The method which has seemed most satisfactory for expressing the relative position of factors, on the theory proposed in the beginning of this paper, is as follows. The unit of 'distance' is taken as a portion of the chromosome of such length that, on the average, one cross-over will occur in it out of every 100 gametes formed. That is, percent of cross-overs is used as an index of distance. In the case of P and M there occurred 109 cross-overs in 405 gametes, a ratio of 26.9 in 100; 26.9, the per cent of cross-overs, is considered as the 'distance' between P and M.

The Linear Arrangement of the Factors

Table 2 shows the proportion of cross-overs in those cases which have been worked out. The detailed results of the crosses involved are

Table 2

FACTORS CONCERNED	PROPORTION OF CROSS-OVERS	PER CENT OF CROSS-OVERS
BCO	193/16287	1.2
BO	2/373	0.5
BP	1464/4551	32.2
BR	115/324	35.5
BM	260/693	37.6
COP	224/748	30.0
COR	1643/4749	34.6
COM	76/161	47.2
OP	247/836	29.4
OR	183/538	34.0
OM	218/404	54.0
CR	236/829	28.5
CM	112/333	33.6
B(C, O)	214/21736	1.0
(C, O)P	471/1584	29.7
(C, O)R	2062/6116	33.7
(C, O)M	406/898	45.2
PR	17/573	3.0
PM	109/405	26.9

given at the end of this paper. The 16287 cases for B and CO are from Dexter ('12). Inasmuch as C and O are completely linked I have added the numbers for C, for O, and for C and O taken together, giving the total results in the lines beginning (C, O) P, B (C, O), etc., and have used these figures, instead of the individual C, O, or CO results, in my calculations. The fractions in the column marked 'proportion of cross-overs' represent the number of cross-overs (numerator) to total available gametes (denominator).

As will be explained later, one is more likely to obtain accurate figures for distances if those distances are short, i.e., if the association is strong. For

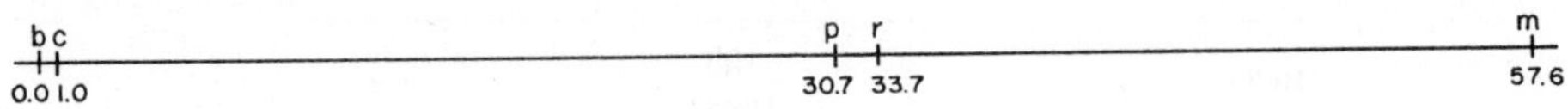

Diagram 1

this reason I shall, in so far as possible, use the percent of cross-overs between adjacent points in mapping out the distances between the various factors. Thus, B (C, O), (C, O) P, PR, and PM form the basis of diagram 1. The figures on the diagram represent calculated distances from B.

Of course there is no knowing whether or not these distances as drawn represent the actual relative spacial distances apart of the factors. Thus the distance CP may in reality be shorter than the distance BC, but what we do know is that a break is far more likely to come between C and P than between B and C. Hence, either CP is a long space, or else it is for some reason a weak one. The point I wish to make here is that we have no means of knowing that the chromosomes are of uniform strength, and if there are strong or weak places, then that will prevent our diagram from representing actual relative distances—but, I think, will not detract from its value as a diagram.

Just how far our theory stands the test is shown by table 3, giving ob-

Table 3

FACTORS	CALCULATED DISTANCE	OBSERVED PER CENT OF CROSS-OVERS
BP....................	30.7	32.2
BR....................	33.7	35.5
BM....................	57.6	37.6
(C, O)R................	32.7	33.7
(C, O)M................	56.6	45.2

served per cent of cross-overs, and distances as calculated from the figures given in the diagram of the chromosome. Table 3 includes all pairs of factors given in table 2 but not used in the preparation of the diagram.

It will be noticed at once that the long distances, BM, and (C, O)M, give smaller percent of cross-overs than the calculation calls for. This is a point which was to be expected, and will be discussed later. For the present we may dismiss it with the statement that it is probably due to the occurrence of two breaks in the same chromosome, or 'double crossing over.' But in the case of the shorter distances the correspondence with expectation is perhaps as close as was to be expected with the small numbers that are available. Thus, BP is 3.2 less than BR, the difference expected being 3.0. (C, O)R is less than BR by 1.8 instead of by 1.0. It has actually been found possible to predict the strength of association between two factors by this method, fair approximations having been given for BR and for certain combinations involving factors not treated in this paper, before the crosses were made.

Double Crossing Over

On the chiasmatype hypothesis it will sometimes happen, as shown by Dexter ('12) and intimated by Morgan ('11 d) that a section of, say, maternal chromosome will come to have paternal elements at both ends, and perhaps more maternal segments beyond these. Now if this can happen it introduces a complication into the results. Thus, if a break occurs between B and P, and another between P and M, then, unless we can follow P also, there will be no evidence of crossing over between B and M, and the fly hatched from the resulting gamete will be placed in the non-cross-over class, though in reality he represents two cross-overs. In order to see if double crossing over really does occur it is necessary to use three or more sex-linked allelomorphic pairs in the same experiment. Such cases have been reported by Morgan ('11 d) and Morgan and Cattell ('12) for the factors B, CO, and R. They made such crosses as long gray red by miniature yellow white, and long yellow red by miniature gray white, etc. The details and analyses are given in the original papers, and for our present purpose it is only the flies that are available for observations on double crossing over that are of interest. Table 4 gives a graphical representation of what happened in the 10495 cases.

Double crossing over does then occur, but it is to be noted that the occurrence of the break between B and CO tends to prevent that between CO and R (or vice versa). Thus where B and CO did not separate, the gametic ratio for CO and R was about 1 to 2, but in the cases where B and CO did separate it was about 1 to 6.5.

Three similar cases from my own results, though done on a smaller scale, are given in the table at the end of this paper. The results are represented in tables 5, 6 and 7.

It will be noted that here also the evidence, so far as it goes, indicated that the occurrence of one cross-over makes another one less likely to occur in the same gamete. In the case of BOPR there was an opportunity for triple crossing over, but it did not occur. Of course, on the view here presented

Table 4

NO CROSSING OVER	SINGLE CROSSING OVER		DOUBLE CROSSING OVER
B CO R	B CO R	B CO R	B CO R
6972	3454	60	9

Table 5

NO CROSSING	SINGLE CROSSING OVER		DOUBLE CROSSING OVER
O P R	O P R	O P R	O P R
194	102	11	1

Table 6

NO CROSSING	SINGLE CROSSING OVER		DOUBLE CROSSING OVER
B O M	B O M	B O M	B O M
278	160	1	0

Table 7

B O P R	B O P R	B O P R	B O P R	B O P R	B O P R	B O P R	B O P R
393	203	19	6	2	1	1	0

there is no reason why it should not occur, if enough flies were raised. An examination of the figures will show that it was not to be expected in such small numbers as are here given. So far as I know there is, at present, no evidence that triple crossing over takes place, but it seems highly probable that it will be shown to occur.[2]

Unfortunately, in none of the four cases given above are two comparatively long distances involved, and in only one are there enough figures to form a fair basis for calculation, so that it seems as yet hardly possible to determine how much effect double crossing over has in pulling down the observed percent of cross-overs in the case of BM and (C, O)M. Whether or not this effect is partly counter-balanced by triple crossing over must also remain unsettled as yet. Work now under way should furnish answers to both these questions.

[2] A case of triple crossing over within the distance CR was observed after this paper went to press.

Possible Objections to These Results

It will be noted that there appears to be some variation in coupling strength. Thus, I found (CO)R to be 36.7; Morgan and Cattell obtained the result 33.9; for OR I got 34.0, and for CR, 28.5. The standard error for the difference between (CO)R (all figures) and CR is 1.84 percent, which means that a difference of 5.5 percent is probably significant (Yule '11, p. 264). The observed difference is 6.1 percent, showing that there is some complication present. Similarly, BM gave 37.6, while OM gave 54.0—and BOM gave 36.7 for BM, and 36.5 for OM. There is obviously some complication in these cases, but I am inclined to think that the disturbing factor

Table 8

(The meaning of the phrase "proportion of cross-overs" is given on p. 6)

BO. P_1: gray eosin ♀ × yellow red ♂
F_1: gray red ♀ × gray eosin ♂
F_2: ♀♀, g.r. 241, g.e. 196
♂♂, g.r. 0, g.e. 176, y.r. 195, y.e. 2

Proportion of cross-overs, $\frac{2}{373}$

BP. P_1: gray red ♀ × yellow vermilion ♂
F_1: gray red ♀ × gray red ♂
F_2: ♀♀, g.r. 98;
♂♂, g.r. 59, g.v. 16, y.r. 24, y.v. 33
Back cross, F_1 gray red ♀♀ from above × yellow vermilion ♂♂
F_2: ♀♀, g.r. 31, g.v. 11, y.r. 12, y.v. 41
♂♂, g.r. 23, g.v. 13, y.r. 8, y.v. 21
P_1: gray vermilion ♀ × yellow red ♂
F_1: gray red ♀ × gray vermilion ♂
F_2: ♀♀, g.r. 199, g.v. 182
♂♂, g.r. 54, g.v. 149, y.r. 119, y.v. 41
P_1: yellow vermilion ♀ × gray red ♂
F_1: gray red ♀ × yellow vermilion ♂
F_2: ♀♀, g.r. 472, g.v. 240, y.r. 213, y.v. 414
♂♂, g.r. 385, g.v. 186, y.r. 189, y.v. 324
F_1: gray vermilion × yellow red (sexes not recorded)
F_1: gray red ♀♀. These were mated to yellow vermilion ♂♂ of other stock
F_2: ♀♀, g.r. 50, g.v. 96, y.r. 68, y.v. 41
♂♂, g.r. 44, g.v. 105, y.r. 86, y.v. 47

Proportion of cross-overs, adding ♀♀ from BOPR (below), $\frac{1464}{4551}$

BR. P_1: miniature yellow ♀ × long gray ♂
F_1: long gray ♀ × miniature yellow ♂

Table 8 (continued)

F_2: ♀♀ l.g. 14, l.y. 2, m.g. 7, m.y. 6;
♂♂ l.g. 10, l.y. 1, m.g. 6, m.y. 8.
P_1: long yellow ♀ × miniature gray ♂
F_1: long gray ♀ × long yellow ♂
F_2: ♀♀, l.g. 148, l.y. 130
♂♂, l.g. 51, l.y. 82, m.g. 89, m.y. 48

Proportion of cross-overs, $\frac{115}{324}$

BM. P_1: long yellow ♀ × rudimentary gray ♂
F_1: long gray ♀ × long yellow ♂
F_1: ♀♀, l.g. 591, l.y. 549
♂♂, l.g. 228, l.y. 371, r.g. 20, r.y. 3
P_1: long gray ♀ × rudimentary yellow ♂
F_1: long gray ♀ × long gray ♂
F_2: ♀♀, l.g. 152
♂♂, l.g. 42, l.y. 29, r.g. 0, r.y. 0

Proportion of cross-overs, $\frac{260}{693}$

COP. P_1: vermilion ♀ × white ♂
F_1: red ♀ × vermilion ♂
F_2: ♀♀, r. 320, v. 294
♂♂, r. 86, v. 206, w. 211
(7 of the vermilion ♀♀ known from tests to be CC, 2 known to be Cc. 7 white ♂♂ Pp, 2 pp.)
Back cross, F_1 red ♀♀ from above × white ♂♂, gave
F_2: ♀♀, r. 195, w. 227,
♂♂, r. 66, v. 164, w. 184
Out cross, F_1 ♀♀ as above × white ♂♂ recessive in P, gave
F_2: ♀♀, r. 35, v. 65, w. 98
♂♂, r. 33, v. 75, w. 95

Proportion of cross-overs, $\frac{224}{748}$

COR. P_1: miniature white ♀ × long red ♂
F_1: long red ♀ × miniature white ♂
F_2: ♀♀, l.r. 193, l.w. 109, m.r. 124, m.w. 208
♂♂, l.r. 202, l.w. 114, m.r. 123, m.w. 174
P_1: long white ♀ × miniature red ♂
F_1: long red ♀ × long white ♂
F_2: ♀♀ l.r. 194, l.w. 160
♂♂ l.r. 52, l.w. 124, m.r. 97, m.w. 41

Proportion of cross-overs, $\frac{563}{1561}$; or, adding such available figures from Morgan ('11 d) and Morgan and Cattell ('12) as are not complicated by the presence of yellow or brown flies, $\frac{1643}{4749}$

Table 8 (continued)

COM. P_1: long white ♀ × rudimentary red ♂
F_1: long red ♀ × long white ♂
F_2: ♀♀, l.r. 157, l.w. 127
♂♂, l.r. 74, l.w. 82, ru.r. 3, ru.w. 2

Proportion of cross-overs, $\frac{76}{161}$

OP. P_1: black red ♀ × black eosin-vermilion ♂
F_1: black red ♀ × black red ♂
F_2: (all black), ♀♀, r. 885
♂♂, r. 321, v. 125, e. 122, e.-v. 268

Proportion of cross-overs, $\frac{247}{836}$

OR. P_1: long red ♀ × miniature eosin ♂
F_1: long red ♀ × long red ♂
F_2: ♀♀, l.r. 408
♂♂, l.r. 145, l.e. 67, m.r. 70, m.e. 100
P_1: long eosin ♀ × miniature red ♂
F_1: long red ♀ × long eosin ♂
F_2: ♀♀, l.r. 100, l.e. 95
♂♂, l.r. 27, l.e. 54, m.r. 56, m.e. 19

Proportion of cross-overs, $\frac{183}{538}$

OM. P_1: long eosin ♀ × rudimentary red ♂
F_1: long red ♀ × long eosin ♂
F_2: ♀♀, l.r. 368, l.e. 266
♂♂, l.r. 194, l.e. 146, ru.r. 40, ru.e. 24

Proportion of cross-overs, $\frac{218}{404}$

CR. P_1: long white ♀ × miniature eosin ♂
F_1: long eosin ♀ × long white ♂
F_2: ♀♀, l.e. 185, l.w. 205
♂♂, l.e. 54, l.w. 147, m.e. 149, m.w. 42
P_1: long eosin ♀ × miniature white ♂
F_1: long eosin ♀ × long eosin ♂
F_2: ♀♀, l.e. 527
♂♂, l.e. 169, l.w. 85, m.e. 55, m.w. 128

Proportion of cross-overs, $\frac{236}{829}$

CM. P_1: long white ♀ × rudimentary eosin ♂
F_1: long eosin ♀ × long white ♂
F_2: ♀♀, l.e. 328, l.w. 371
♂♂, l.e. 112, l.w. 217, ru.e. 4, ru.w. 0

Proportion of cross-overs, $\frac{112}{333}$

Table 8 (continued)

PR. P_1: long vermilion (yellow) ♀ × miniature red (yellow) ♂
F_1: long red yellow ♀ × long vermilion yellow ♂
F_2: (all y.) ♀♀, l.r. 138, l.v. 110
♂♂, l.r. 8, l.v. 117, m.r. 97, m.v. 1
P_1: long vermilion (gray) ♀ × miniature red ♂
F_1: long red ♀ × long vermilion ♂
F_2: ♀♀, l.r. 116, l.v. 110
♂♂, l.r. 2, l.v. 81, m.r. 96, m.v. 1
P_1: miniature red ♀ × long vermilion ♂
F_1: long red ♀ × miniature red ♂
F_1: ♀♀, l.r. 45, m.r. 49
♂♂, l.r. 1, l.v. 27, m.r. 26, m.v. 0
F_1 long red ♀♀ from above × miniature red ♂♂ of other stock, gave
F_2: ♀♀, l.r. 74, m.r. 52
♂♂, l.r. 3, l.v. 66, m.r. 46, m.v. 1

Proportion of cross-overs, $\frac{17}{573}$

PM. P_1: long vermilion ♀ × rudimentary red ♂
F_1: long red ♀ × long vermilion ♂
F_2: ♀♀, l.r. 451, l.v. 417
♂♂, l.r. 105, l.v. 316, ru.r. 33, ru.v. 4

Proportion of cross-overs, $\frac{109}{405}$

OPR. P_1: long vermilion ♀ × miniature eosin ♂
F_1: long red ♀ × long vermilion ♂
F_2: ♀♀, l.r. 205, l.v. 182
♂♂, l.r. 1, l.v. 109, l.e. 8, l.e.-v. 53, m.r. 49, m.v. 3, m.e. 85, m.e.-v. 0

BOM. P_1: long red yellow ♀ × rudimentary eosin gray ♂
F_1: long red gray ♀ × long red yellow ♂
F_2: ♀♀, l.r.g. 530, l.r.y. 453
♂♂, l.r.g. 1, l.r.y. 274, l.e.g. 156, l.e.y. 0, ru.r.g. 0, ru.r.y. 4, ru.e.g. 4, ru.e.y. 0

BOPR. P_1: long vermilion brown ♀ × miniature eosin black ♂
F_1: long red black ♀ × long vermilion brown ♂
F_2: ♀♀, l.r.bl. 305, l.r.br. 113, l.v.bl. 162, l.v.br. 256
♂♂, l.r.bl. 0, l.r.br. 2, l.v.bl. 3, l.v.br. 185, l.e.bl. 9, l.e.br. 0, l.e.-v.bl. 127, l.e.-v.br. 0, m.r.bl. 1, m.r.br. 76, m.v.bl. 1, m.v.br. 10, m.e.bl. 208, m.e.br. 3, m.e.-v.bl. 0, m.e.-v.br. 0

discussed below (viability) will explain this. However, experiments are now under way to test the effect of certain external conditions on coupling strength. It will be seen that on the whole when large numbers are obtained in different experiments and are averaged, a fairly consistent scheme results. Final judgment on this matter must, however, be withheld until the subject can be followed up by further experiments.

Another point which should be considered in this connection is the effect of differences in viability. In the case of P and M, used above as an illustration, the rudimentary winged flies are much less likely to develop than are the longs. Now if the viability of red and vermilion is different, then the longs do not give a fair measure of the linkage, and the rudimentaries, being present in such small numbers, do not even up the matter. It is probable that there is no serious error due to this cause except in the case of rudimentary crosses, since the two sides will tend to even up, unless one is very much less viable than the other, and this is true only in the case of rudimentary. It is worth noting that the only serious disagreements between observation and calculation occur in the case of rudimentary crosses (BM, and (CO)M). Certain data of Morgan's now in print, and further work already planned, will probably throw considerable light on the question of the position and behavior of this factor M.

Summary

It has been found possible to arrange six sex-linked factors in Drosophila in a linear series, using the number of cross-overs per 100 cases as an index of the distance between any two factors. This scheme gives consistent results, in the main.

A source of error in predicting the strength of association between untried factors is found in double crossing over. The occurrence of this phenomenon is demonstrated, and it is shown not to occur as often as would be expected from a purely mathematical point of view, but the conditions governing its frequency are as yet not worked out.

These results are explained on the basis of Morgan's application of Janssens' chiasmatype hypothesis to associative inheritance. They form a new argument in favor of the chromosome view of inheritance, since they strongly indicate that the factors investigated are arranged in a linear series, at least mathematically.

LITERATURE CITED

BOVERI, T., 1902: Ueber mehrpolige Mitosen als Mittel zur Analyse des Zellkerns. Verh, Phys.-Med. Ges. Würzburg., N.F., Bd. 35, p. 67.

DEXTER, J. S., 1912: On coupling of certain sex-linked characters in Drosophila. Biol. Bull., vol. 23, p. 183.

JANSSENS, F. A., 1909: La théorie de la chiasmatypie. La Cellule, tom. 25, p. 389.

LOCK, R. H., 1906: Recent progress in the study of variation, heredity, and evolution. London and New York.

McClung, C. E., 1902: The accessory chromosome—sex determinant? Biol. Bull., vol. 3, p. 43.
Morgan, T. H., 1910: Sex-limited inheritance in Drosophila. Science, n.s., vol. 32, p. 120.
1910 a: The method of inheritance of two sex-limited characters in the same animal. Proc. Soc. Exp. Biol. Med., vol. 8, p. 17.
1911: The application of the conception of pure lines to sex-limited inheritance and to sexual dimorphism. Amer. Nat., vol. 45, p. 65.
1911 a: The origin of nine wing mutations in Drosophila. Science, n.s., vol. 33, p. 496.
1911 b: The origin of five mutations in eye color in Drosophila and their modes of inheritance. Science, n.s., vol. 33, p. 534.
1911 c: Random segregation versus coupling in Mendelian inheritance. Science, n.s., vol. 34, p. 384.
1911 d: An attempt to analyze the constitution of the chromosomes on the basis of sex-limited inheritance in Drosophila. Jour. Exp. Zoöl., vol. 11, p. 365.
Morgan, T. H. and Cattell, E., 1912: Data for the study of sex-linked inheritance in Drosophila. Jour. Exp. Zoöl., vol. 13, p. 79.
Stevens, N. M., 1905: Studies in spermatogenesis with special reference to the 'accessory chromosome.' Carnegie Inst. Washington, publ. 36.
1908: A study of the germ-cells of certain Diptera. Jour. Exp. Zoöl., vol. 5, p. 359.
Sutton, W. S., 1902: On the morphology of the chromosome group in Brachystola magna. Biol. Bull., vol. 4, p. 39.
Wilson, E. B., 1905: The behavior of the idiochromosomes in Hemiptera. Jour. Exp. Zoöl., vol. 2, p. 371.
1906: The sexual differences of the chromosome-groups in Hemiptera, with some considerations on the determination and inheritance of sex. Jour. Exp. Zoöl., vol. 3, p. 1.
Yule, G. U., 1911: An introduction to the theory of statistics. London.

Author's Notes, 1961

The terminology for the genes used is very different from that now current; the following list of equivalents may be useful in correlating this paper with later literature:

1913	B	b	C	c	O	o	P	p	R	r	M	m
Current	y^+	y	w^+	w	w^+	w^e	v^+	v	m^+	m	r^+	r

McClung's suggestion as to the relation of the X chromosome to sex-determination was made in 1901 (Anat. Anz. 20: 220–226) in a preliminary note on the material in the longer paper of 1902 which is cited in the text.

Here, and in other early papers on the genetics of Drosophila, occurs the statement that "the male has only one sex-chromosome." This was due to a misinterpretation of the work of Stevens. She described and figured the chromosomes as now known, including an XY pair, but we were misled especially by one of her figures (number 61). This confusion was not cleared up until the cytological studies of Metz and of Bridges.

There is an obvious arithmetical error in the discussion of Table 1. The total number of males is 458, not 405 as stated, and the "per cent of cross-overs" is 23.8, not 26.9. This mistake is incorporated in Tables 2 and 8, in Diagram I, and in the calculated values of Table 3.

2

THE HIMALAYAN RABBIT CASE, WITH SOME CONSIDERATIONS ON MULTIPLE ALLELOMORPHS †

It has been shown by Castle ('06, '09), Hurst ('06) and Punnett ('12) that the Himalayan pattern in rabbits behaves as a simple recessive to self color, and as a simple dominant to albino. Thus, as Punnett points out, we might suppose self to be the double dominant, Himalayan a recessive in one factor, and albino a double recessive. But, to use Punnett's words:

> The F_2 from self × albino should consequently contain Himalayans as well as true albinos. But among the large number of animals reared from such matings no Himalayans have hitherto been recorded, and for the present the relations between these various forms remain obscure.

If we suppose that albino may be either the second single recessive or the double recessive we avoid this difficulty, but are then unable to explain why albino × Himalayan should not, at least occasionally, produce selfs by recombination.

Now it seems to me that the facts of the case are fitted equally well by either of two hypotheses. In the first place, we may consider, as above, that Himalayan is a single recessive and albino a double recessive—if we suppose the two factors concerned to be completely linked. The gametic (not zygotic) constitution of the three types would then be represented thus, *C* being the color producer and *S* the factor changing Himalayan to self.

Self —*CS*
Himalayan—*Cs*
Albino —*cs*

If *C* and *S* be completely linked no *cS* individual can be obtained, and *CS* × *cs* would give no *Cs* in F_2.

On the other hand, we may consider the factor for self as allelomorphic to that for Himalayan pattern, and also to that for albinism. Then the three pure types might be represented thus (zygotic formulæ):

† Reprinted from The American Naturalist, Vol. 47, 1913, pp. 234–238.

Self —*SS*
Himalayan—*HH*
Albino —*AA*

S, H, and *A* being allelomorphic each to itself or to either of the others, the crosses would result thus:

Self —*SS*
Himalayan—*HH*

F_1 *SH*—self

$$F_2 \left\{ \begin{array}{l} \left.\begin{array}{l} SS \\ SH \\ SH \end{array}\right\} \text{3 self} \\ HH\text{—1 Himalayan} \end{array}\right.$$

Self —*SS*
Albino —*AA*

F_1 *SA*—self

$$F_2 \left\{ \begin{array}{l} \left.\begin{array}{l} SS \\ SA \\ SA \end{array}\right\} \text{3 self} \\ AA\text{—1 albino} \end{array}\right.$$

Himalayan—*HH*
Albino —*AA*

F_1 *HA*—Himalayan

$$F_2 \left\{ \begin{array}{l} \left.\begin{array}{l} HH \\ HA \\ HA \end{array}\right\} \text{3 Himalayan} \\ AA\text{—1 albino} \end{array}\right.$$

An explanation similar to the second one above has been given by de Meijere ('10) for Jacobson's results with *Papilio Memnon.* The evidence on this case is, however, very incomplete, and there are complications due to sex. Either triple allelomorphs or complete coupling would seem to cover the facts as we have them at present. Shull ('11) has also used a system of three allelomorphs for a case in *Lychnis dioica.* I shall refer to this case again.

It will be seen that triple allelomorphs may be substituted for complete coupling as an explanation of any case where only three of the four combinations possible on the complete coupling scheme are known. But if we have the double dominant, both single recessives, and the double recessive, then triple allelomorphism will no longer work. Thus, if a race of albino rabbits is discovered which produces self when mated to Himalayan, complete linkage will be the most likely explanation of the case.

There are certain other cases which fulfil the above requirements. Emerson ('11) has reported a case in beans (green leaves—green pods, green leaves—

yellow pods, and yellow leaves—yellow pods are the three races concerned). The similar cases of complete linkage reported for corn by East and by Emerson are probably more easily explainable by linkage than by multiple allelomorphs, as, at least in some cases, all four possible races are found. Baur ('12) has a case in *Aquilegia,* where three types of leaves are found—green, variegated (green and yellowish green), and yellowish green. These behave toward each other in a manner exactly similar to that of the self, Himalayan and albino rabbits. Finally, Morgan ('12) has reported a case in *Drosophila ampelophila.* Red eye is a dominant to eosin and to white, and eosin is also a dominant to white. No two types ever give the third when crossed, either in F_1 or in F_2. The explanation which has been given in beans, columbines and flies has been that of two allelomorphic pairs, completely linked to each other.

The question as to which of these views is the more probable is closely bound up with the presence and absence hypothesis. On a strict application of this idea there is of course no possibility of more than two members of any given allelomorphic group. The presence and absence hypothesis as a universal principle has been criticized by Morgan ('13) in a recent paper, on what seem to me very strong grounds. It seems very unlikely that protoplasm (chromatin?) is such a simple substance that the only possible change in a given unit (molecule?) involves the loss of that unit. On the other hand, if a slight change takes place in a chemically complex gene, is it necessary to suppose that its allelomorphic relations must be upset? That very slight changes in the constitution of a gene might easily affect its behavior in ontogeny will, I think, be readily granted.

It is to be noted that in all the cases cited above the supposed three allelomorphs have similar ontogenetic effects. Thus the three in rabbits, in *Aquilegia,* and in *Papilio* all affect the distribution of pigment (and, in *Papilio,* also the shape of the wings), those in *Lychnis* the sex, those in beans the production of the same color in different organs, and those in *Drosophila* the production of different colors in the same organ. This may perhaps seem to be in favor of the view that we have here different modifications of the same gene, rather than two distinct genes and their absences.

The history of the red-white-eosin group of eye colors in *Drosophila* is interesting when considered from the viewpoint of the presence and absence hypothesis. The first white-eyed fly arose as a mutant in red stock. On presence and absence it must have been caused by the simultaneous loss of two factors, which were called *C* and *O* by Morgan. Then, in white-eyed stock there appeared an eosin-eyed fly. Here the factor called *O,* just lost, must have been put back again. Finally, in one of my own cultures, eosin has given rise to white by mutation.[1] In both these latter cases the flies had

[1] After this paper went to press it was pointed out to me by Mr. H. J. Muller that there is another possible explanation of this case, which does not involve mutation from eosin to white. This interpretation can not be entered into until certain phenomena observed by Mr. C. B. Bridges have been more fully investigated.

miniature wings, and in the white-to-eosin case they also had black body color. These characters give a check on the results, and make it extremely unlikely that any contamination had occurred. Further evidence to this effect is given, in the eosin-to-white case, by the fact that the mutant fly was one of 127 obtained from a single pair, all of her brothers and sisters being of the expected classes (half of the females heterozygous for white), as were likewise the flies from six sister pairs.[2]

The presence and absence hypothesis involving a dropping out of whole genes or addition of entirely new ones does not offer as simple an explanation of this case as does the conception that we have here a relatively unstable gene, which does not drop out entirely, but undergoes various changes, that from white to eosin being reversible.

It should be noted that Shull ('11) has reported a case of what he calls reversible mutation in the sex-determining factor in *Lychnis,* which is very similar to the above red-eosin-white case. He has adopted a system of triple allelomorphs to explain it, though admitting that complete linkage will also cover the facts. He has also considered the bearing of the case upon the presence and absence hypothesis and upon the nature of mutation, reaching conclusions somewhat similar to those given above.

LITERATURE CITED

BAUR, E., 1912: Vererbungs- und Bastardierungsversuche mit *Antirrhinum.* II. Faktorenkoppelung. *Zts. ind. Abst.-Vererb.-Lehre,* 6, p. 201.

CASTLE, W. E., 1906. Heredity of Coat Characters in Guinea-pigs and Rabbits. Carnegie Inst. Wash., publ. 23.

———, (and others), 1909: Studies of Inheritance in Rabbits. Carnegie Inst. Wash., publ. 114.

EMERSON, R. A., 1911: Genetic Correlation and Spurious Allelomorphism in Maize. *Ann. Rept. Nebr. Agr. Sta.,* 24, p. 58.

HURST, C. C., 1906: Mendelian Characters in Plants and Animals. Rep. Conf. *Gen., Roy. Hort. Soc.* London, p. 114.

DE MEIJERE, J. C. H., 1910: Uber Jacobsons Züchtungsversuche bezüglich des Polymorphismus von *Papilio Memmon* L. ♀, und über die Vererbung sekundärer Geschlechtsmerkmale. *Zts. ind. Abst.-Vererb.-Lehre,* 3, p. 161.

MORGAN, T. H., 1912: Further Experiments with Mutations in Eye Color in *Drosophila. Jour. Acad. Nat. Sci. Phila.,* 15, p. 323.

———, 1913: Factors and Unit Characters in Mendelian Heredity. *Am. Nat.,* 47, p. 5.

PUNNETT, R. C., 1912: Inheritance of Coat-color in Rabbits. *Jour. Genet.,* 2, p. 221.

SHULL, G. H., 1911: Reversible Sex-mutants in *Lychnis dioica. Bot. Gaz.,* 42, p. 329.

[2] There has been at least one case where eosin has seemed to arise as a mutant in red stock, but as there was no other character to serve as a check against contamination, the case does not carry very great weight.

Author's Notes, 1961

The alternative interpretation referred to in the first footnote was that of equational non-disjunction, just then under study by Bridges. As implied in the footnote, the reported case of mutation from eosin to white was in fact spurious.

Here, and in paper no. 3, the older name *ampelophila* is used for what is now known as *Drosophila melanogaster*.

3

A THIRD GROUP OF LINKED GENES IN *DROSOPHILA AMPELOPHILA* †

The existence of a group of sex-linked genes in *Drosophila ampelophila* which are linked to each other in different degrees has been demonstrated by Morgan in numerous papers ('10, '11, '12) and by Sturtevant ('13). The fact that the black and vestigial factors, which are not sex-linked, are linked to each other, was reported by Morgan and Lynch ('12) and by Morgan ('12), and these genes were considered as lying in the "second chromosome." The present paper presents evidence showing the existence of still another group of genes, which are located in the "third chromosome."

The pink-eyed fly, first described by Morgan ('11), has been shown (Morgan, '11, '12) to behave as an ordinary Mendelian recessive to the normal red. That this factor is independent of those in the second chromosome is shown by the following experiments:

Red black ♀ × Pink gray ♂ ↘↗ red gray ♂ and ♀ →

red gray	113
red black	52
pink gray	47
pink black	22

Red gray ♀ × Pink black ♂ ↘↗ red gray ♂ and ♀ →

red gray	279
red black	96
pink gray	107
pink black	40

F_1 red gray ♀ from above × Pink black ♂ ↘↗

red gray	96
red black	111
pink gray	67
pink black	77

Vestigial red ♀ × Long pink ♂ ↘↗ long red ♂ and ♀ →

long red	717
long pink	268
vestigial red	136
vestigial pink	50

Except for disturbances due to differences in viability these are obviously 9 : 3 : 3 : 1 ratios, except the third, which is 1 : 1 : 1 : 1, as expected. Evidently pink is not linked either to black or to vestigial.

† Reprinted from Science, Vol. 37, 1913, pp. 990–992.

In the balloon-winged stock (see Morgan, '11) there appeared a very dark mutant, which has been named ebony. This fly is darker and more shiny than the black fly, and has a slightly greenish, smoky color when young. It behaves as a simple Mendelian recessive to the normal gray color, though the heterozygous form is often somewhat darker than the wild fly. The females of this race are often sterile, and seldom produce many offspring, but the males breed freely when mated with other females, and produce many offspring. That the gene concerned is not in the second chromosome is indicated by the following experiment:

Vestigial gray ♀ × Long ebony ♂ → long gray ♂ and ♀ →

long gray	268
long ebony	79
vestigial gray	94
vestigial ebony	24

This is a simple 9 : 3 : 3 : 1 ratio, except that the viability of the ebonies is not quite as good as that of the grays. Ebony has also been shown to be independent of two other wing factors in the second chromosome—curved and balloon.

When pink-eyed flies were crossed to those with ebony body color, the following result was obtained:

Gray pink × Ebony red → gray red →

gray red	3764
gray pink	1369
ebony red	1112
ebony pink	0

This count includes a few F_3 flies derived from F_2 red grays mated together, but also includes large numbers of F_2's from each of the reciprocal crosses—ebony ♀ × pink ♂ and pink ♀ × ebony ♂. In the later generations from these crosses the writer was able to obtain a few ebony pink flies, and a pure stock of this combination was then made up. When these flies were used for "coupling" experiments the result was as follows:

Ebony pink × Gray red → gray red →

gray red	272
gray pink	12
ebony red	10
ebony pink	65

The result indicates that the two genes involved are really linked in the ordinary sense. In order to get more exact data doubly heterozygous flies have been crossed to double recessives (ebony pink). Owing to the difficulty of breeding ebony females it has not yet been possible to make any extensive tests of the amount of crossing over in males, but the 88 flies so far raised from doubly heterozygous males by ebony pink females have none of them been cross-overs. Similar tests of doubly heterozygous females have disclosed the existence of much variability in the amount of crossing over, the extreme values so far obtained being about 1 percent, on the one hand, and about 27 percent, on the other. This matter is being further investigated, and it seems probable, from results already obtained, that external conditions play a large part of it. The detailed counts are reserved until this side of the question is more thoroughly worked out, but the totals are given below, because they show beyond question that linkage occurs. When females that showed "coupling" (in Bateson's ('06) sense)—*i. e.,* in which red and gray, pink and ebony were linked—were tested by mating to ebony pink males, the result was as follows:

Cross		Offspring	Number
Gray red (heterozygous) ♀	↘	gray red	880
×		gray pink	86
Ebony pink ♂	↗	ebony red	65
		ebony pink	691

Tests of females showing Batesonian "repulsion" (having red linked with ebony, pink with gray—Bateson and Punnett, '11) gave the following result:

Cross		Offspring	Number
Gray red (heterozygous) ♀	↘	gray red	3
×		gray pink	601
Ebony pink ♂	↗	ebony red	584
		ebony pink	4

This gives a total of 158 cross-overs out of 2,914 flies, but the absolute proportion is not very significant, because of the variability mentioned above. That rather strong linkage exists, however, there can be no doubt.

Author's Note, 1961

The variability in linkage that was found here was later found to be due to heterogeneity for an inversion, brought in by the ebony stock.

4

EXPERIMENTS ON SEX RECOGNITION AND THE PROBLEM OF SEXUAL SELECTION IN *DROSOPHILA* †

Much has been written on the subject of sexual selection since Darwin first developed the theory, and many remarkable observations have been recorded. There has, however, been very little experimental work in this field. Darwin and those who have followed him have obtained much of their evidence from the insects, and within this group some of the most striking cases of elaborate mating habits have been reported in the Diptera, and here too there is to be found a most remarkable array of secondary sexual characters. Perhaps the most extreme case is that of the *Elaphomyia* described by Wallace, in which the male has long, hornlike processes arising from his head, which are absent in the female. The families of Platypezidae, Dolichopodidae, and Empididae are especially rich in secondary sexual characters, which occur in the legs, wings, antennae, face, or other parts. In the two latter families some very curious observations on mating habits have been recorded (see especially Poulton's account ('13) of Hamm's work on Empididae).[1]

Some of the best experimental evidence in favor of sexual selection is that obtained by Lutz (1911), who worked with an abnormal wing venation in *Drosophila ampelophila* which he found to be strongly selected against. To Dr. Lutz is due the suggestion that the mutants in this fly obtained by Morgan would form excellent material for the study of the problem. The method of some of the experiments was also suggested by Lutz. I took up the matter at the suggestion of Professor Morgan, to whom I am greatly indebted for most of the material and for his encouragement and criticisms. Several discussions of the matter with Dr. Lutz have also been very helpful. Some of the work was done at the Cold Spring Harbor Laboratory in the summer of 1911, and I wish to express my appreciation of the interest in the experiment shown by Dr. C. B. Davenport at that time.

† Reprinted from The Journal of Animal Behavior, Vol. 5, 1915, pp. 351–366.

[1] I have myself observed courtship in a few Dolichopodidae and in a number of other Diptera. These observations will be published in full later.

Courtship and Mating

Most of my observations on courtship in *Drosophila* have been made upon *D. ampelophila* Loew. In general the process is very similiar in *D. busckii* Coq., but differs in several respects in *D. amoena* Loew, *D. repleta* Woll., and *D. funebris* Fabr. The first and most noticeable act in courtship occurs when the male, being near the female, extends one wing at about right angles to his body, and vibrates it for a few seconds. The wing is then returned to the normal position and the process is repeated, usually with the other wing. But between times there is a scissors-like movement of the wings repeated several times. This vibrating of the wings is often repeated many times, and may be done in any position relative to the female, though the male always faces her. Usually, in fact, he swings quickly around her in a semicircle once or oftener during the process. Soon the male begins to protrude his genitalia and, if the female remains quiet, to lick her posterior end. Some white matter now protrudes from her ovipositor, and other males in the same vial are usually observed to become excited now and begin courting; indicating odor as a cause of sexual excitement. If the female runs or flies away the male is excited, moves his wings jerkily, and walks around rapidly, but seems unable to follow the female accurately or to locate her quickly. The penis is directed forward by bending up the abdomen underneath, towards the thorax, and is jerked toward the female (the male always standing facing her at this stage), but not always toward her genitalia, as I have seen it strike her in the eye.[2] If it does strike the mark the male mounts on the female's back, between her wings. Mounting never takes place until after the actual copulation has occurred, in which respect *Drosophila* differs from some related flies (e.g., Muscidae, Anthomyidae, Sepsidae, Borboridae, and Ephydridae, so far as my observations go). In these forms the male flies and lights on the female, after which copulation may or may not take place, probably depending upon the way the female responds.

Berlese ('02) and Hewett ('08) find that in the house fly the final step in copulation is taken by the female, which inserts the ovipositor into the genital opening of the male. I have not been able to verify this for *Drosophila,* but it is probably true here also. In *Drosophila,* as in some other related flies (I have examined a few Anthomyids and *Sepsis violacea*), the ovipositor enters the male opening, instead of the penis entering the female duct. But I cannot state positively that the female inserts it instead of the male drawing it in by means of his genital armature. In any case, it is certain that the female is not entirely a passive agent. The time required for copulation to take place depends largely upon whether she stands quietly and allows the male to pair, or moves away when he begins to court. In the latter case

[2] The male in this case, however, had white eyes, and so was perhaps blind. Normally the aim is accurate.

very active males have been seen to pair while the female was walking away, but this is exceptional. Occasionally a female seems to frighten off a male by spreading her wings and moving quickly toward him.[3] When this happens he moves off, and does not so far as I have seen, then pair with that female, although she has been known to pair with another male within a few minutes afterward. That this is really a threat on the part of the female seems likely from observations of fighting between males. If two males are courting the same female they often grow very excited, especially if she is unwilling to stay quiet. In such cases they may sometimes be seen to spread their wings, run at each other, and apparently butt heads. One of them soon gives up and runs away. If the other then runs at him again within the next few minutes he usually makes off without showing fight.

The time occupied by the process of courtship varies greatly with the age and condition of the flies and with the temperature. Copulation may occur within a few seconds after the flies are put together, with little preliminary courtship. While experimenting with flies about 3 or 4 days old, which had never been allowed to pair, I have found that 20 minutes or a little less is about the average length of time before copulation occurs. The flies may remain *in copula* for only a few seconds, but so far as my observation goes this occurs only after a prolonged courtship in which the female has seemed unwilling to mate, and is, I think, due to the male not getting a good hold. I do not know whether or not such pairings are successful. Ordinarily copulation lasts about 20 minutes.[4]

Experiments on Other Insects

There is a considerable body of evidence relating to the question of sex recognition in the Arthropods. A short review of the subject and a bibliography are given by Chidester ('11), so I shall confine myself here to the evidence dealing with insects. There is evidence from several groups of insects, but most of it points the same way. Sex recognition at a distance is by smell, but the actual process of copulation depends upon the sense of touch.

In the Orthoptera, Stockard ('08) has observed the male walking stick to pair with the detached abdomen of a female which was fastened to a stick with wires for legs.

In Coleoptera, Féré ('98) finds that male cockchafers do not pair if the antennae are removed. Males will sometimes pair with males, provided the latter have just paired with females, or have been artifically impregnated with female juices.

[3] Compare Howard's ('02, pp. 141 and 145) accounts of Asilid and Empidid flies eating males which were courting them.

[4] In 20 cases the duration of copulation was timed with the following result (minutes): ½, 5, 16, 17, 18, 18, 19, 20, 21, 21, 21, 21, 22, 22, 23, 24, 25, 26, 27, 33.

Mast ('12) shows that in the firefly the males find the females by means of the flashing lights, signals being made by both sexes. Tower ('06) reports that in *Leptinotarsa* males normally never try to actually copulate with males, but that if the antennae are removed or painted with shellac they will try to pair with any individual they happen to touch. If this individual is a female pairing will occur. If the abdomens of females are removed the males are attracted by them, though not by the wings, head and thorax. Females with their abdomens coated with shellac are not attarctive to the males.

Féré ('98a) found that in the silkworm moth males will pair with other males, which have not been given a chance to get the female odor, if the latter males are sluggish, as after removal of the antennae. Males without antennae will copulate. Kellogg ('07) has since reported that males without antennae find the females only by chance, while normal males go straight to them. If one antenna be removed the male travels in a circular path instead of going toward the female. If a male without antennae happens to touch a female he immediately shows strong sexual excitement, such as normal males show when brought near females. Scent glands on the abdomen of the female were shown to be the seat of the olfactory attraction, for when the abdomens were removed males were attracted by them and not by the rest of the female. Blackening the eyes produced no change in the behavior.

Mayer (1900) performed numerous experiments on the moth *Callosamia Promethea.* He showed that it is odor that attracts the males, and that this odor comes from the abdomen of the female. Interchanging of wings indicated that the marked sexual dimorphism in color, which occurs in this species, has no selective value.

Kirkland (1896)[5] showed that in the Gypsy moth odor is again the main element concerned, but the wings of the female, as well as her abdomen, have an exciting odor. Mayer and Soule (1906) tried experiments on this form, which they supposed indicated that normal females discriminated against males without wings. I find, by applying Yule's (1911) formula for the standard error of the difference, that the difference between the per cent of times winged males paired without resistance and the per cent that wingless males paired without resistance (the measure of sexual selection used by these authors) is almost exactly three times the standard error. This means that the result is not conclusive. Mayer and Soule blinded females and found this apparent discrimination against wingless males to disappear, but the per cent of resistance dropped, as a whole, from 46% in the normal females to 27% in the blinded ones. This may mean that the blinded ones were too greatly disturbed by the blinding to pay much attention to what male mated with them. While I am inclined to suspect that there is some odor connected with the male's wings, still, as stated above, it is not certain that any effect at all is produced by removing the wings of the male. Further evidence against

[5] I have not seen this paper, but make the statement on the authority of Mayer and Soule (1906).

the importance of sight is furnished by the fact that females did not discriminate against males with wings painted in unusual colors.

Federley (1911) gives evidence indicating that odor is a strong sexual stimulator in the moth *Pygaera,* and may even cause males to attempt to pair with the bars of the cage in which they are confined. Entomological literature contains many accounts of the remarkable ability of male moths to locate females through smell. A few of the more striking cases are given by Washburn ('09, pp. 87 and 88.)

Sex Recognition in Drosophila

It was noted above that when an egg protrudes from the ovipositor of a female any males in the same bottle usually become excited and begin courting. When a female is killed eggs are likely to protrude, and males are often seen vigorously courting females which have been freshly killed, even though the killing agent was so strong smelling a substance as ether. In such cases it is the posterior end of the abdomen which seems to be the chief focus of attraction. The attraction may last for at least 30 minutes after death, and probably longer. I have not as yet been able to cause males to copulate with dead females, probably because the female normally takes an active part, as indicated above.

If a male which has just paired is placed with a male which has been isolated from females for several days he may be courted by that male, though this is not frequent. If two males are kept in solitary confinement for four or five days and are then put together they often court each other, in one case even the doubling up of the abdomen being seen, and the wing movement (*not* fighting) being almost invariably observed. Such mutual courtship between males has not been seen under other conditions, but a dead male which had not been with females for two days before killing has been seen to be courted by other males. This courting, however, seems more likely to occur when juices from female abdomens are put on the dead males. In *Psychoda* sp. courtship of males by males seems to be very frequent, and often copulation is attempted. The genitalia even become attached, and stay so for several seconds. Males have been seen to mount males in *Fucellia marina, Sepsis violacea,* and *Sarcophaga* sp.

In order to test what part is played by the female juices mentioned above I placed some on bits of filter paper and put these in the same vial with males. The males paid no attention to the paper, and showed no signs of excitement, though that they were sexually ripe was shown by the fact that they courted and paired with females when they were put in with them. This experiment has been repeated several times with the same result. When a female is thoroughly crushed and the remains heaped up in a little pile the males will sometimes court slightly, but as a general rule, the more mashed the female is, the less excited the males become. No great importance is to be placed upon this latter experiment, as the result is complicated by the presence of so many other body juices. This objection, however, will hardly hold in the case of the filter paper impregnated with female juices. The objection that not enough odor was present may perhaps be justifiable in that case. But neither of these objections would seem to apply to the following experiment. Two females were placed in a small, dark, cloth bag and this was put in a vial with a few males. The males were close to the females, which they could not see or touch, but should be able to smell. This experiment was done three times and in no case did the males show any signs of sexual excitement, though in all three cases they did court immediately afterwards when given females in the usual way.

It has been shown by Barrows ('07) that the sense organs for smell in *Drosophila,* in so far as one may judge from reactions to food substances, are located in the terminal antennal joints. For this reason I was led to perform the following experiments, in an effort to determine the part played by the sense of smell in courtship and copulation.

The antennae were removed from several males,[6] which were after several days placed with virgin females. Such males are very sluggish, and it was therefore not surprising that no courtship was observed. However, one such male was found copulating, and at least three of them left offspring after the operation. One of the three had white eyes, and was therefore probably also blind. Courtship has been observed in two males from which the antennal aristae had been removed. A normal male has been seen copulating with a female from which the antennae had been removed, and this pairing resulted in the production of offspring.

The above experiments failed to give any evidence demonstrating that smell alone can cause males to show signs of courtship. Another series of experiments, now to be described, has indicated, however, that smell may be a secondary factor in causing sexual excitement. Males and virgin females were isolated for three or four days and were then placed, in pairs, in clean vials. The length of time before copulation occurred was recorded; and, after it had taken place, the flies were removed and a new pair was placed in the same vial. The only difference between the two sets was that one lot was in clean vials, the other was in vials in which copulation had just occurred. In these experiments an equal number of each sort was done on each day (placing a third pair in some vials when necessary in order to get an equal number), and all flies used on a given day were as nearly the same age and size as practicable, and were from the same culture. Table I gives the result of the experiments. It shows that in the first five or six minutes the flies in the used vials are more likely to copulate than are those in the clean vials; between ten and twenty minutes there are more copulations in the

[6] I have found that antennae may easily be removed without using the complex method described by Barrows, if the operation be performed on very young flies.

Table I

Minutes before Copulation	Number of times observed	
	1st pair	2nd pair
1– 3	13	22
4– 6	12	22
7– 9	5	7
10–12	7	6
13–15	11	2
16–18	7	5
19–21	8	5
22–24	4	3
25–27	5	3
28–30	2	1
31–33	7	1
34–36	2	2
37–39	0	2
40 and more, including failures.	24	25

clean vials, and after that the two series are parallel. Apparently there are a certain number of pairs that are nearly ready to mate, and these will mate more quickly if another copulation has just occurred in the same vial. But if much courtship is to be required before copulation, then the effect of smell is negligible. It is not clear from these results whether the effect is produced upon the males, the females, or both.

In order to test whether sight is the sense which stimulates the male I carried out the following experiment. Two vials having the same bore were filled with cotton up to about three centimeters from the mouths, and then placed with their mouths together. A thin glass cover-slip was then placed between them, and a male fly was placed in one vial, a female in the other. These flies had been isolated for several days, and were sexually ripe. The male could see the female, but could not touch or smell her. Once or twice they met "head on" with only the cover-slip between them, but the male showed no signs of recognition. After thirty minutes the female was let in the vial containing the male; and the vials were left in position. He courted her within two minutes, and paired in five minutes. The experiment was repeated with the same result, except that copulation now occurred in three minutes after putting the flies together.

It is also evident that sight is not necessary for pairing from the facts that Drosophila breeds freely in the dark (see Payne, '11), and that pure stock of white-eyed flies (which may be blind) has been kept for many generations.

In *Fucellia marina* and in *Musca domestica* males seem to see their mates from some distance and fly directly to them, lighting on the back of the mate. Here sight would seem to play considerable part, and in *Fucellia* still more convincing evidence was obtained from observation of violent courtship of a fly which was separated by thick glass from the courter. Not only did this male court the other fly, but when she walked around rather rapidly, he very accurately followed her on the other side of the glass. This has been observed twice.

Since the wings play such a conspicuous part in courtship I was led to try the effect of cutting them off. Two males of the same age were used, the wings of one being cut off at the base. These were kept in the same vial for several days, and then placed with a virgin female several days old. The vial was then watched until copulation took place. This has been done 125 times, using different individuals each time. The normal male paired 72 times, the clipped male 53 times. As in all the similar experiments here reported care was taken to have the competing flies as nearly the same age and size as possible. The result indicates that there is very slight, if any, selection against the clipped males. It seemed possible that courtship made the female ready to copulate, but that she would then mate with either male. To test this hypothesis another series of experiments was carried out. A single pair of flies was placed in each vial, using each day an equal number of normal and of clipped males. The length of time before copulation was recorded in each case. A preliminary account of this experiment has been published by Morgan ('13); but the data used there have been discarded, since two serious sources of error had not then been recognized. It was not realized that the time before copulation might be influenced by a previous copulation having occurred in the same vial; and sufficient precautions against drying were not taken—a very important factor. A new series of experiments, in which these points were controlled, gave the results shown in table II. As a matter of fact, however, these data are very similar to the discarded series.

Table II

Minutes before Copulation	Number of times observed	
	Normal ♂	Clipped ♂
0– 3	15	4
4– 6	19	9
7– 9	10	7
10–12	15	5
13–15	3	7
16–18	8	4
19–21	3	10
22–24	3	4
25–27	9	4
28–30	2	2
31–33	2	2
34–36	0	4
37–39	1	2
40–42	3	2
43–45	2	1
46 and over, including failures.	43	73

This table seems to justify the suspicion that led to the experiment. Had the females discriminated against the clipped males to the extent shown above when both kinds were present, the normal males would certainly have appeared at a greater advantage. That the result was not due to less activity on the part of the clipped males is indicated both by the contests described above and by the following observations. In some of the experiments recorded in Table III the number of minutes before courtship began

Table III

Minutes before Courtship	Normal	Clipped
1– 2	12	9
3– 4	2	3
5– 6	1	2
7– 8	1	2
9–10	0	1
11–12	1	0
13 and over	7	7

was observed. Table III indicates that the clipped males began courting as quickly as did the normals.

From these experiments it seems certain that the wings are of value in courtship; but the effect probably is to produce sexual excitement in the female, rather than to cause her to select a male that uses his wings. No "choice" is involved; but, as pointed out by Watson ('14, p. 173), the effect, in nature, would be strongly in favor of the normal male.

It seems probable that touch is of considerable importance in the sexual process, and all my observations are consistent with that view, but I have no direct experimental evidence to that effect. It is not possible to get evidence from *Drosophila* such as Kellogg obtained from *Bombyx,* because the flies are more active, and less easily sexually excited, than is the silkworm moth.

Two kinds of experiments have been carried out in an effort to find out what part of the body is responsible for causing sexual excitement. A female without an abdomen, but alive and active, was placed with males that had been isolated from other females for four days. She was vigorously courted.

Three gynandromorphs have been tested to determine their sexual behavior. None showed any certain indications of male behavior, but all were vigorously courted by males. Of these three gynandromorphs the external characters were as follows: (A) All female, except one side of the head, which was male; (B) female on one side of the whole body, male on the other side; (C) female, except the genitalia, which were male. It is doubtful what conclusion, if any, is to be drawn from these few observations.

Sexual Selection—Artificial Abnormalities

The male of *Drosophila ampelophila* bears a small comb on his front metatarsi, a secondary sexual character not found in most species of the genus.[7] Lutz ('11) has shown that the removal of this comb has no effect upon the availability of males for copulation. That is, the females were not influenced by the sex-comb in their choice of mates.

It sometimes seems to be difficult for the male to get the wings of the female out of his way so that he can mount her. For this reason I carried out an experiment the converse of one recorded above; using two virgin females of the same age, one with normal wings, the other with wings removed. These were put with a normal male. In 52 successful trials the normal female was paired with 25 times, the clipped one 27. Again it would seem as though clipping the wings has very little if any effect.

Sexual Selection—Mutants

Lutz ('11) found that certain slight abnormalities in wing venation were selected against both by normal and abnormal flies of both sexes. The abnormal flies were not noticeably different from the normals in behavior, so that it seems quite unlikely that the results were due to a difference in the activity of the two types. Moreover, if, let us say, the normal male was more active than the abnormal, so that he would be more likely to pair first, it would seem that the normal female would not be so easily paired with as the abnormal; or *vice versa,* if we suppose the abnormal female to be less active and therefore more likely to be paired with, it is hard to see why the abnormal male should be at a disadvantage. Further evidence bearing out this view of the effect of differences in activity will be given below. Lutz suggested that it seems unlikely that sight could have any influence, and that perhaps there is some unpleasant smell correlated with the abnormality, this being the basis of selection.

I have conducted a series of experiments upon some of Morgan's *Drosophila* mutants, in an effort to find if there was any selective mating in connection with them. The following mutants were used:

(1) White eyes. This form was first described by Morgan ('10). There is no color in the eye. This probably means that the fly is blind, and it has often been observed to be less strongly phototactic than the wild fly. White-eyed flies are also less active and vigorous than normals.

(2) Yellow body color. Described by Morgan ('11). The whole body of

[7] So far as my observation goes it is present only in *D. ampelophila, D. confusa, D. obscura,* and three or four undescribed species.

this fly is lighter than that of the normals, and there is a distinct yellow color to the wings. Sight seems to be normal, but again the flies are not so active as are the normals.

(3) Curved wings. The main interest of this form in this connection is that the wings are always held extended, in somewhat the same position as that of the courting males. The flies are perhaps a little less active than the normals.

(4) Vermilion eyes. Described first by Morgan ('11 a). These flies differ from the normals in that the eyes are of a brighter and less intense red. Their vigor and activity is very little, if at all, inferior to that of the normals.

Some of the flies used had yellow bodies and white eyes, and in one experiment a few vermilion eyed yellow black colored females were used. This latter combination was at the time considered one of the weakest stocks in the laboratory, and was used for that reason.

The method of the experiment was as described above for the wing clipping experiment. For instance, a white eyed female would be given her choice between a red male and a white one of the same age and size. Then the experimenter simply watched until pairing was seen, and the flies were then thrown away. The following table (IV) gives the results obtained.

Table IV

"Chooser"	"Chosen"	Number of cases
Red vs. white eyes. (Normal body color.)		
Red ♂	Red ♀	54
	White ♀	82
White ♂	Red ♀	40
	White ♀	93
Red ♀	Red ♂	53
	White ♂	14
White ♀	Red ♂	62
	White ♂	19
Gray (normal) vs. yellow body color. (Red eyes.)		
Gray ♂	Gray ♀	25
	Yellow ♀	31
Yellow ♂	Gray ♀	12
	Yellow ♀	30
Gray ♀	Gray ♂	60
	Yellow ♂	12
Yellow ♀	Gray ♂	25
	Yellow ♂	8
Vermilion Black-yellow ♀	Red-gray ♂	13
	White-gray ♂	1
Gray and yellow body colors. (White eyes.)		
Gray ♂	Gray ♀	11
	Yellow ♀	4

Table IV (continued)

Gray ♀	Gray ♂	21
	Yellow ♂	3
Red and white eyes. (Yellow body color.)		
Red ♂	Red ♀	3
	White ♀	4
White ♂	Red ♀	9
	White ♀	9
Red ♀	Red ♂	9
	White ♂	2
White ♀	Red ♂	21
	White ♂	1
Red and vermilion eyes. (Gray body color.)		
Red ♂	Red ♀	7
	Vermilion ♀	5
Vermilion ♂	Red ♀	4
	Vermilion ♀	4
Red ♀	Red ♂	11
	Vermilion ♂	14
Long and curved wings. (Other characters normal.)		
Long ♂	Long ♀	10
	Curved ♀	13
Curved ♂	Long ♀	5
	Curved ♀	3
Long ♀	Long ♂	14
	Curved ♂	4
Curved ♀	Long ♂	9
	Curved ♂	5

It appears that sexual selection is not involved in any of these cases. The impression gained from observation of over 1,000 "contests" of this sort is that the outcome is not a matter of choice. A female, in the great majority of cases, seems to allow the first active amorous male that comes along to pair with her; or, if she is disinclined to mate, resists all males apparently indifferently. When a male is sexually excited he pairs with the first female he finds which will allow him to. As a result, the more active and vigorous males are likely to win their contests, and the greater the difference in vigor, the greater the proportion of times the better male wins. This is just what the results show to happen. So far as the small numbers show, the vermilion male is at no disadvantage and the curved male is not quite able to hold his own, while it is certain that the yellow and white are far behind their normal opponents.

In the converse case, a less vigorous female will be less likely to resist or escape from the male successfully if she be disinclined to mate. But this influence should have less effect on the result than the one discussed above, since both females will often be willing to mate and will not try to escape. This again agrees exactly with the facts as given above.

The two results of unequal vigor discussed above seem to me quite adequate to explain all the results obtained. There is no evidence of any "choice" on the part either of males or of females. Unlike the abnormal venation studied by Lutz, these mutants probably have no significance from the point of view of sexual selection, in the narrower sense of that term.

Summary

Experiments indicate that sight is not essential in sex recognition in *Drosophlia.* The olfactory and tactile senses are probably both concerned, as in most other insects.

The wings of the male play a conspicuous part in normal courtship. Experiments with males from which the wings had been removed indicated that the function of these organs in courtship is the production of sexual excitement in the female.

Experiments were carried out with four mutants (white eyes, vermilion eyes, yellow body color, and curved wings), involving the observation of 839 contested matings.

As a result of these experiments it seems probable that the four characters involved have no selective value, except in so far as results from the fact that at least two of these classes are less active than normals.

In general it is probable that, in *Drosophila,* neither sex exercises any "choice" in the selection of a mate. A female that is ready to mate will accept any male, and a male that is ready to mate will do so with the first female that will allow him to.

LITERATURE CITED

BARROWS, W. M., 1907: The reactions of the pomace fly, Drosophila ampelophila, to odorous substances. *Jour. Exper. Zool.,* IV.

BERLESE, A., 1902: L'accoppiamento della Mosca domestica. *Rev. Patol. Veg.,* IX.

CHIDESTER, F. E., 1911: The mating habits of four species of the Brachyura. *Biol. Bull.,* XXI.

DARWIN, C., 1874: The descent of man and selection in relation to sex. 2nd edition.

FEDERLEY, H., 1911: Vererbungsstudien an der Lepidopteren-gattung Pygaera. *Arch. Rass. u. Ges.-Biol.,* VIII.

FÉRÉ, C., 1898: Expériences relatives aux rapports homosexuels chez les hammetons. *C. R. Soc. Biol.,* V.

1898 a: Expériences relatives a l'instinct sexuel chez le bombyx du murier. *C. R. Soc. Biol.,* V.

HEWETT, C. G., 1908: Structure, development, and bionomics of the housefly. *Quart. Jour. Micr. Sci.,* LII.

HOWARD, L. O., 1902: The insect book, New York.
KELLOGG, V. L., 1907: Some silkworm moth reflexes. *Biol. Bull.,* XII.
KIRKLAND, G. H., 1896: Assembling of the gypsy moth. Gypsy Moth Report, Mass. Board. Agr.
LUTZ, F. E., 1911: Experiments with Drosophila ampelophila concerning evolution. *Carnegie Inst. Wash. publ.* 143.
MAST, S. O., 1912: Behavior of fire-flies (Photinus pyralis) with special reference to the problem of orientation. *Jour. Animal Behav.,* II.
MAYER, A. G., 1900: On the mating instinct in moths. *Ann. Mag. Nat.* H., V.
MAYER, A. G. and C. G. SOULE. 1906: Some reactions of caterpillars and moths. *Jour. Exper. Zool.,* III.
MORGAN, T. H., 1910: Sex-limited inheritance in Drosophila. *Science,* XXXII.
1911: The origin of nine wing-mutations in Drosophila. *Science,* XXXIII.
1911 a: The origin of five mutations in eye-color in Drosophila, etc. *Science,* XXXIII.
1913: Heredity and sex. New York.
PAYNE, F., 1911: Drosophila ampelophila bred in the dark for sixty-nine generations. *Biol. Bull.,* XXI.
POULTON, E. B., 1913: Empidae and their prey in relation to courtship. *Entom. Mag.,* XXIV.
STOCKARD, C. R., 1908: Habits, reactions and mating instincts of the walking stick, Aplopus mayeri. *Carnegie Inst. Wash. publ.* 103.
TOWER, W. L., 1906: An investigation of evolution in Chrysomelid beetles of the genus Leptinotarsa. *Carnegie Inst. Wash. publ.* 48.
WASHBURN, M. F., 1909: The animal mind. New York.
WATSON, J. B., 1914: Behavior. New York.
YULE, G. U., 1911: An introduction to the theory of statistics. London.

Author's Note, 1961

The three species referred to in footnote 7 are all misnamed. They are now known as *D. melanogaster, D. affinis,* and *D. pseudoobscura.* Later studies have shown that sex-combs are present in numerous representatives of the *melanogaster* and *obscura* species groups.

5

GENETIC FACTORS AFFECTING THE STRENGTH OF LINKAGE IN *DROSOPHILA* †

In September, 1913, a wild female Drosophila of a stock from Liverpool, Nova Scotia, was crossed to a male bearing the second chromosome mutant characters vestigial and speck. A single daughter of this mating was tested, and gave no crossovers among 99 offspring, though vestigial and speck usually show about 37% crossing over. This strain has since been bred in very large numbers, and the experiments are being continued; but it has seemed advisable to report briefly on some of the results obtained.[1]

It has become clear that the original result was due to something in the second chromosome derived from the Nova Scotia female. Two of her granddaughters and all of her later descendants that were known, from linkage, to have received the second chromosome in question gave similar results; but those descendants known not to have received it gave the 'usual' result. This was true whether the chromosome came from the mother (as above) or from the father. And when other second chromosome genes were substituted for vestigial and speck, unusual ratios still appeared; but for any combination the result was relatively constant, and the combined 'unusual' results give as consistent a scheme as the 'usual' results. The first column of table 1 shows the percentage of crossing over in the ordinary females for various combinations of second chromosome genes.[2] The second column gives similar data for tests of females that carry a Nova Scotia chromosome. Figures 1 and 2 are chromosome maps constructed on the basis of these two columns.

A large number of these experiments have been made with females having one Nova Scotia chromosome, and its mate bearing the mutant genes black, purple, and curved. Numerous tests have been made in order to determine the nature of the crossovers produced. It has been found that those crossovers that receive the part of the original Nova Scotia chromosome lying to

† Reprinted from the Proceedings of the National Academy of Sciences, Vol. 3, 1917, pp. 555–558.

[1] Some of the early results were reported at the 1913 meeting of the American Naturalists, and brief references have been published by me (1915) and by Muller (1916).

[2] Much of this data has not hitherto been published. It has been collected mainly by Dr. C. B. Bridges, to whom I am indebted for permission to use it.

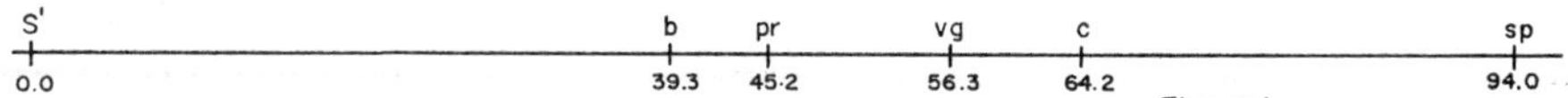

Figure 1

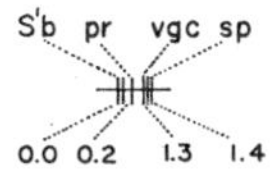

Figure 2

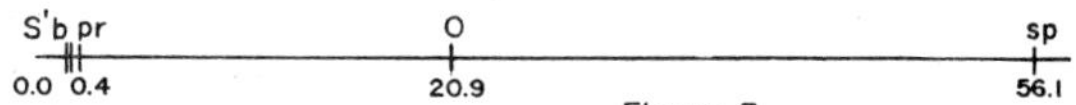

Figure 3

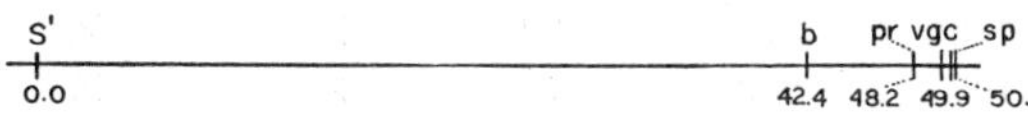

Figure 4

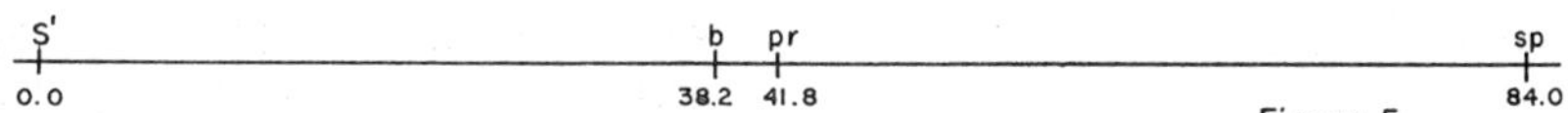

Figure 5

Figure 6

the left of the purple locus still give unusual results and transmit the peculiarity to all of their descendants that receive this piece. The results produced by such females are not, however, the same as those produced when all the Nova Scotia chromosome is present, as is shown in column 3 of table 1 and in figure 3. When the other piece of the Nova Scotia chromosome, lying to the right of purple (and, in many experiments, that part of it that is also to the left of speck) is tested, a still different result is obtained, as shown in column 4 and figure 4.

It follows from these results that the original Nova Scotia chromosome contained at least two factors causing reduced crossing over, and each affecting chiefly the region in which it lies. We may call these two factors 'C_{IIl}' and 'C_{IIr}' (C for crossing over, after Muller,[3] II to distinguish them from similar factors affecting other chromosomes, *l* and *r* for left hand and right hand).

By appropriate matings it has been possible to obtain females homozygous for C_{IIr}, and a surprising result has been observed: the effect produced by C_{IIr} when heterozygous disappears when it becomes homozygous (see columns 5 and 6, figures 5 and 6). There can be no doubt of this result, as it has been obtained repeatedly, and has often been checked by tests of the offspring. It has also been paralleled more recently by results obtained by Muller[3] (1916) and by me[4] with C_{III}. Similar tests of homozygous C_{IIl} have

---

[3] Muller, H. J., *Amer. Nat.,* Lancaster, Pa., **50,** 1916, (193, 284, 350, 421).

[4] Sturtevant, A. H., *Zs. Abst. Vererb.,* **13,** 1915 (234).

Table 1

GENES	PERCENTAGE OF CROSSING OVER					
	Normal	C_{III} C_{IIr} / —	C_{III} / —	C_{IIr} / —	C_{IIr} / C_{IIr}	C_{III} C_{IIr} / C_{IIr}
Star black	39.3	0.0	0.0	42.4	38.2	0.3*
Star purple	40.2	0.0	×	45.4	40.6	0.3*
Star curved	46.5	×	25.9	47.3	×	×
Star speck	48.3	0.4	49.0	46.0	49.0	47.4
Black purple	5.9	0.2	0.4	5.8	3.6	0.1
Black vestigial	16.9	1.2	×	8.9	×	×
Black curved	22.7	1.3	25.6	7.5	×	×
Black speck	49.3	1.1	48.8	9.3	43.5	47.0
Purple vestigial	11.1	1.2	×	0.6	×	×
Purple curved	19.0	1.1	20.5	1.7	×	×
Purple speck	46.5	1.2	47.4	2.8	42.2	47.4
Vestigial speck	36.7	0.2	×	0.0	×	×
Curved speck	29.8	0.1	35.2	0.2	×	×
†Total length	94.0	1.4	56.1	50.1	84.0	47.8

* These few (6) crossovers are doubtful. None of them were tested; and there is apt to be a small percentage of error in classifying Star flies.

† Calculated as Star black + black purple + purple curved + curved speck in all but the last two columns, where purple speck is used. In no case further corrected for double crossing over.

not been possible since no chromosome has yet been obtained that contains both that factor and any mutant factor within its "sphere of influence."

Backcross tests have shown that males of all the above types with respect to C_{III} and C_{IIr} resemble "normal" males in giving no crossovers at all.

Included in the above tables are a considerable number of data involving three or more loci at once; and these agree with the table in showing clearly that the linear order of the factors established for the usual second chromosome is unchanged by the factors under discussion. The amount of crossing over is altered, often markedly, and not usually proportionately in different regions; but the factors keep their same sequence. This result serves to emphasize the importance of considering the distances on chromosome maps as only diagrammatic, not as representing actually proportionate distances between the genes, although actual distance is evidently an important factor influencing the end result. It does not, I think, in any way weaken the case for the chromosome hypothesis, but merely shows, together with the results of Bridges[5] (1915) and Plough[6] (1917), that any chromosome map is available for purposes of numerical prediction only when the conditions under which it was made are duplicated.

[5] Bridges, C. B., *J. Exp. Zool.*, Wistar Inst. Philadelphia, **19,** 1915 (1).

[6] Plough, H. H., these PROCEEDINGS, **3,** 1917 (553–555).

Author's Note, 1961

These two "factors," C_{IIl} and C_{IIr}, are now known to be inversions, currently named In(2L)NS and In(2R)NS. The argument in the last paragraph needs revision accordingly (see paper no. 15).

6

A PARALLEL MUTATION IN *DROSOPHILA FUNEBRIS* †

A mutant of *Drosophila funebris* Fabr. has recently appeared that is so strikingly similar to a well-known mutant of *D. melanogaster* Meig. (*ampelophila* Loew) that there can be little doubt that the same mutation has occurred independently in the two species. The new form, called notch, agrees with the notch *melanogaster* in at least eight different respects, as will appear below.

Origin.—A female *funebris* of a stock from Mitchell, S. D., was mated to a male of a stock from New York City. The descendants were mated in pairs for several generations, and no variations were observed except an occasional fly with one of the anterior scutellar bristles missing. Such flies were found also in the uncrossed New York stock. In the line under consideration selection was carried on, in an attempt to increase the percentage of such flies, but no marked result was obtained. In F_5 one pair (5201) produced 35 normal females, 34 notch females, and 36 normal males. The sex ratio here is significant, since an excess of males is more frequent than an excess of females in this species. The pair from which the parents of 5201 came produced 19 females and 31 males, which is not an unusual excess when complete counts are not obtained. In *D. funebris* the males usually emerge in a little less time than the females. This relation is just the reverse of that found in *D. melanogaster*. Evidently the female parent of 5201 was genetically notch. She was not observed to be abnormal, and had been destroyed when her offspring began to emerge. It seems probable that she did not have notched wings, but she may well have had the characteristic veins and acrostichal hairs, since these would more easily have been overlooked.

Description.—Notch *melanogaster* is characterized by having the wings somewhat nicked, more especially at the apical posterior corner. But this character is somewhat variable, being often unlike in the two wings of the same female, and sometimes even entirely absent.[1]

† Reprinted from Science, Vol. 48, 1918, pp. 72–73.

[1] See Morgan, 1917, "The Theory of the Gene," *Amer. Nat.*, 51, for figure and a discussion of this variability.

In addition the eyes are often smaller than those of the wild-type flies and somewhat roughened.[2]

Furthermore the veins of notch are somewhat thickened, more especially the apical portions of the second and fifth longitudinal veins. This character is the most invariable and convenient index of the presence of the notch gene. The anterior scutellar bristles of notch are often doubled. The acrostichal hairs are more numerous than those of the wild-type fly, and are irregularly arranged, instead of being in eight fairly definite rows.[3] The notch gene thus produces an unusually large number of morphological peculiarities.

Notch *funebris* agrees in all of the above respects. The wings are nicked in the same way, but are often asymmetrical and sometimes normal; the eyes are often small and roughened; the wing veins are thickened even more than those of notch *melanogaster,* the second and fifth being affected most, and this character being again the most convenient and reliable for purposes of classification; the anterior scutellar bristles are often doubled, in spite of the fact that notch arose in a family selected for the absence of these bristles; the acrostichal hairs are irregularly arranged, but differ from those of notch *melanogaster* in being entirely wanting on each side in a narrow band just inside the dorsocentral row.

The unusual features of notch in *melanogaster* are not limited to its morphological nature. Notch is one of the few dominant mutant genes, and in addition is sex-linked and has a recessive lethal effect. The result is that a notch female gives equal numbers of wild-type and notch daughters and of wild-type sons. Notch males never appear. This is the only known dominant sex-linked gene that is also lethal—except *funebris* notch. We have seen that the original notch culture, 5201, gave the characteristics 1: 1: 1: 0 ratio; and this has been repeated by the notch females produced in that culture, both when mated to their brothers and when mated to unrelated wild-type males.[4]

The striking parallel between these two mutants makes it highly probable that they represent the same genetic change. This view is strengthened by the fact that notch is one of the most frequent mutations in *melanogaster* (known to have occurred seven times), and might therefore be expected to be one likely to occur in another species.

Summary.—Notch *melanogaster* and notch *funebris* agree in the following respects:

1. Wings usually irregularly nicked at tip.
2. Certain veins thickened.

[2] Bridges has shown that notch is probably an allelomorph of the roughened eye known as facet. Metz and Bridges, 1917, "Incompatibility of Mutant Races in *Drosophila*," *Proc. Nat. Acad. Sci.*, 3.

[3] The peculiarity of the acrostichal hairs was not observed here until it was looked for after notch *funebris* was found to have unusual acrostichals.

[4] It is theoretically possible that *funebris* notch is not sex-linked, but that the gene is dominant in females, lethal in males. This can be determined by finding gynandromorphs, or by finding other sex-linked genes and observing their linkage to notch.

3. Eyes often small and roughened.
4. Acrostichal hairs not in definite rows.
5. Anterior scutellar bristles often doubled.
6. Character is dominant.
7. Gene has a recessive lethal effect.
8. Gene is sex-linked in *melanogaster,* almost certainly so in *funebris.*
9. Mutation is one of the most frequent in *melanogaster,* and the first certain one in *funebris.*

7

INTERSEXES IN *DROSOPHILA SIMULANS* †

On the first day of January, 1920, a stock of *Drosophila simulans* Sturtevant [1] from Rochester, Minn., was found to contain intersexual individuals. Over 200 such intersexual specimens from this stock and derivatives of it have now been examined. About a dozen of them have been dissected and about the same number have been cleared in KOH and examined in balsam. All these specimens apparently belong to a single type. Male and female parts are both present, as will appear from the following table.

The intersexes are sterile, inasmuch as their gonads are almost, if not quite, absent. Their sexual behavior seems to agree best with that of the normal females. They are courted by males, but mating has not been seen.

	Males	Females	Intersexes
Sex combs on fore tarsi	Present	Absent	Absent
Number of dorsal abdominal tergites	5	7	7
Ovipositor	Absent	Present	Present
Spermathecæ	None	2	2
Penis	Present	Absent	Absent
First genital tergite	Present	Absent	Present
Anal plates	Lateral	Dorsal and ventral	Lateral
Claspers	Present	Absent	Present
Tip of abdomen	Black	Banded	Black
Gonads	Testes	Ovaries	Very minute if present

Crosses of normals from the intersex stock have made it possible to study the character. The intersexes are modified females—*i.e.,* they have two X-chromosomes. This is shown by the fact that in cultures in which half of the males show sex-linked recessive characters but all the females are wild-type, the intersexes never show these sex-linked characters. This relation has been found to hold true for three sex-linked characters that are not closely

† Reprinted from Science, Vol. 51, 1920, pp. 325–327.

[1] For a description of this species see *Psyche* (1919), 26, p. 153.

linked to each other; and the intersex gene itself has been found not to be sex-linked (see below). Therefore the relation just noted can not be due to linkage between the intersex gene and the sex-linked genes in question.

Numerous crosses of the intersex stock to unrelated stocks have never given intersexes in F_1, but have frequently produced them in F_2. The intersex character is therefore recessive.

Pair matings that have produced intersexes have given a total of 510 ♀ :165 intersex: 754 ♂. There is an excess of males, but this is evidently a 3:1 ratio of females to intersexes, indicating not only that the gene is recessive but also that it is not sex-linked. The final proof of the latter point has been obtained through the discovery that the intersex gene is linked to the autosomal recessive gene for "plum" eye-color. Three F_1 pairs from a mating between the intersex stock and the plum stock have given in F_2:

Females		Intersexes		Males	
Wild-type	Plum	Wild-type	Plum	Wild-type	Plum
198	91	87	0	293	65

The absence of the intersex plum class shows that the two genes are linked; and plum is known to be an autosomal recessive.

It has been shown by Morgan and Bridges[2] that individuals of *D. melanogaster*[3] that are partly male and partly female are produced, though only rarely, by most stocks. These "gynandromorphs" have been shown, by genetic evidence, to have two X-chromosomes in their female parts and only one X in their male parts. They are sex mosaics, and each part develops as it would in a whole animal of the same genetic constitution. There is strong evidence that the intersexes described here are not of this nature. The male and female parts in them probably both possess two X-chromosomes. This has been shown as follows. A total of 104 intersexes have been produced by females heterozygous for the sex-linked gene for "yellow" hairs and bristles. Half of these intersexes—about 50—must then themselves have been heterozygous for yellow. If the intersexes are really gynandromorphs, the male parts at the posterior end of the abdomen should have contained a single X-chromosome, and in about half of the specimens that were heterozygous for yellow (*i.e.,* in about 25 individuals) this should have been the yellow-bearing X. As Morgan and Bridges have shown, these parts should then have borne yellow hairs and bristles. The 104 intersexes were all carefully examined for this point, and *none of them had yellow male parts.*

We may conclude that the intersexes are females, modified by a recessive autosomal mutant gene that causes them to show male parts, though these parts themselves still have two X-chromosomes. The normal sex-determining

[2] Carnegie Inst. Washington (1919), publ. 278, pp. 3–122.

[3] I have unpublished data on exactly similar cases in *D. simulans* itself.

mechanism is not affected at all, but the end result is modified by a gene that is not even in the sex chromosomes. It has been assumed by Goldschmidt,[4] Hertwig,[5] Banta,[6] and others working with intersexes that in their animals the normal sex-determining mechanism itself was failing to function as usual. The present example shows that such an assumption can not be accepted without proof.

[4] *Proc. Nat. Acad. Sci.* (1916), 2, pp. 53–58; *Jour. Exper. Zool.* (1917), 22, pp. 593–611, and elsewhere.

[5] *Biol. Zentralbl.* (1912), 32, pp. 65–111, and elsewhere.

[6] *Proc. Nat. Acad. Sci.* (1916), 2, pp. 578–583, and (1918), 4, pp. 373–379.

8

THE VERMILION GENE AND GYNANDROMORPHISM †

Morgan and Bridges (1919, Carnegie Inst. Wash. publ. 278) have recently described and discussed a large number of gynandromorphs of *Drosophila melanogaster*. They conclude that all female parts in gynandromorphs of this animal contain two X-chromosomes, and that all male parts contain only one X. The peculiarities of a given part are thus due to its own constitution, and are not dependent on the rest of the body for their differentiation. The same principle was found to hold for the characters determined by the sex-linked genes, which are carried by the X-chromosomes.

I have recently obtained evidence indicating that the sex-linked character vermilion forms an exception to this rule. A gynandromorph was obtained in which the male parts showed the sex-linked characters scute, echinus, cut, garnet, and forked. These male parts included the whole head, in which region effects of all five of these sex-linked genes could be identified with certainty. Only one of these genes was present in the mother of the gynandromorph, which was heterozygous for the sex-linked genes eosin, ruby, forked, "vermilion lethal," and perhaps for vermilion. All five of them were known to be present in the single X of the father of the gynandromorph, which must therefore have been the X present in the male parts of the gynandromorph itself. But the X of the father was known to carry also the gene for vermilion, and the eyes of the gynandromorph were not vermilion. The not-vermilion color was, then, apparently determined not by the genetic constitution of the eye-pigment itself, but by that of some other portion of the body.

I have obtained two other gynandromorphs in which vermilion seems to have reacted in the same fashion; but these are not so certain as the case described. One of them can be accounted for by the rare type of "double-nucleus" gynandromorphism described by Morgan and Bridges, and in the other the eye-color, while not vermilion, was still not quite normal.

Among the large number of gynandromorphs described by Morgan and Bridges there are only three in which the male parts were known to be

† Reprinted from the Proceedings of the Society for Experimental Biology and Medicine, Vol. 17, 1920, pp. 70–71.

genetically vermilion while the female parts were not genetically vermilion, and in which these male parts included eye tissue. In all of them, however, eosin eye color was also present with the vermilion. Whether the eosin is responsible for the difference between these cases and the one reported here, or whether that difference is due to a difference in the genetic constitution of some other part of the body cannot be determined as yet.

9

GENETIC STUDIES ON *DROSOPHILA SIMULANS*

I. INTRODUCTION. HYBRIDS WITH *DROSOPHILA MELANOGASTER*[1] †

Introduction

Many attempts have been made to hybridize different species of Drosophila, but hitherto all but the combination to be described here have been unsuccessful. The present case is itself only to be considered a partial success, since the hybrids have so far all been completely sterile. Furthermore, since the two parent species are extremely similar and probably have identical chromosome groups, the data that are to be obtained from the study do not throw as much light as might be wished on many of the problems concerning the nature of the specific differences found in the genus Drosophila. Nevertheless, the investigation has led to interesting results bearing on such subjects as interspecific sterility, parallel mutations, chromosome maps, and sex determination.

Description of Drosophila simulans

Drosophila simulans STURTEVANT (1919) is very similar to the well-known *D. melanogaster* MEIGEN (*ampelophila* LOEW), and has been confused with it until recently. There is only one satisfactory character by which the two species may always be separated; viz., the external male genitalia (figures 1 and 2). In living or relaxed pinned material the posterior process (**P**) of the genital tergite appears like a clam-shell in *D. simulans,* and like a small hook in *D. melanogaster*. The eggs (figures 3 and 4) are also characteristically different, but are more variable than the male genitalia, besides being less convenient to obtain. In addition to these characters, the cheeks are a little narrower and the eyes a little larger in *simulans,* and in general the newly hatched specimens have a more definite dark streak down the dorsal surface of the thorax (mesonotum) than has *melanogaster*. With a little

[1] Contribution from the CARNEGIE INSTITUTION OF WASHINGTON.

† Reprinted from Genetics, Vol. 5, 1920, pp. 488–500.

experience one can usually separate most of the individuals in a mixed stock by the use of these characters. In addition there are other still more indefinite differences that are observable when large series are compared. *Drosophila simulans* averages somewhat stouter in abdominal shape and darker in color than *melanogaster*. In mixed laboratory cultures *simulans* is apt to be smaller, though in pure cultures or in wild material there is no appreciable size difference.

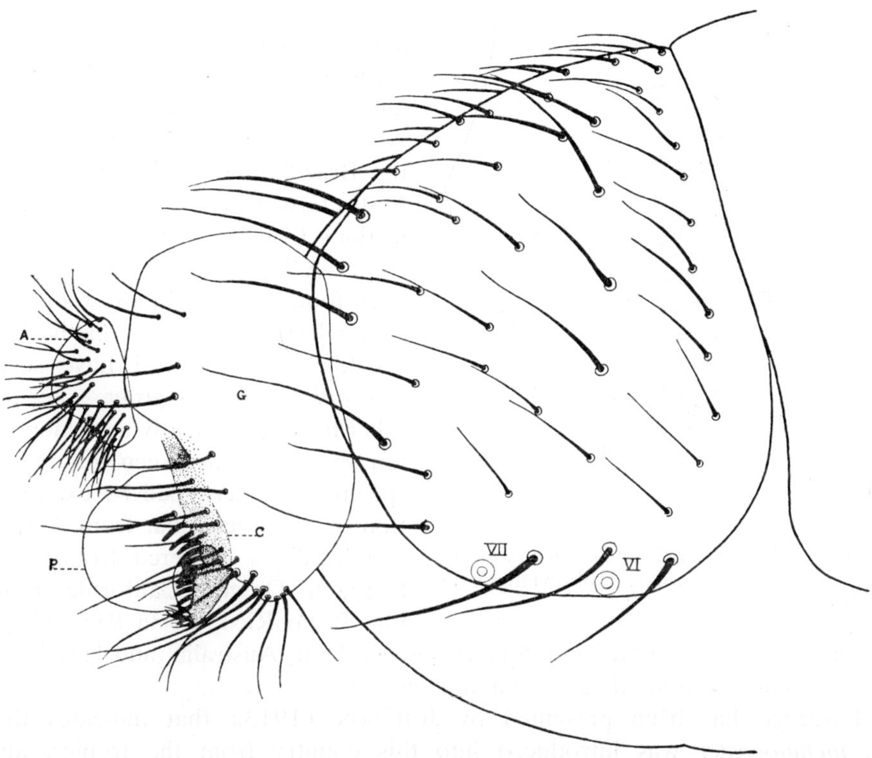

Figure 1. External male genitalia of *Drosophila simulans*. From a specimen cleared and mounted in balsam. A, anal plate; C, clasper; G, first genital tergite; P, posterior process of genital tergite; VI, VII, abdominal spiracles on fifth abdominal tergite.

Though experience will give one the impression that the two forms are distinct and should be classed as "good species," they are so close that a technical description of *D. melanogaster* or a drawing of the style familiar to geneticists would pass equally well for *D. simulans*. So far as I have been able to discover, the mating and breeding habits and the length of the life-cycle are identical. Dr. C. W. METZ reports, after a preliminary study, that the chromosome groups are probably also identical. I am unable to distinguish the larvae, the pupae, or the female genitalia. The structure of the male genitalia and the sterility of the hybrids remain as the only valid grounds for making a specific distinction; but I think there can be no question that these grounds are sufficient.

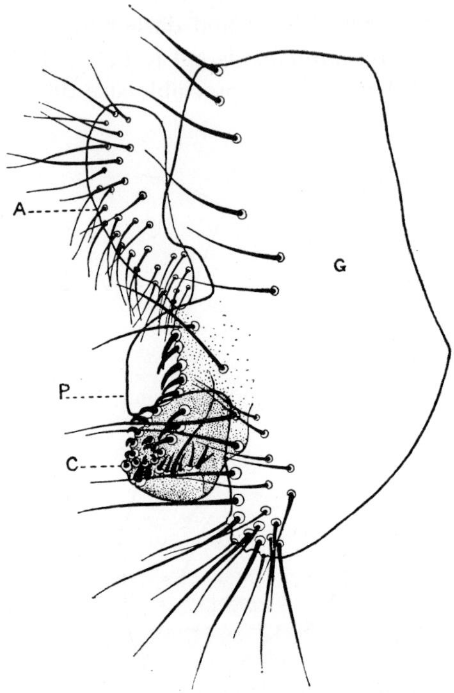

Figure 2. External male genitalia of *Drosophila melanogaster*. Abbreviations as in Figure 1.

Distribution of Drosophila simulans

I described *D. simulans* from stock collected at Lakeland, Florida. I have also seen specimens from New Hampshire, New York, Minnesota, Virginia, Georgia, Alabama, Costa Rica, Panama, and the Brazilian states São Paulo and Matto Grosso. Rather extensive collections from the Pacific coast and from the West Indies have so far not been found to contain any *simulans,* and it has not appeared in the Old World material that I have seen. *Drosophila melanogaster,* on the other hand, is evidently a cosmopolitan species. I have seen undoubted specimens, determined by examination of male genitalia, from American localities scattered from Nova Scotia to Oregon, California, Florida, Porto Rico, Costa Rica, Chile, and Brazil; and from Holland, Spain, New Zealand, Australia, and Hawaii. It is also recorded from Africa, India, and elsewhere.

Evidence has been presented by JOHNSON (1913) that indicates that *D. melanogaster* was introduced into this country from the tropics, and became common soon after 1870. The arguments advanced by JOHNSON are equally applicable to *simulans,* so far as they indicate the rarity or absence of the species before 1870, since the two forms have practically identical

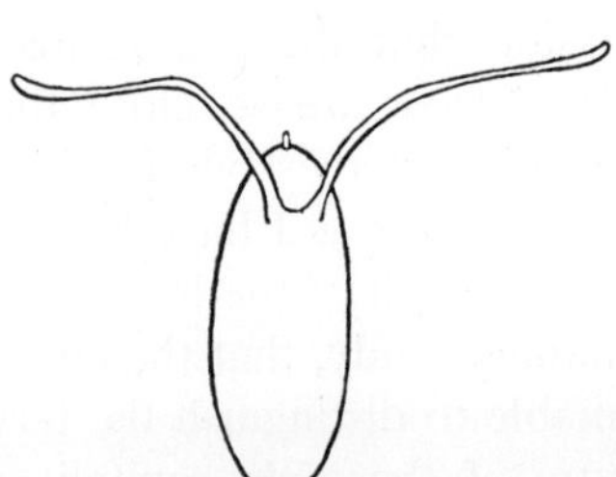

Figure 3. Egg of *Drosophila simulans.*

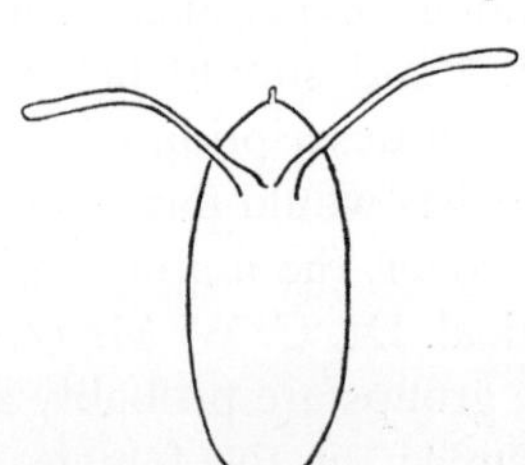

Figure 4. Egg of *Drosophila melanogaster.*

habits, and are so similar in appearance that even when JOHNSON wrote they had never been distinguished. The period at which *simulans* was introduced remains doubtful. It seems certain that QUACKENBUSH (1910) had specimens of it, derived from a stock collected at Woods Hole, Massachusetts, in 1908, and this is the earliest record known to me. The earliest specimen that I have actually seen was collected in Florida in 1912. It is quite probable, however, that an examination of large series of earlier material would show the species to have been established much earlier. The present distribution, outlined above, indicates that the species must have come from South or Central America.

Discovery of Drosophila simulans

At present *Drosophila simulans* and *D. melanogaster* are about equally common around New York City, in Florida, in Alabama, and in southwestern Brazil. Since the two are distinguishable only with some practice, and since a fresh wild stock usually contains both, it seems very surprising that *simulans* was not accidentally discovered long ago in the course of genetic experiments with *melanogaster*. It must have been in the laboratory many times, but seems to have led to unexpected results only two or three times. The curious "unisexual broods" recorded by QUACKENBUSH (1910) were surely hybrid broods. Professor E. A. ANDREWS reported in correspondence some years ago that he had had difficulty in crossing a mutant race of *melanogaster* to a wild race collected at Baltimore, Md., and it now seems probable that the wild race was *simulans*. Finally, Mr. A. M. BROWN, in the COLUMBIA laboratory, obtained anomalous results with a wild race collected by me in Alabama, which ultimately led to the recognition of *simulans*. Mr. BROWN found that males of this race (now known definitely to have been *simulans*), when crossed to mutant races of *melanogaster,* gave only sterile female offspring.

Because of these results observed by Mr. BROWN, I asked Dr. C. W. METZ to send me a number of wild races from Lakeland, Florida, in the spring of 1919. These were tested by crossing to *melanogaster,* and about half of them were found to give unisexual broods. It was noticed that stocks that gave this result looked somewhat different from ordinary *melanogaster,* and had different eggs. Later, Dr. C. B. BRIDGES discovered the striking difference in the male genitalia. Not until many experiments had been made with the Florida race was it noticed that *simulans* also occurs in many other places. I have since shown by crosses, both to the Florida race and to *melanogaster,* that several New York races and one each from New Hampshire and from Minnesota, behave in the same way as does the Florida race. There can now be no question that the male genitalia, by which these later races were first identified, are a certain index as to the genetic behavior.

I have kept mixed stocks of the two species for several months, using the ordinary laboratory technique for keeping stocks of *melanogaster,* and have found that both species may persist together for at least five months. But in such mixed stocks *melanogaster* is usually far more numerous, and the few individuals of *simulans* are almost always small and pale. When the culture becomes old and dry *melanogaster* seems to be unfavorably affected more quickly, so that the percentage of *simulans* rises. Greater resistance to such unfavorable conditions may perhaps account for the fact that *simulans* maintains itself in nature in spite of the fact that, under favorable conditions, it is outdone by *melanogaster*. In the small numbers and small size of the *simulans* specimens from mixed stocks we apparently have the reasons why it was not discovered sooner, for when one selects flies for mating he usually picks out large vigorous ones, and such a procedure would almost always lead to the selection of *melanogaster*. Furthermore, even if *simulans* was occasionally selected it would usually fail to cross, and the absence of offspring would simply be ascribed to sterility.

Drosophila melanogaster ♀ × *D. simulans* ♂

This cross was first made by QUACKENBUSH (1910), to judge from his account. It was also made by Mr. A. M. BROWN, as stated above. I first obtained it in May, 1919, and have since repeated it many times. If the mother is an ordinary XX individual the result is that *only female offspring are produced.* Hundreds of offspring have been obtained from matings of this type, without the production of males.[2] These hybrid females are completely sterile (with *simulans, melanogaster, obscura,* or hybrid males) and have rudimentary gonads, as QUACKENBUSH first showed.

These females always have the sex-linked characters expected of females from the cross. If the mother carries recessive sex-linked genes that are not present in the father, the corresponding recessive characters are never present in the hybrid daughters. This is true for all the recessive mutant characters tested, 12 in number. If sex-linked recessives are present in the *simulans* father, these also do not appear in the hybrid females. Three such have been tested. It follows that the hybrid females have two X chromosomes, and are not transformed males.

It has been shown by BRIDGES (1916) that females of *melanogaster* that have a Y chromosome (XXY) produce exceptional offspring of two types, due to the production of XX and Y eggs, respectively. The first (XX) type of eggs, if fertilized by Y sperm, produce daughters that get all their sex-linked genes from their mother, and are therefore matroclinous exceptions

[2] The results to be presented below indicate that an occasional male, due to primary non-disjunction, is to be expected. One male obtained in the cross involving scarlet *simulans* (to be described in part III of these studies) may have been of this nature.

to the usual behavior of sex-linked characters. The second (Y) type of eggs, if fertilized by X sperm, give rise to sons that get their sex-linked genes from their father, and are thus patroclinous exceptions to the usual behavior of sex-linked characters. In order to further analyze the sex-ratio of the hybrid families, I have crossed XXY *melanogaster* females by *simulans* males. The result has been that *only the regular daughters and the exceptional sons are produced.* The other two classes, regular sons and exceptional daughters, are entirely absent. Two experiments of this type have given the same result, and have produced a total of 128 regular females and 59 exceptional males.[3]

Hybrid males produced in this way are sterile, and have rudimentary testes. They are in no way distinguishable from the hybrid males obtained from the reciprocal mating. They have the sex-linked characters of the father, and are identified as non-disjunctional exceptions by this fact.

Drosophila simulans ♀ × *D. melanogaster* ♂

The reciprocal hybrids are more difficult to obtain. They have so far been produced most often from yellow mothers, except in one mating made by QUACKENBUSH. This mating has been made most often by Prof. T. H. MORGAN (1920), but I have made it successfully a few times. Usually the *only offspring produced are regular sons,* but MORGAN records having sometimes obtained also a few regular daughters, especially as the cultures became old. I have on one occasion used *simulans* females from a stock known to carry extra Y's, and obtained a hybrid female that was undoubtedly a non-disjunctional exception.[4]

As in the case of the reciprocal mating, all these hybrids were completely sterile.

Analysis of Results

The results of the four types of interspecific matings may be tabulated as follows:

melanogaster ♀ (XX)	× *simulans* ♂	= regular ♀♀.
melanogaster ♀ (XXY)	× *simulans* ♂	= regular ♀♀ + exceptional ♂♂.
simulans ♀ (XX)	× *melanogaster* ♂	= regular ♂♂ + a few regular ♀♀.
simulans ♀ (XXY)	× *melanogaster* ♂	= regular ♂♂ + exceptional ♀♀.

[3] The large number of exceptions is due to the fact that a strain of "high non-disjunction" was used, and does not indicate a higher viability of the exceptional males.

[4] The nature of this female was certain because she had the characteristic appearance of a hybrid, and because she was known to have received a *melanogaster* third chromosome but not a *melanogaster* X chromosome.

Regular females will probably occur in the last mating in small numbers, as in the third mating. It is possible that other classes may appear, but none of them, aside from the exceptional males of the last mating, can survive in nearly as large a percentage of cases as do the types noted.

Analysis of these results indicates that the egg cytoplasm must play a part in determining what types survive. We may represent this cytoplasm by the symbol C, and denote the origin of it and of the chromosomes by sub-letters (m for *melanogaster,* s for *simulans*). The types investigated may then be tabulated as follows:[5]

♀	♂
$C_mX_mX_s$,—lives	$C_mX_mY_s$,—dies
$C_mX_mX_sY_m$,—lives	$C_mX_mY_mY_s$,—dies
$C_mX_mX_mY_s$,—dies	$C_mX_sY_m$,—lives
$C_sX_mX_s$,—usually dies	$C_sX_sY_m$,—lives
$C_sX_mX_sY_s$,—usually dies	
$C_sX_sX_sY_m$,—lives	

The Y would not be expected to affect the result, and may probably be disregarded. The above formulae then simplify as follows:

♀	♂
$C_mX_mX_s$,—lives	C_mX_m,—dies
$C_mX_mX_m$,—dies	C_mX_s,—lives
$C_sX_mX_s$,—usually dies	C_sX_s,—lives
$C_sX_sX_s$,—lives	

This table presents many anomalous points. The difference between the two kinds of regular females (C_m X_mX_s and C_s X_mX_s) can only be due to a difference in the cytoplasm of the mother, since the chromosomes of these two types must be identical.[6]

That hybrid females should develop in *melanogaster* cytoplasm only in case they carry a *simulans* X is surprising. This result must be due in part to the autosomes (or to the *simulans* Y), since C_m X_mX_m, that dies as a hybrid, is the formula of the normal *melanogaster* female. In general, it appears that hybrids develop only if they carry a *simulans* X, but that in the presence of *simulans* cytoplasm a *melanogaster* X usually inhibits development even though a *simulans* X is also present.

Fate of C_mX_m Males

It has been assumed in the preceeding discussion that the missing classes of hybrids die. It has been shown, by a consideration of the distribu-

[5] The autosomes are alike in all cases (one member of each pair from each parent) and have therefore been left out of the formulae.

[6] It is, of course, probable that the nature of this egg cytoplasm is determined in the first instance by the chromosomes of the female in which the eggs develop.

tion of the sex-linked characters, that they are not changed over into any of the other classes. The possibility of selective fertilization is the only remaining alternative to the assumption that the individuals belonging to the missing classes are formed but die before they reach the adult stage. The evidence now to be presented shows that in the hybrid cultures from *melanogaster* ♀ × *simulans* ♂, many individuals die in the larval stage; and there is every reason to believe that these represent the missing males.

An eosin vermilion forked *melanogaster* female was mated to a forked *simulans* male. Larvae were observed in about five days, and the pair was removed to a fresh vial. A small amount of food was supplied, and the eggs laid on it were counted and removed, individually, to fresh food. Forty-seven eggs collected in this way gave rise to 14 adults, all of them hybrid forked females as expected. An attempt was made to count the pupa-cases after the adults had all emerged, but this was not found possible. It was clear from the attempt, however, that most of the pupae gave rise to adults, since only a few unhatched puparia could be discovered. Evidently, then, there is heavy mortality in the egg or larval stage. Observation of the isolated eggs indicated that larvae hatched from most of them. A few days after hatching it was noticeable that the larvae were of two sizes;—one type seemed to have grown normally, while the other was much smaller. It had previously been observed that a few larvae were usually present in a hybrid culture at the time adults ceased emerging, and that a number of hybrid larvae died without pupating. These observations indicate that the male larvae grow slowly and never pupate. Their history has not yet been followed in detail, nor has the fate of the other missing classes of hybrids been determined.

Sex-Ratios of Other Species Hybrids

Numerous examples of anomalous sex-ratios among species hybrids are on record. These are evidently due to different causes, as will appear from an examination of a few examples.

When the domestic cow is fertilized by the male bison, the male offspring have a hump, derived from the bison, which is so large that it usually prevents their being born alive. The female hybrids lack this hump, and are born normally (BOYD 1914, etc.). Here, then, an abnormal sex-ratio is brought about at the time of birth by causes external to the hybrid itself,—i.e., by the structure of the mother.

GUYER (1909) has collected data on species hybrids among certain birds, and has shown that there is a decided excess of males in F_1. Although it is possible that some of the individuals were in reality females that had developed male plumage as a result of imperfect development of their ovaries, the evidence indicates a real excess of males. RIDDLE (1916) has recorded the reverse relation (excess of females) for the cross *Streptopelia risoria* × *S. alba* (species of doves) under certain conditions. The cross involves a sex-linked gene, and should give only dark males and white females. Examination of RIDDLE's data for the sex and color of the hybrids shows that, contrary to RIDDLE's own conclusion, the excess of females is not due to "changing over" of males. The female hybrids failed to receive a maternal Z Chromosome and would therefore be expected to be females. The data also show that there is not a high enough mortality of embryos to account for the deficiency of males. The only remaining alternative explanation is that, under the conditions of the experiment, the Z chromosome went to a polar body at reduction oftener than it remained in the egg.

GOLDSCHMIDT (1916, 1917, 1919), HARRISON (1919), and others working with Lepidopterous hybrids have often obtained broods that were largely or entirely of one sex. HARRISON has shown that in at least one instance there is not a high enough mortality to account for the results; and both authors have presented evidence indicating strongly that the results are due to a "changing over" of the sex of half the individuals. Crosses in the same groups often produce intersexes, that are more or less intermediate between males and females. These intersexes commonly replace one of the expected sexes, while the other is present in the expected proportions. Intersexuality may be very slight or more and more marked. The unisexual broods evidently represent the final stage where one sex has been completely changed over into the other. The nature of this change is not yet satisfactorily explained; but it is clear that the anomalous sex-ratios are due to causes different from those operating in RIDDLE's doves or in the cow-bison hybrids. The intersexes will be discussed again, in connection with the *simulans* intersexes (part III of these studies).

The *melanogaster-simulans* hybrid sex-ratio evidently represents still a fourth type. It is not due to a changing over of sex, as shown by the distribution of the sex-linked genes. It is not due to a preferential maturation of the egg, since the female Drosophila is not heterozygous for sex. It is not due to external mechanical difficulties, since both types of larvae emerge from the eggs. The cause is an internal one, but acts merely by killing certain classes of larvae.

Description of Hybrids

In general, *melanogaster-simulans* hybrids, of all types that have been observed, are intermediate between the parent species in those characters in which the parents differ. The male genitalia are shown in figure 5. The width of the cheeks is also intermediate, as is the streak in the dorsal surface of the thorax. Since the hybrid females do not produce eggs, their egg-filaments cannot be examined.

In several respects the hybrids differ from both parents. The complete sterility of both sexes has already been referred to. It is connected with the rudimentary condition of the gonads. The ovaries especially are very minute, and in the testes no sperm have been seen. The hybrids (more especially the females) frequently have fewer dorso-central and scutellar bristles than the number (4 of each sort) that is normal for each parent species. All of these bristles are sometimes missing, though in both pure species individuals with extra bristles are more frequent than those with bristles missing, and I have never seen wild specimens in which more than two were missing. Other bristles on the thorax (noto-pleurals, supra-alars, post-alars) are also often missing in the hybrids but not in the pure species. Another peculiarity that is not infrequent among the hybrids (though less common than the missing bristles) is a broken or missing cross-vein. In the same category belongs abnormally banded abdomen. Both of these characters are very rare in wild individuals of each parent species. None of these characteristics of the hybrids suggests any related species of fly known to me.

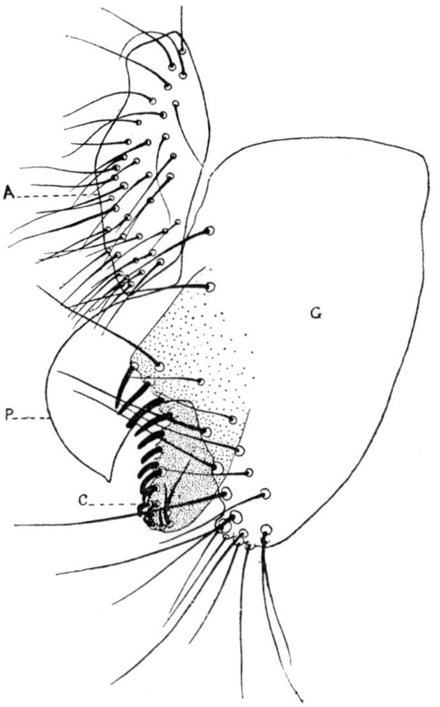

Figure 5. External male genitalia of hybrid male. A, anal plate; C, clasper; G, first genital tergite; P, posterior process of the genital tergite. The upper part of the first genital tergite was broken off in this specimen; this part of the plate is normally shaped as in the two parent species.

These characters that appear in the hybrids but not in either parent species must be due to complemental genes carried by the two species. Since they are evidently dominant genes, it may be surmised that F_2, were it possible to obtain it, would include individuals with numerous other new characters, due to recessive complementary genes.

Sexual Selection

A small series of observations indicates strongly that females of *melanogaster* or of *simulans* are far more likely to mate with males of their own species than with those of the other species. This is probably the reason why so many cross-matings fail to produce offspring. Both kinds of males will court females of either species, apparently indiscriminately; and that it is the females that are responsible for the small percentage of successful cross-matings is also indicated by the fact that normal *simulans* females have seldom been successfully crossed to *melanogaster* males. QUACKENBUSH used a female that had been subjected to heat, and nearly all later crosses have been from yellow females. I have shown (STURTEVANT 1915) that yellow females of *melanogaster* are more easily mated with by *melanogaster* males than are wild-type females.

Sexual selection, then, is one means whereby the two species are kept from crossing. It would, however, probably be ineffective if fertile hybrids were produced when cross-mating does occur.

Hybrids, of both sexes, have been seen to mate with both parent species. No sufficient data are at hand to show what preferences, if any, are shown by the hybrids, or how readily the parent species will mate with them.

Summary

1. *Drosophila simulans* is a species closely similar to *D. melanogaster* in structure, habits, and chromosome group.

2. The two species may be crossed, though many cross-matings fail, apparently because females do not readily allow males of the other species to mate with them.

3. *Drosophila melanogaster* ♀ × *D. simulans* ♂ produces only female offspring, unless the mother is XXY in constitution. In the latter case non-disjunction occurs, and regular daughters and exceptional sons are produced.

4. *Drosophila simulans* ♀ × *D. melanogaster* ♂ usually produces only male offspring, but females sometimes come in small numbers. One exceptional female, due to non-disjunction, has also appeared.

5. The regular females from these two matings are of the same chromosome constitution, but in one case they live and in the other they usually fail to emerge. This relation can only be accounted for by supposing that the egg cytoplasm influences the result.

6. Hybrids survive only if they carry a *simulans* X. In the presence of *simulans* egg cytoplasm and a *melanogaster* X, survival is not usual even if a *simulans* X is also present.

7. In the cross of *melanogaster* ♀ × *simulans* ♂, evidence is presented indicating that the regular males die in the larval stage.

8. The sex of the hybrids corresponds in all cases to their chromosome constitution. XX is female, X is male.

9. Hybrids of all classes are completely sterile, and have rudimentary gonads.

10. The hybrids are intermediate between the parent species in all respects in which the latter differ.

11. The hybrids often show some characters (absence of certain bristles, missing or broken cross-vein, abnormally banded abdomen) not present in either pure species.

LITERATURE CITED

BOYD, M. M., 1914: Crossing bison and cattle. Jour. Heredity **5:** 189–197.

BRIDGES, C. B., 1916: Non-disjunction as proof of the chromosome theory of heredity. Genetics **1:** 1–52, 107–163.

GOLDSCHMIDT, R., 1916: Experimental intersexuality and the sex-problem. Amer. Nat. **50:** 705–718.

1917: A further contribution to the theory of sex. Jour. Exper. Zool. **22:** 593–611.

1919: Intersexualität und Geschlechtsbestimmung. Biol. Zentralbl. **39:** 498–512.

GUYER, M. F., 1909: On the sex of hybrid birds. Biol. Bull. **16:** 193–198.

HARRISON, J. W. H., 1919: Studies in the hybrid Bistoninae. IV. Concerning the sex and related problems. Jour. Genetics **9:** 1–38.

JOHNSON, C. W., 1913: The distribution of some species of Drosophila. Psyche **20:** 202–204.

MORGAN, T. H., 1920: Variations in the secondary sexual characters of the fiddler crab. Amer. Nat. **54:** 220–246.

QUACKENBUSH, L. S., 1910: Unisexual broods of Drosophila. Science **32:** 183–185.

RIDDLE, O., 1916: Sex-control and known correlations in pigeons. Amer. Nat. **50:** 385–410.

STURTEVANT, A. H., 1915: Experiments on sex-recognition and the problem of sexual selection in Drosophila. Jour. Anim. Behav. **5:** 351–366.

1919: A new species closely resembling *Drosophila melanogaster*. Psyche **26:** 153–155.

10

A CASE OF REARRANGEMENT OF GENES IN *DROSOPHILA* † [1]

Seven mutant genes of Drosophila simulans have been shown to be allelomorphic to previously known mutant genes of D. melanogaster.[2] Five of these lie in the X-chromosome, and a study of their linkage relations was shown to indicate that the sequence of the five loci concerned is the same in both species, and that the percentages of crossing over in comparable regions, while not identical, is still not very different. The other two allelomorphic mutant genes, scarlet and peach, lie about 3 units apart in the third chromosome of melanogaster; in simulans they lie in the same chromosome (which is thus identified as the third one), but they were found to be at least 45 units apart.

More recently two more mutant genes of simulans that lie in the third chromosome have been studied. One of these, dachs,[3] lies to the left of scarlet; the other, deltoid,[4] lies between scarlet and peach. The latter, since it makes possible the detection of a portion of the double crossovers, has resulted in a more accurate determination of the scarlet peach distance. The map based on the linkage relations of these four loci (not corrected for unobservable double crossing over) is shown in figure 1.

Both of the new mutant types, dachs and deltoid, resemble previously known mutant types in D. melanogaster. Dachs in melanogaster lies in the second chromosome; it is accordingly not surprising that tests have shown it not to be allelomorphic to dachs simulans. Deltoid resembles delta melanogaster. Since both genes are dominant, the usual test of allelomorphism could not be applied; but each has also a recessive lethal effect, and crosses of delta melanogaster by deltoid simulans have shown that the hybrids that receive both mutant genes do not develop. It follows that the two genes are allelomorphic. The map of the melanogaster third chromosome, including

† Reprinted from the Proceedings of the National Academy of Sciences, Vol. 7, 1921, pp. 235–237.

[1] Contribution from the Carnegie Institution of Washington.

[2] Sturtevant, A. H., *Genetics,* **6,** 1921 (63, 179).

[3] Discovered by Prof. T. H. Morgan.

[4] Discovered by Dr. C. B. Bridges.

the known ends and the three loci occupied by parallel mutations, is shown in figure 2.[5]

A comparison of figures 1 and 2 show that *the three identical loci are not in the same sequence in the two species.*

Mr. D. E. Lancefield has obtained evidence suggesting a similar rearrangement of genes in the X-chromosome of D. obscura. His results were obtained before those here reported, but are not yet published. Since D. obscura has not yet been crossed with any other species, the evidence for identity of loci is not conclusive in this case.

The only analogous case so far reported appears to be that briefly described by Bridges[6] under the name of "vermilion duplication." In this case a section from the middle of an X-chromosome of D. melanogaster appears to have broken loose and attached itself to the left-hand end of a normal X-chromosome. Weinstein[7] has shown that such an occurrence might lead to a change of sequence of identical loci such as is here reported. If we suppose that the simulans third chromosome was originally constituted as is that of melanogaster, the situation as we now find it may be supposed to have arisen as follows: A section, including the peach locus, broke loose and attached itself near the right-hand end of a normal third chromosome. After this condition had become established the peach locus near the middle of the chromosome mutated or became "deficient," so that in effect the peach locus was moved to the right end of the chromosome. Such an interpretation will account for the observed facts.[8]

There is, however, another possible method whereby the same result might be supposed to have been brought about, viz., by the simple inversion of a section of a normal chromosome. Such an accident seems not unlikely to occur at the stage of crossing over. If we suppose a chromosome to occasionally have a "buckle" at a crossing over point, it is conceivable that crossing over might be followed by fusion of the broken ends in such a way as to bring about an inversion of a section of chromosome.

Either of the two suppositions discussed will account for the observed results, but they should lead to different relations for other loci in the same chromosome; it is hoped that further work will lead to the discovery of additional parallel mutations, so that the maps may be studied in more detail.

If an inversion of the kind suggested above occurred within a species, then individuals bearing one normal chromosome and one chromosome with an inverted section would probably show no crossing over in the region in question, since it seems probable that synapsis in this region would be abnormal or absent. It would also be not surprising if crossing over in adjacent

[5] This map is based on the more extensive one published by Bridges in these PROCEEDINGS.

[6] Bridges, C. B., *J. Gen Physiol.,* **1,** 1919 (645).

[7] Weinstein, A., these PROCEEDINGS, **1,** 1921 (625).

[8] It is, of course, possible to invert this interpretation by supposing the simulans situation to be the original one.

regions was decreased. But individuals homozygous for the inverted section would be expected to show free crossing over again, since there should now be no difficulty at synapsis.

The relations indicated are those that have actually been found in the cases of the two "crossover genes" in melanogaster known as C_{IIl}[9] *and* C_{IIr}.[10] These "genes" both cause, in individuals heterozygous for them, the disappearance of crossing over in the immediate regions where the "genes" themselves lie, and a considerable reduction of crossing over in neighboring regions. In individuals homozygous for either of these "genes," however, the percentage of crossing over rises to (or beyond) that found in "normal" individuals. Experiments are now under way in an attempt to determine if these "genes" are simply inverted chromosome sections, but it will probably be a long task to definitely settle the matter.

The demonstration of a change in sequence of identical loci that is here reported makes the identification of parallel mutations in species that cannot be crossed even more difficult than it has previously seemed; for identity of sequence in a group of identical loci now appears not to be necessarily expected.

0.0	dachs
14.2	scarlet
39.0	deltoid
76.5	peach

FIGURE 1

0.0	roughoid
41.5	scarlet
44.5	peach
63.5	delta
101.0	minute-23

FIGURE 2

[9] Muller, H. J., *Amer. Nat.*, **50,** 1916 (103, 284, 350, 421), and Sturtevant, A. H., *Carnegie Inst. Wash. Publ.,* No. 278, 1919 (305).

[10] Sturtevant, A. H., these PROCEEDINGS, **3,** 1917 (555), and *Carnegie Inst Wash. Publ.,* No. 278, 1919 (305).

11

REVERSE MUTATION OF THE BAR GENE CORRELATED WITH CROSSING OVER †

The bar gene of Drosophila melanogaster reverts to its normal or wild-type allelomorph in from thirty to sixty individuals per hundred thousand (May 1917, Zeleny 1918, 1921, 1923). This is a far higher frequency of mutation than is shown by any other locus in Drosophila. We have now found that bar reversion shows a relation to crossing over in the bar region.

Zeleny (1921) found that bar reversions usually occur singly, so that it is not probable that the mutation occurs far back in the history of the germ-cells. It has been the experience of all who have studied bar stocks that in any given specimen the gonads and the eyes are of the same constitution—that is, that a bar-eyed fly breeds as a bar, and a round-eyed fly breeds as a not-bar. These facts further indicate that the bar reversion occurs either late in the history of the germ-cells, or else before the first cleavage division.

If reversion occurs in the germ-cells of both sexes with equal frequency homozygous stocks should produce twice as many heterozygous bar females as wild-type males; if it occurs only in females the number of heterozygous females produced should be equal to the number of wild-type males; if it occurs only in males the reverted offspring should all be heterozygous females. Zeleny (1923) has obtained from homozygous bar stocks a total of fifty-two heterozygous bar females and fifty wild-type males, thus indicating that reversion occurs only (or chiefly) in the germ-cells of females. We can confirm the absence of reversion in males, from extensive experiments in which bar males were mated to not-bar females, without the production of a single reversion.

In an attempt to determine whether the reversion of bar affects one or both chromosomes of a homozygous female, the following experiment was carried out. Females were made up that were homozygous for bar and for the sex-linked recessive eye-color eosin, and that carried a Y-chromosome. These females were also made up in such a way as to give "high" non-disjunction. Such females were mated to vermilion (not-bar) males. Disregarding gynandromorphs, mutations other than bar reversion and a few anomalous

† In collaboration with T. H. Morgan. Reprinted from Science, Vol. 57, 1923, pp. 746–747.

results probably due to non-virginity or to mutation, the following counts were obtained by Miss E. M. Wallace:

REGULAR OFFSPRING			
red heterozygous bar	eosin bar	red not-bar	eosin not-bar
♀	♂	♀	♂
46518	40303	19	8
Percentage of reversions, .03			

EXCEPTIONAL OFFSPRING		
eosin homozygous bar	vermilion not-bar	eosin heterozygous bar
♀	♂	♀
12223	11440	1
Percentage of reversions, .008		

These results, taken together with the facts previously outlined, at once suggested that bar reversion has something to do with crossing over, since the most striking difference between the regular and exceptional offspring of an XXY female is that the former include almost or quite the normal proportion of crossovers, while the latter are only very rarely crossovers (Bridges 1916). On this view the one eosin heterozygous bar female produced might be interpreted as an "equational" exception (in which case the experiment was inconclusive for the purpose for which it was intended).

The crossover suggestion has now been verified directly in two separate experiments. The two sex-linked recessives forked (f, locus 56.5) and fused (fu, locus 59.5) lie on opposite sides of the bar locus, but are only about three units apart (Bridges 1921). In the first experiment ultra-bar (B^u), an allelomorph of bar described by Zeleny (1920) was used. Zeleny has shown that homozygous ultra-bar stocks give rise to reversions to wild-type with about the same frequency as do homozygous bar stocks. In the present case females of the constitution $\frac{\text{B}}{\text{f B}^u\text{ fu}}$ were tested. Three reverted offspring were produced in about 6,500 flies, thus indicating a reversion frequency for the heterozygote that is of about the same order as the frequency in homozygous bar or homozygous ultra-bar. All three of these reverted individuals arose from crossing over between forked and fused: one was forked not-fused and the other two were not-forked fused. In the second experiment ultra-bar was not used and forked and fused were in opposite chromosomes, *i.e.,* the females tested were of the constitution $\frac{\text{B fu}}{\text{f B}}$. Three reversions have so far appeared—two wild-type males and one forked fused male. Thus all six reversions in these experiments were accompanied by crossing over between forked and fused; yet such crossovers include only about three percent of the total number of individuals examined. It follows that *reversion of bar to normal is associated with crossing over at or near the bar locus.*

12

INHERITANCE OF DIRECTION OF COILING IN LIMNAEA†

A recent paper by Boycott and Diver (1923, Proc. Roy. Soc., 95 B; 207) on the inheritance of dextral and sinistral coiling in the snail Limnaea suggests that this character may give an exceptionally clear illustration of "maternal" inheritance that is nevertheless dependent upon the chromosomes.

These authors find that if a single individual of Limnaea is isolated at an early stage it will reproduce, presumably by self-fertilization. Broods produced in this way are always either wholly dextral or wholly sinistral (with the rare exceptions noted below)—but either type of parent may produce either type of brood. This result agrees with the findings of Mayor (1902) and Crampton (1916) on the viviparous Tahitian land-snail Partula, where a given individual contains in its brood-pouch only one type of young. A sinistral individual may have either sinistral or dextral young—but never both types at once; and the same is true for a dextral mother.

Boycott and Diver have also mated together two individuals, and have reared from such pairs mixed broods, which they report as giving 3 dextral : 1 sinistral or 1 dextral : 1 sinistral. In the absence of numerical data, and in view of the fact that the eggs from the two parents were not separated in these experiments, one may doubt if these ratios are anything more than fortuitous ones due to the two members of the pairs in question producing different types of offspring. If one does interpret these ratios as merely chance ones, it becomes possible to formulate a much simpler interpretation than the one suggested by these authors.

An analysis of the data presented suggests that the case is a simple Mendelian one, with the dextral character dominant, but with the nature of a given individual determined, not by its own constitution but by that of the unreduced egg from which it arose.

This last assumption becomes extremely plausible when it is recalled that it was shown by Crampton and by Kofoid in 1894 that dextral and sinistral snails can be distinguished at least as early as the second cleavage division (perhaps at the first), since the cleavage-pattern of one is the mirror-image of that of the other. A character that appears so early in development might well be expected to be determined by the genes present in the mother—*i.e.,* in the unreduced egg, rather than by the combination present after reduction and

† Reprinted from Science, Vol. 58, 1923, pp. 269–270.

fertilization. Yet the results obtained by Boycott and Diver can not be accounted for unless it is supposed that the sperm does actually produce an effect, though the effect is delayed for one generation.

The hypothesis here suggested may be made clearer by the following elaboration. Let the recessive gene for the sinistral character be represented by *l*, and its dominant allelomorph for the dextral character by *L*. Then any heterozygote, Ll, will produce by self-fertilization three types of offspring —LL, Ll and ll. Since all the eggs contained the gene L before reduction, all these individuals will be dextral in somatic appearance; but the ll individuals will themselves produce only sinistral offspring. If an ll individual of this family mates, as a female, to an LL, the offspring will all be sinistral (since the mother carried no L); but they will be Ll in constitution and will therefore produce only dextral offspring. Further combinations may easily be worked out.

It is probable that dextral snails can not mate with sinistral ones; this being the case one might expect that heterozygous individuals would quickly disappear from the colonies, in which case no such results as recorded would be obtainable. The paper under discussion gives a clue as to why the heterozygotes do not disappear. In families that were expected to be purely sinistral a dextral individual occasionally appeared. If such individuals are due to some environmental cause and are genetically sinistral, they will of necessity mate with dextrals and produce new families of heterozygotes. This interpretation is borne out by Lang's results with Helix, where the occasional cases of reversed symmetry were found not to be inherited at all.

Further data on the case of Limnaea will be awaited with interest, for it seems likely that we shall have here a model case of the Mendelian inheritance of an extremely "fundamental" character, and a character that is impressed on the egg by the mother.

13

THE EFFECTS OF UNEQUAL CROSSING OVER AT THE BAR LOCUS IN *DROSOPHILA* † [1]

Introduction

If one thinks of mutations as being simply inherited changes, it becomes necessary to distinguish changes that involve whole chromosomes (e.g., non-disjunction or tetraploidy), changes that involve several adjacent genes (deficiencies and duplications), and what have been called "point-mutations" or "gene-mutations." Probably this last type includes quite diverse processes. It is therefore important to collect information as to the nature of specific examples of mutation. For this purpose it will commonly be necessary to work with a frequently recurring mutation. Only one frequently mutating gene has hitherto been discovered in Drosophila, namely, bar. Crossing over has proved to be the key to the mutation behavior of bar, as will be shown in the present paper. The case appears not to be, strictly speaking, a point-mutation after all, but a new kind of section-mutation, in which the section concerned is so short as to include only a single known gene, and in which unequal crossing over furnishes the mechanism for bringing about the new types.

Historical

In 1914 Tice (1914) found a single male of *Drosophila melanogaster* that had narrow eyes (see figures 1 and 2). The new type, called bar (or "barred" in the earlier literature), was found to depend on a sex-linked gene located at 57.0 in the X chromosome. It was further found that the bar character is dominant, in the sense that females carrying one bar gene have eyes distinctly different from the wild-type or "round" eye (figure 3). Because of this dominance the type has been extensively used in linkage experiments. May (1917) reported that the bar gene occasionally reverts to normal (figures 4 and 5)—a process that has more recently been extensively studied by Zeleny (1919, 1920, 1921). Zeleny found that the frequency of reversion is variable, but in many stocks is such that about

† Reprinted from Genetics, Vol. 10, 1925, pp. 117–147.
[1] Contribution from the Carnegie Institution of Washington.

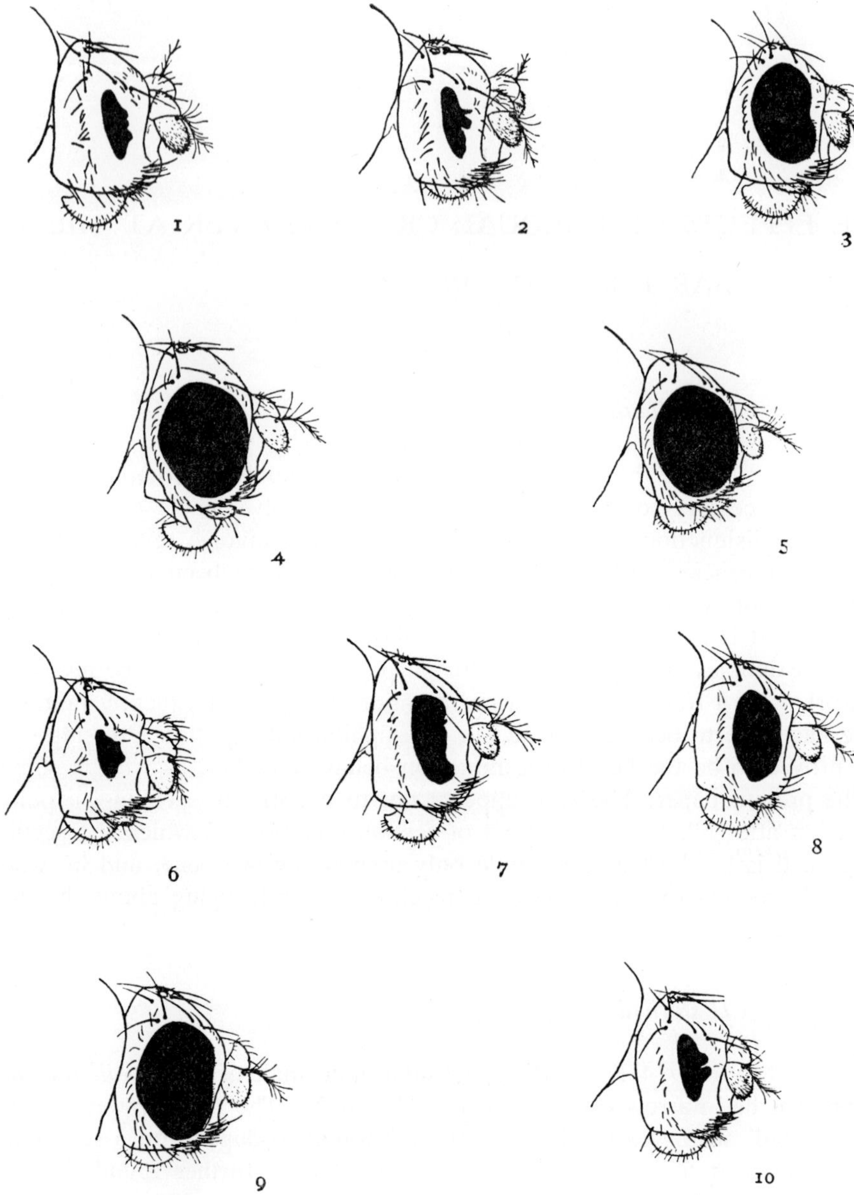

LEGEND FOR PLATE 1

Figure 1. Homozygous bar female.
Figure 2. Bar male.
Figure 3. Bar-over-round female.
Figure 4. Female homozygous for round, that was obtained by reversion.
Figure 5. Male that carries round, obtained by reversion.
Figure 6. Double-bar male.
Figure 7. Homozygous infrabar female.
Figure 8. Infrabar male.
Figure 9. Infrabar-over-round female.
Figure 10. Double-infrabar male.

1 in 1600 offspring from a pure bar stock receives a not-bar, or round, allelomorph. ZELENY also concluded that the reversion probably occurs chiefly (or perhaps exclusively) in females. His argument, based on the sex ratio found among reverted individuals, is not as convincing as the direct tests that will be described in this paper, and which verify his conclusion. ZELENY also found that homozygous bar gives rise to a new and more extreme allelomorph of bar, that he has called "ultra-bar." For reasons that will be developed in this paper, I prefer to call it "double-bar." The eyes of double-bar are distinctly smaller than those of bar (figure 6). ZELENY has shown that the type is more strongly dominant over round than is bar, and also that double-bar is largely dominant over bar.

ZELENY likewise found that homozygous double-bar stocks revert to round with a frequency not very different from that of homozygous bar stocks, and that double-bar occasionally mutates to bar; that is, it can go all the way back to round at one step, or it can give bar, which, in turn, is capable of reverting to round. ZELENY has argued that the three types, round, bar, and double-bar, have the same characteristic properties, regardless of their origin. The round eye of reverted bar is indistinguishable from wild-type; bar derived from double-bar does not differ from the original bar, etc. This point will be considered in more detail in a later section.

STURTEVANT and MORGAN (1923) showed that double-bar over bar[2] also gives rise to round-eyed individuals. They reported three reversions from this combination and three from homozygous bar. In all cases the mothers had been heterozygous for forked,[3] which lies 0.2 units to the left of bar, and for fused [4] which lies 2.5 units to the right of bar. All six reversions represented crossovers between forked and fused, though the total forked fused crossovers constituted less than 3 percent of the number of individuals examined. STURTEVANT and MORGAN also reported that experiments in which bar entered only through the males had failed to give any reversions, though no numerical data were reported. The present paper is based on the results of a more detailed study of the relations first shown by STURTEVANT and MORGAN (1923).

Mutations and Crossing Over

The results from homozygous bar females, that were reported by STURTEVANT and MORGAN, were from females of the constitution $\frac{+B\,f_u}{f\,B+}$.

[2] In this paper the constitution of heterozygotes will be expressed as above in order to avoid circumlocution or indefiniteness. "Double-bar over bar" is to be understood as: "carrying double-bar in one X chromosome and bar in the other X chromosome."

[3] Forked is a recessive bristle modification. Locus 56.8 in the X chromosome (see MORGAN and BRIDGES 1916).

[4] Fused is a recessive venation character. Its locus is at 59.5 in the X chromosome (see MORGAN and BRIDGES 1916).

A more efficient type of experiment is that in which females of the constitution $\frac{+B+}{f\,B\,f_u}$ are mated to forked bar fused males. Table 1 (first row) shows the results obtained from an extensive series of this type. In the second row of table 1 are given the results from mating a few females of the above constitution to forked fused males.

Table 1

$\frac{B}{fBf_u}$ ♀ × fBf_u ♂ (*1st row*) *or* ff_u ♂ (*2nd row*).

TYPE OF MALE USED	BAR								TOTAL	ROUND					DOUBLE-BAR	
	♀				♂					♀		♂			♀	♂
	0*		1		0		1			f_u	f	+	f_u	f	f	f
	+	ff_u	f_u	f	+	ff_u	f_u	f								
fBf_u	5413	3749	94	140	5218	4160	93	124	18991	0	2	1	2	1	1	1
ff_u	374	359	6	6	352	324	10	7	1438	1	0	0	1	0	0	0
Total...	5787	4108	100	146	5570	4484	103	131	20429	1	2	1	3	1	1	1

* In this and the following tables "0" signifies non-crossover classes; "1," classes resulting from crossing over in the first region, etc.

The mutant females that appeared in these experiments were, of course, all heterozygous for whichever allelomorph (bar or round) they received from their fathers, but in tables 1 to 8 this paternal gene is ignored and the females classified according to the maternal gene. In all doubtful cases (including both double-bars), the classification was checked by raising offspring from the mutants, since there is sometimes difficulty in classifying single individuals, a difficulty that can be removed by examination of a series of specimens of a given constitution.

It will be observed that seven of the eight reversions, and also both of the double-bars, occurred in gametes that came from crossing over between forked and fused, though the total fused crossovers constituted only 2.4 percent of the population. The one exceptional case, a wild-type male, is not above suspicion of having arisen through contamination rather than reversion of bar. That he was really round-eyed was proved by tests. Only 4 other exceptions to the rule that mutation in this locus is accompanied by forked-fused crossing over have been met with in the work here reported. These will be discussed separately later.

On the basis of these results we may formulate the working hypothesis that both reversion and the production of double-bar are due to *unequal crossing over*. If we suppose that, in a female $\frac{+B+}{f\,B\,f_u}$, crossing over occurs in such a way that the respective points of interchange lie to the left of the bar locus in one chromosome, but to the right of it in the other one, there will result chromosomes of the constitution $fBB+$ and $+f_u$ (or $f+$ and $+BBf_u$). The hypothesis is that reverted round is simply *no-bar,* and that double-bar is *BB,*—this being the reason for abandoning ZELENY's name, ultra-bar.

This hypothesis makes reverted round and double-bar complementary crossovers, and they should accordingly be produced with equal frequency. Table 1 agrees with ZELENY's more extensive data in showing that round is apparently far more frequent than double-bar; but such a result was to be expected for two reasons. Double-bar is not as viable as round, so that fewer of the double-bar mutant individuals would be expected to survive; and double-bar is not always clearly distinguishable from bar, so that some mutant individuals are probably overlooked, while it is not likely that any reversion is overlooked through difficulty of classification.

The double-bar over bar experiments reported by STURTEVANT and MORGAN (1923) can be interpreted in the same way: the reversion is here due to unequal crossing over just as in homozygous bar. In the earlier cultures of the experiments previously reported, only the reversions were classified for forked and for fused. Two of the reversions were in such incompletely classified cultures. Table 2, including all the double-bar over bar data for which complete counts are available, contains one of the previously reported reversions and one new one. This table includes only the male offspring, since the *BB* derived from the fathers rendered the classification of the females uncertain.

The process of unequal crossing over might be expected to give rise to triple-bar from the females that are double-bar over bar. No individual that could be so identified was obtained, though several specimens with very small eyes were tested. All those that were fertile proved to be double-bar. Apparently triple-bar is either inviable or sterile. This problem will be discussed again below.

Table 2

$$\frac{B}{fBBf_u} \text{♀} \times fBBf_u \text{♂}.$$

B OR BB ♂					ROUND ♂
0		1			
+	ff_u	f_u	f	TOTAL	f
3933	2741	62	73	6809	2

My own experiments with homozygous double-bar have not yielded any mutations, probably because I have found this type hard to breed and have therefore not obtained large numbers of offspring. It may be recalled that ZELENY has obtained both bar and round from such females, but not in experiments in which forked and fused were present. In table 3, showing the data I have obtained, only males are recorded, for the same reason as in table 2, and also because the females could not be classified for fused.

Table 3

$\frac{fBB}{BBf_u}$ ♀ × fBB ♂.

BB♂				
0		1		
f_u	f	+	ff_u	TOTAL
627	956	37	24	1644

In the case of double-bar over round, bar should be produced by any crossover between the two bars of the double-bar chromosome. This event might seem less unlikely to occur than the type of unequal crossing over invoked in the preceding experiments; and both of the chromosomes resulting from such crossing over should yield bar, whereas in the preceding cases, a given crossing over must always have yielded chromosomes bearing two different kinds of bar allelomorphs. It is accordingly in agreement with the hypothesis that tables 4 and 5 show a higher percentage of mutation than do tables 1 and 2. Two non-disjunctional individuals (a ff_u♂ and a $\frac{+\,+\,+}{f\,BB\,f_u}$ ♀) have been omitted from table 5.

Table 4

$\frac{BB}{ff_u}$ ♀ × ff_u ♂.

	DOUBLE-BAR AND ROUND								
	0		1		2			BAR	
SEX	BB	ff_u	f_u	fBB	BBf_u	f	TOTAL	f_u	f
♀	1174	784	1	1	18	30	2018	0	1
♂	993	943	1	1	14	19	1971	1	1

No mutations are to be expected from females that are heterozygous for bar and for round $\left(\frac{B}{+}\right)$, since crossing over can not produce any new combination. The results of the tests are shown in tables 6 and 7, and are in agreement with this expectation. One non-disjunctional female $\left(\frac{+\,+\,+}{f\,B\,f_u}\right)$ was also produced in this series.[5]

[5] This series also produced one female that was wild-type in appearance. Such a female would be either a double crossover—which is not at all probable in such a short

Table 5

$\frac{+}{fBBf_u}$ ♀ × ff_u ♂.

	DOUBLE-BAR AND ROUND							BAR	
	0		1		2				
SEX	+	$fBBf_u$	BBf_u	f	f_u	fBB	TOTAL	f_u	f
♀	464	392	0	1	14	23	894	1	0
♂	495	406	0	0	6	10	917	1	1

L. V. MORGAN (1922) has described a race of *D. melanogaster* in which the two X chromosomes of the female are attached to each other, so that such a female gives 100 percent non-disjunction. This race, and others separate in origin but having the same peculiarity of attached X's, have been used to test the mutability of bar in the male. If a round-eyed female with attached X's is crossed to a bar male, all the sons get their X chromo-

Table 6

$\frac{+}{fBf_u}$ ♀ × ff_u ♂.

	BAR						
	0		1		2		
SEX	+	fBf_u	Bf_u	f	f_u	fB	TOTAL
♀	331	276	0	0	12	13	632
♂	316	265	0	1	10	11	603

somes from the bar father and accordingly furnish a direct test of the mutability of bar in males.[6] A total of 10,079 bar males has been observed from such matings, with no rounds or double-bars. There was, however, one other male that had an eye intermediate between bar and round. This new type, called infrabar (figures 7, 8, 9) has been shown to represent a new allelomorph of bar. Its somatic appearance will be described in more detail later in this paper.

Tests soon showed that infrabar behaved as a single unit in inheritance. Bar can not be recovered from it, and it shows the same linkage relations

chromosome section, or a non-crossover reversion of bar. This female was mated to an unrelated bar male, and produced 124 bar-over-round daughters and 96 round-eyed sons; but none of the sons showed either forked or fused. This result must mean that the exceptional female was due to contamination.

[6] Occasionally the attached X's separate, and a regular son carrying a maternal X is produced. In the present series of experiments this source of error was eliminated by having the two maternal X's differ from the paternal one in at least one other mutant gene besides bar.

Table 7

$\frac{B}{ff_u}$ ♀ × ff_u ♂.

	BAR						
	0		1		2		
SEX	B	ff_u	f_u	fB	Bf_u	f	TOTAL
♀	1792	1356	3	7	36	41	3235
♂	1523	1334	4	3	26	36	2926

as bar. The convincing proof that it represents a modification of the bar gene will appear below.

Homozygous infrabar behaves like bar in that it reverts to normal, and also produces a new and more extreme type, double-infrabar (figure 10), analogous to double-bar. As shown in table 8, both these events are again associated with forked-fused crossing over.

Of the three double-infrabar individuals, one was sterile, and one was accidentally lost, but the constitution of the other was established by breeding tests. The type was also obtained in another experiment; and its appearance and mutation behavior will be described later. This series shows the same excess of rounds over double forms as did the bar series, and the explanation is doubtless the same. The difficulty of separation is even greater here, owing to a greater variability in eye size (see below).

Table 8

$\frac{B^i}{fB^if_u}$ ♀ × fB^if_u ♂.

	INFRABAR					ROUND		B^iB^i	
	0		1						
SEX	+	ff_u	f_u	f	TOTAL	f_u	f	f_u	f
♀	4607	3606	94	142	8449	0	5	0	1
♂	4341	3905	97	106	8449	2	11	2	0

Bar-over-infrabar females of three different constitutions have been tested. They have produced rounds, and also a new double type that had eyes only very slightly larger than those of double-bar. These must evidently have bar and infrabar in the same chromosome. Tables 9, 10 and 11 show these results. In most cases the fathers of the cultures in table 9 did not carry fused; for this reason the females, among which no mutants were detected, are not listed. The forked fused mutant male was sterile, but the constitution of the other was tested. All the experiments here described, on "bar-infrabar" (figure 10), concern this mutant gene. The stability of this new type in the

Table 9

$\frac{fB^i}{Bf_u}$ ♀ × *various* ♂♂.

BAR AND INFRABAR MALES					DOUBLE TYPE	
0		1			♂	
f_u	f	+	ff_u	TOTAL	+	ff_u
623	682	26	18	1349	1	1

male has also been tested. Matings of bar-infrabar males to attached-X females (also differing in at least one other sex-linked gene) have produced 9042 non-disjunctional sons,—all of them bar-infrabar.

Table 10

$\frac{B^i}{fBf_u}$ ♀ × fBf_u ♂.

	BAR AND INFRABAR					
	0		1			
SEX	+	ff_u	f_u	f	TOTAL	MUTANTS
♀	497	415	10	18	940	0
♂	505	476	15	16	1012	0

The double-type male of table 11, which resembled those from table 9, was tested. His descendants are discussed below under the name "infrabar-bar."

Table 11

$\frac{B}{f\,B^i\,f_u}$ ♀ × $f\,B^i\,f_u$ ♂

	BAR AND INFRABAR					ROUND		DOUBLE TYPE
	0		1					
SEX	+	ff_u	f_u	f	TOTAL	f_u	f	f
♀	2445	2203	59	74	4781	0	1	0
♂	2365	2347	64	71	4847	1	0	1

These new double types have made it possible to devise crucial tests of the theory of unequal crossing over, which may now be described. The first test made was that of bar-infrabar over round. Tables 12 and 13 show the results obtained. In addition one culture of this series gave two wild-type females, one in the first count and another in the last count, ten days later. These females give rise to the same difficulties as did the wild-type obtained in the series reported in table 7. One of them was tested, and gave unexpected results. The mother had carried vermilion in her forked fused X, and the father had also been vermilion forked fused. The tested exceptional female was found to carry one wild-type X and one forked fused X, but did not carry vermilion at all. Under the circumstances it is open to question whether these two exceptional individuals were not due to contamination. They will be discussed again later.

Table 12

$\frac{fBB^i}{f_u}$ ♀ × ff_u ♂.

	BAR-INFRABAR AND ROUND							BAR	INFRABAR
	0		1		2				
SEX	f_u	fBB^i	BB^i	ff_u	+	fBB^if_u	TOTAL	ff_u	+
♀	303	394	4	0	14	8	723	0	1
♂	321	308	0	2	7	3	641	1	1

Tables 12 and 13 show that bar and infrabar can both be recovered from bar-infrabar. It appears then that in the double form the individual elements maintain their identity. Even more important, however, is the

Table 13

$\frac{BB^i}{ff_u}$ ♀ × ff_u ♂.

	BAR-INFRABAR AND ROUND							BAR	INFRABAR
	0		1		2				
SEX	BB^i	ff_u	f_u	fBB^i	BB^if_u	f	TOTAL	f_u	f
♀	2549	1684	3	1	43	74	4354	0	0
♂	2322	1857	5	8	48	63	4303	2	4

indication that they maintain their sequence in the chromosome. As shown in table 9, the bar-infrabar first came from the combination $\frac{f\,B^i}{B\,f_u}$ as a not-forked not-fused male. If the two elements of double forms are arranged in the same linear series as the rest of the genes, this result must mean that the bar now lies to the left of the infrabar. This supposition is entirely borne out by tables 12 and 13, which are experiments of the usual type used to establish sequence of genes. All the 9 single types recovered agree with the supposed sequence.

Table 14

$\frac{BB^i}{fB^if_u}$ ♀ × fB^if_u ♂.

	BB^i AND B^i							B^iB^i
	0		1		2			
SEX	BB^i	fB^if_u	B^if_u	fBB^i	BB^if_u	fB^i	TOTAL	f
♀	850	702	3	3	17	28	1603	0
♂	867	860	4	2	22	23	1778	1

Bar-infrabar has also been tested against infrabar (table 14).

Here again there is an opportunity for the production of a triple form,—but since the corresponding round did not appear, its absence is not significant. The one double-infrabar confirms the sequence of the component parts of bar-infrabar. It was tested, and all the results from double-infrabar reported below were obtained from flies descended from it.

Examination of the data in table 11 shows that the double form obtained there must have had infrabar to the left of bar; that is, it was infrabar-bar instead of bar-infrabar. The tests made with it appear in table 15.

Table 15

$\frac{fB^iB}{f_u}$ ♀ × ff_u ♂.

	INFRABAR-BAR AND ROUND							BAR	INFRABAR
	0		1		2				
SEX	f_u	fB^iB	B^iB	ff_u	+	fB^iBf_u	TOTAL	+	ff_u
♀	483	469	1	1	15	15	984	2	0
♂	478	408	1	9	17	15	928	0	1

These results show, in fact, that the sequence is infrabar-bar as supposed. Except for this difference the mating is the same as in table 12. It will be observed that the two not-forked not-fused single types in that table were both infrabar, while the two obtained here were bar; the one forked fused there was bar, here it was infrabar. There is a total of 13 single-type mutants in tables 12 to 15, all of them agreeing in indicating that the two elements of double types maintain not only their individuality but their sequence.

The double-infrabar obtained in table 14 has been tested against round (tables 16 and 17).

The results in these tables show that double-infrabar was correctly identified, and that it behaves as was to be expected, giving infrabar by both kinds of crossing over,—just as double-bar gives bar in both cases.

Table 16

$\frac{fB^iB^i}{f_u}$ ♀ × ff_u ♂.

	B^iB^i AND ROUND							B^i
	0		1		2			
SEX	f_u	fB^iB^i	B^iB^i	ff_u	+	$fB^iB^if_u$	TOTAL	+
♀	538	629	1	1	12	17	1198	0
♂	574	505	1	1	13	7	1101	2

Table 17

$\frac{B^iB^i}{f\ f_u}$ ♀ × ff_u ♂.

	DOUBLE INFRABAR AND ROUND							INFRABAR	
	0		1		2				
SEX	B^iB^i	ff_u	f_u	fB^iB^i	$B^iB^if_u$	f	TOTAL	f_u	f
♀	1664	1404	3	3	39	37	3150	1	1
♂	1449	1193	3	3	29	42	2719	1	2

Double-infrabar over infrabar has also been tested, in the hope of obtaining triple-infrabar (table 18).

Table 18

$\frac{B^iB^i}{fB^if_u}$ ♀ × fB^if_u ♂.

	DOUBLE INFRABAR AND INFRABAR*					ROUND
	0		1			
SEX	+	ff_u	f_u	f	TOTAL	f
♀	851	777	25	31	1684	1
♂	665	750	26	29	1470	0

* The double-infrabar and infrabar flies are not entered separately. In the original counts they were separated; but the classification is uncertain at times.

No triple-infrabar was detected; but its absence is not surprising, since the corresponding round occurred only once, and since it is not at all sure that the triple form could be distinguished from the double one.

One further type of female was tested, namely, bar-infrabar over infrabar-bar (table 19).

The one mutant infrabar obtained in the series is in agreement with the sequence in which bar and infrabar were supposed to lie in the two X chromosomes of the mother; and this mutant is also the only one yet obtained from a mother carrying a double-type allelomorph in each X, where forked

Table 19

$\frac{BB^i}{fB^iBf_u}$ ♀ × fB^iBf_u ♂ (*1st row*), *or* ff_u ♂ (*2nd row*).

	BB^i AND B^iB									B^i
	♀				♂					♂
	0		1		0		1			
	+	ff_u	f_u	f	+	ff_u	f_u	f	TOTAL	f
1*	...	...	.	..	95	74	3	4	176	0
2	428	335	8	15	341	319	7	16	1469	1

* Females not counted in the mating to B^iB male.

and fused were heterozygous. It therefore serves to complete the demonstration of the relation of crossing over (between forked and fused) to mutation in the bar locus.

The experiment of table 19 was, however, planned for another purpose. It will be seen that in the mother, which was $\frac{B\ B^i}{B^i\ B}$, *equal* crossing over might give rise to new types, namely, double-bar and double-infrabar. The first could not be distinguished, in somatic appearance, from the unmutated double types (BB^i and B^iB); but the double-infrabar should be readily detected. Such an individual would be forked. It may accordingly be concluded that none of the 35 forked (not-fused) offspring represented equal crossing over between the halves of the two double-type bar allelomorphs present. It therefore seems probable that crossing over of this kind is not much, if any, more frequent than is that between the two elements of a double-type allelomorph when the other chromosome carries round (tables 4, 5, 12, 13, 15, 16).

Several of the above tables agree with a small series of infrabar over round, heterozygous for forked and for fused, in showing that infrabar lies between forked and fused. It must clearly be either an allelomorph of bar, or bar plus a modifier that lies near bar.

The experiments with bar-infrabar and with infrabar-bar show that these two types both contain infrabar as a unit distinct from bar.

Since bar-infrabar was produced by an unequal crossover that occurred very close to the left of infrabar, it becomes unlikely that a modifier can lie on that side of the bar locus; infrabar-bar furnishes similar evidence that there is no modifier to the right. All the evidence thus indicates clearly that the infrabar gene is really a modification of the bar gene itself.

Frequency of Bar Mutations

The data presented in tables 1 to 19 have been examined in an attempt to formulate some general statements as to the relative frequency of the various types of mutation in the bar locus. It is probable that homozygous double types, and double over single show the lowest frequencies of mutation, and that double type over round shows the highest. Both these results might have been expected. There is, however, so much variability among crosses of the same general nature that these conclusions must be accepted with caution. For example, the two largest series are those from homozygous bar (20,438 offspring) and from homozygous infrabar (16,918 offspring). The mechanical conditions should be alike in the two cases, since both represent homozygous single types. Yet from the first there appeared 0.03 percent of reversions, or 1 in each 2920 offspring; from the second there were 0.11 percent, or 1 in 940 offspring. In view of such unexplained differences as this, and in view of the statistical difficulty of determining probable errors for such small percentages, it does not seem profitable to discuss further this aspect of the data, except to note that mutation frequency does not appear to be correlated with frequency of forked-fused crossing over.

The Crossover Values for Forked, Bar and Fused

The experiments recorded in tables 1 to 19 include by far the largest series of data yet accumulated for the crossover values of the three loci, forked, bar and fused. These are summarized in tables 20 and 21. In these tables all the mutant individuals have been omitted. Their inclusion

Table 20

fBf_u crossing over.

TYPE OF FEMALE TESTED	NON-CROSSOVERS	CROSSOVERS		TOTAL
		Region 1	Region 2	
Single type over round	7,193	18	185	7,396
Double type over round	22,937	52	603	23,592
Double type over single	3,279	12	90	3,381
Total................	33,409	82	878	34,369
Percentage...........	97.21	0.24	2.55	

Table 21

ff_u crossing over.

TYPE OF FEMALE TESTED	NON-CROSSOVERS	CROSSOVERS	TOTAL
Single type over single.......	48,966	1290	50,256
Double type over single......	9,717	246	9,963
Double type over double.....	3,175	114	3,289
Total.....................	61,858	1650	63,508
Percentage................	97.40	2.60	

would not have affected any of the values appreciably. In table 22 the data already published on these loci are included with the above, in order to arrive at final "map values" for the three loci.

On the basis of these data, it seems best to map forked 0.2 unit to the left of bar, that is, at 56.8; and fused 2.5 units to the right of bar, at 59.5, thus making the forked-fused interval 2.7 units.

Facet Number

It has been shown by ZELENY and MATTOON (1915), MAY (1917) and ZELENY (1922) that selection for number of facets is effective in isolating lines of bar flies with high facet numbers or with low. Though no detailed genetic analysis has yet been reported, there is abundant evidence in

Table 22

Total linkage data.

LOCI	SOURCE OF DATA	CROSSOVERS	TOTAL	PERCENT
fB	MORGAN and BRIDGES 1916	8	1,706	0.5
	BRIDGES 1917	5	980	0.5
	Table 20	82	34,369	0.2
	Total	95	37,055	0.26
ff_u	WEINSTEIN 1918	200	8,298	2.4
	Table 21	1650	63,508	2.6
	Total "primary" data	1850	71,806	2.58
	MORGAN and BRIDGES 1916	37	1,201	3.1
	Table 20	960	34,369	2.8
	Grand total	2847	107,376	2.65
Bf_u	MORGAN and BRIDGES 1916	222	8,768	2.5
	BRIDGES 1917	46	1,401	3.3
	Table 20	878	34,369	2.5
	Total	1146	44,538	2.57

these papers that ordinary bar stocks are heterogeneous for modifiers (not in the bar locus) that affect facet number. This is also the impression I have gained from extensive but less exact studies, with numerous crosses involving bar.

Another source of variability in facet number is temperature. It was shown by SEYSTER (1919) that high temperature decreases the facet number of bar, and this relation has been studied in great detail by KRAFKA (1920), ZELENY (1923), A. H. HERSH (1924) and R. K. HERSH (1924). These observers have shown that the effect is present, though in varying degree, in double-bar, round, and in various heterozygotes, as well as in bar. Although these studies furnish essential data for any complete analysis of the mode of action of the bar series of allelomorphs upon development, they need not be further discussed here.

The evidence just reviewed indicated that it would be necessary to get stocks as nearly uniform for modifiers as possible, and also to control the temperature, if any reliable data were to be collected as to the relative effectiveness on facet-number of the various combinations of bar and infrabar. Accordingly, a female that was $\frac{fBf_u}{fBB}$,[7] was mated to a round-eyed vermilion male from vermilion stock. The descendants from this mating were inbred (brother-sister pair matings) for seven generations. In each generation a female heterozygous for two of the three bar allelomorphs concerned was mated to a male carrying the third allelomorph. The line was made homozygous for forked; but both vermilion and fused were eliminated. No other selection was practiced. A female of the third inbred generation was mated to an infrabar male, and a daughter that was infrabar over round was mated to a male from the fourth generation. For three more generations the infrabar series was crossed to the inbred line. After the eighth generation the pedigrees are somewhat more complex, but as close inbreeding as was compatible with maintaining four allelomorphic sex-linked genes was continued for six more generations before the facet counts were begun. The other types studied (bar-infrabar, reverted bar, etc.) were all crossed to the inbred stock just described, at least five times (mostly using females to allow crossing over and get as much of the X's uniform as possible) before being used in the counts. This procedure should have resulted in making the various stocks practically alike with respect to modifying genes, and the results obtained are sufficiently consistent to indicate that there was no heterogeneity in major modifiers, though it is still possible to interpret some of the minor differences observed as being due to uneliminated diversity in modifiers.

The temperature control used was not very exact, but maximum-minimum daily records show that 25°C was maintained to within about ±1°C, and even these deviations were probably of short duration. For the main body of the experiments it has not proved possible to detect any systematic effect of the fluctuations in temperature that did occur. The few experiments in which such an effect is perhaps present will be specified when described.

The facets of the smaller eyes,—up to and including homozygous bar,—were counted directly under the binocular microscope, usually on etherized flies, but in some cases on alcoholic specimens. The eyes larger than this were not found to be workable by this method. Such specimens were killed

[7] Derived from the experiments of table 1, so that the *BB* of the following discussion is known to have come from the *BB* male of that table, and to have been derived from homozygous bar by forked-fused crossing over.

and cleared in KOH. The surface of each eye was then removed and mounted on a slide. By the aid of a camera lucida a drawing was made, representing each facet by a dot, and these dots were then counted, each dot being marked by a check as counted. In all cases the right eye alone was used.

The main series of data is shown in table 23.

The table shows that homozygous infrabar is about like bar over round in facet number, but the two types can be separated by a peculiarity common

Table 23

Facet number of flies carrying various combinations of bar allelomorphs.

		BB	BB^i	B^iB^i	*B*	B^i	+
	n	47					
BB	M	24.96 ± .30					
	σ	3.03 ± .21					
	n	20	51				
BB^i	M	24.10 ± .35	26.69 ± .29				
	σ	2.36 ± .25	3.07 ± .21				
	n	25	32	27			
B^iB^i	M	26.68 ± .29	27.91 ± .36	38.19 ± .71			
	σ	2.13 ± .20	2.98 ± .25	5.48 ± .50			
	n	51	86	28	50		
B	M	36.43 ± .30	37.01 ± .25	38.29 ± .59	68.12 ± 1.09		
	σ	3.18 ± .21	3.49 ± .18	4.64 ± .42	11.15 ± .77		
	n	30	53	3	30	18	
B^i	M	41.77 ± .41	37.79 ± .52	138.2	73.53 ± 1.29	348.4 ± 12.4	
	σ	3.34 ± .29	5.62 ± .37	——	10.52 ± .92	78.0 ± 8.8	
	n	161	70	27	25	25	25
+	M	45.42 ± .24	50.46 ± .40	200.2 ± 8.6	358.4 ± 7.9	716.4 ± 3.4	779.4 ± 4.1
	σ	4.57 ± .17	4.93 ± .28	66.7 ± 6.1	58.2 ± 5.6	24.9 ± 2.4	30.6 ± 2.9
	n	261	247	40	40	25	25
♂	M	29.02 ± .17	29.68 ± .15	45.98 ± 1.19	91.03 ± 1.76	478.1 ± 13.7	738.8 ± 6.5
	σ	4.02 ± .12	3.94 ± .10	11.13 ± .84	16.61 ± 1.25	101.8 ± 9.7	48.1 ± 4.6

to all the larger infrabar and double-infrabar types, namely, a roughened appearance of the eye, due to irregularities in the rows of facets. This peculiarity is not present in bar eyes, and is almost completely recessive in bar over infrabar. In infrabar over round (which is not far from round in facet number) the roughness is variable in extent, and may be not at all evident,—in which case the type can not be distinguished with certainty from homozygous round. In other stocks, where the modifiers are different, it often happens that infrabar over round is regularly conspicuously roughened and is easily distinguishable from round. This roughness of the eyes may be taken as evidence that the infrabar gene is qualitatively different from bar, rather than being merely a fraction of bar.

The table shows that in general when bar and infrabar are both present in an individual, the infrabar produces almost as great an effect in reducing facet number as would another bar, though in the absence of bar the infrabar is far less effective. For example, $\frac{B}{B} = 68$, $\frac{B^i}{B^i} = 348$; but $\frac{B}{B^i} = 74$. And in general, BB^i is practically as effective as BB throughout the table. In two cases the observed differences, though surely not significant, indicate that BB^i is more effective than BB (that is, the combinations with BB and with B^i). Similar relations are shown in other parts of the table.

The most striking relation shown by table 23 is that the relative positions of identical genes affect their action on facet number. There are three similar comparisons to be made:

$$\frac{B}{B} = 68.1 \textit{ versus } \frac{BB}{+} = 45.4$$

$$\frac{B}{B^i} = 73.5 \textit{ versus } \frac{BB^i}{+} = 50.5$$

$$\frac{B^i}{B^i} = 292.6^{8} \textit{ versus } \frac{B^iB^i}{+} = 200.2$$

Since the bar allelomorphs are to be thought of as inhibitors of facet development, it will be seen that this comparison indicates that *two bar allelomorphs lying in the same chromosome are more effective than are the same two allelomorphs when they lie in opposite chromosomes.*

Such an unexpected result must of course be checked up carefully. Only two possibilities of avoiding the above conclusion seem open. The results are due to (1) accidental differences in temperature or modifiers; or (2) the round allelomorph of bar brings about a reduction in facet number just as does bar. Both of these possibilities can be eliminated, as the following paragraphs will show.

There is no temperature effect, since in each case the cultures were reared side by side; and in the case of $\frac{B}{B}$ *versus* $\frac{BB}{+}$, several different tests all gave the same type of result.

If the results are due to accidental modifiers it is scarcely conceivable that these should lie anywhere but near the bar locus, because of the inbreeding to which the stocks have been subjected. As will be shown below, an-

[8] This value for $\frac{B^i}{B^i}$ is different from the one of table 23. It is based on a series reared at the same time as the $\frac{B^iB^i}{+}$ with which it is here compared. The difference between this value and that of the table is probably due to temperature.

other bar chromosome (derived by mutation from the inbred *BB* of this strain) has been found to give results sufficiently close to the bar used here so that the conclusions as to the effect of position must apply also to the new bar. And two other not-bar ("round") chromosomes have been found to give substantially the same result as the one employed here (see below). These facts eliminate the possibility of explaining the result as due to accidental genetic or environmental differences.

The second possible escape from the conclusion as to the effect of position lies in the assumption of an effect produced by the round allelomorph. This has been tested by determining the effect on facet number of reverted bar and reverted infrabar. Round obtained by reversion from homozygous bar or infrabar stocks cannot carry a normal allelomorph on the view advanced in this paper, unless such an allelomorph is already present in the parent stocks. But such a gene is, almost by definition, not an allelomorph of bar; and in any case cannot be supposed to produce the effects here under discussion, since the $\frac{B}{B}$ and the $\frac{BB}{+}$ would both carry it.

Two different rounds by reversion have been introduced into the inbred stocks, by the same method as used for BB^i and B^iB^i. Care was taken never to use flies carrying one of these reverted rounds in the same culture with the old round of the inbred stock, so that it is certain that these new rounds are really due to reversion,—if not from the supposed source, then from some of the allelomorphs within the inbred stock, since a new reversion may have occurred during the process of getting the desired modifiers into the reverted round stocks. One of the reverted rounds, called "rev. *B*," came from the homozygous bar experiments of table 1; the other, called "rev. B^i," came from the homozygous infrabar experiments of table 8. The results of these tests are shown in table 24.

Table 24

Facet counts to test nature of reversions.

	CHROMOSOME TESTED AGAINST		
	BB	BB^i	Y (i.e., ♂)
Wild-type (control)	45.42 ± .24	50.46 ± .40	738.8 ± 6.5
Rev. *B*..........	45.63 ± .45	45.48 ± .47	....
Rev. B^i.........	43.92 ± .26	46.13 ± .36	704.4 ± 8.0

These data suggest that the reversions,—especially the more thoroughly tested one from infrabar,—may give slightly smaller eyes than does the wild-type allelomorph. Certainly they do not give larger eyes. And still more certainly, with *BB* or BB^i they do not give as large eyes as do the corresponding types $\frac{B}{B}$ and $\frac{B}{B^i}$. These data eliminate the second possible method of explaining away the effect of the position of the bar allelomorphs on facet number. The conclusion must stand as stated on p. 137.

It seems probable that such an influence of the relative position of genes on their effectiveness in development may be interpreted in terms of diffusion and localized regions of activity in the cell. This idea is, however, scarcely worth elaborating until more evidence is obtained. It may, however, be pointed out that there is another possible application of the hypothesis of a position effect.

It has been shown by BRIDGES (1921) that in triploid individuals the recessive genes brown, plexus and speck do not become dominant to their normal allelomorphs even when two recessive allelomorphs and one dominant are present. Unpublished data collected by BRIDGES (in part verified in an independently arisen series of triploids in my own experiments) show that this relation is a general one for allelomorphs that do not produce an obviously intermediate diploid heterozygote. But BRIDGES (BRIDGES and MORGAN 1923) has shown that a different relation may occur, even for some of the same genes that show the former relation in triploids. In the example referred to, a portion of the second chromosome, carrying the normal allelomorph of plexus (among other genes) has become attached to a chromosome III. It is possible, therefore, to obtain individuals with two complete second chromosomes, each of these carrying the plexus gene, while a normal allelomorph of plexus is present in the section attached to a chromosome III. Such individuals are plexus in appearance,—none as extreme as those in a pure plexus stock, but far more like such a stock than like the ordinary triploids carrying two plexus and one not-plexus allelomorphs. While it is possible that the difference here is due to a different "balance" of modifying genes in the extra section of chromosome II, it seems likely that the effective agent is a difference in position. It is to be remembered that in Drosophila the homologous chromosomes lie closely apposed in somatic divisions (see METZ 1916), so that there is probably a real difference in relative positions in the two cases.

Besides the comparison of round (ultimately from vermilion stock) with round obtained by reversion from bar and from infrabar, two other derived types have been compared (as to facet number) with the corresponding original types.

An infrabar from bar-infrabar over round (table 13) was introduced into the inbred strain by six successive backcrosses, and was then compared with the old inbred infrabar, both types being made heterozygous for double-bar. The control (old infrabar) value is rather lower than the value given in table 23, probably because the temperature ran slightly higher.

Since the difference between the two means is about 4.5 times its own probable error, it is probably significant, but more extensive data will be required to establish this point.

A female of the inbred series that was double-bar over round, mated to a round male, gave rise to one bar male by mutation. This male was mated to double-bar-over-round females, and the resulting double-bar-over-bar daughters were compared with double-bar over the old inbred bar, derived from cultures made up at the same time and put side by side. Here again there is a slight difference from the value of table 23, perhaps due to a temperature difference.

The difference between the means is slightly over 3 times its probable error, and may be considered as doubtfully significant. Here again, more data are needed.

It may be pointed out that in tables 25 and 26 the derived type is presumably the larger in both cases. It is possible that this result is to be correlated with that recorded in table 24, namely, that round by reversion is perhaps smaller than wild-type. Both relations are consistent with the view that there exists a normal allelomorph of bar that has an effect on facet number opposite to, but much weaker than, that of bar; for both of the derived single types tested were from double type over round, so that these

Table 25

Facet counts from $\frac{BB}{B^i}$.

	NUMBER	MEAN	σ	DIFFERENCE OF MEANS
Old B^i..........	40	34.73 ± .35	3.24 ± .24	2.05 ± .46
New B^i..........	36	36.79 ± .31	2.76 ± .22	

Table 26

Facet counts from $\frac{BB}{B}$.

	NUMBER	MEAN	σ	DIFFERENCE OF MEANS
Old B..........	31	33.10 ± .33	2.76 ± .24	1.77 ± .56
New B..........	30	34.87 ± .45	3.65 ± .32	

derived forms may really be single type plus normal allelomorph. This possibility requires further experimental investigation.

Are Mutations in General Due to Unequal Crossing Over?

One of the first problems raised by the discovery of the nature of bar reversion is as to how widespread may be the phenomenon of unequal crossing over. One direct test has been attempted, making use of MULLER's method of testing for the frequency of occurrence of new lethal mutations.

Females were made up that carried one wild-type X chromosome and one X with the mutant genes scute (locus 0.0), echinus (5.5), crossveinless (13.7), cut (20.0), vermilion (33.0), garnet (44.4), and forked (56.8). Such females were mated to males carrying all the mutant genes named. In such matings it is possible to detect practically all the crossing over that occurs in the X chromosome, except that to the right of forked (about 13 units). Counts were made from individual females, in order to make sure that they carried no lethals. Forty-one wild-type daughters (non-crossovers) were tested from such matings, to see if the non-crossover X chromosomes carried new lethals. The only lethal that occurred was in a paternal (i.e., scute echinus crossveinless cut vermilion garnet forked) chromosome. Thirty-eight double-crossover daughters and one triple-crossover (i.e., a total of 79 crossings over) were also tested; and again no lethals occurred in any of the maternally derived chromosomes, though there were two doubtful cases of new lethals in the chromosomes derived from the multiple-recessive fathers. While this experiment was done on a small scale, it furnishes no indication that crossover chromosomes are more likely to contain new lethals than are non-crossovers.

There is, however, another kind of evidence that argues against any general applicability of the unequal-crossing-over explanation of mutation, namely, the cases in which mutations can be shown to have occurred in the X chromosomes of males, since it may be taken as established that crossing over does not occur between the X and the Y of a male.

We have seen earlier in this paper that infrabar arose from bar in a male, and that its later behavior was not in agreement with the view that it represented a quantitative change in the bar gene, as it should if due to unequal crossing over. I have also obtained yellow, a fused allelomorph, and a lozenge allelomorph under the same circumstances, namely, from mothers with attached X's and in experiments where known sex-linked genes were present, so that breaking apart of the attached X's was known not to have occurred.

Unpublished data are available for 5 other cases of the same sort, either from attached-X or from "high-non-disjunction" mothers, as follows: rudimentary (C. B. BRIDGES), a dusky allelomorph (C. B. BRIDGES), a sable allelomorph (E. M. WALLACE), white (L. V. MORGAN), and a new lozenge allelomorph from lozenge (C. B. BRIDGES). In all of these cases, as in that of infrabar, the mutant type first appeared as a single male.

MULLER (1920) reported the occurrence of white as a "somatic" mutation in a male. From a stock in which white was not present he obtained a male with one wild-type eye and one white eye. This male also transmitted white to some (all that were tested) of his daughters. In the same paper MULLER described briefly a mosaic male that was partly yellow, and transmitted the new character to his offspring. Doctor BRIDGES informs me that he has a similar (unpublished) record for yellow. MOHR (1923 a) reports a similar case for a singed allelomorph, though here some of the X-bearing sperms carried singed, while others did not. I have observed two other such cases,—both in *D. simulans*. The mutant types dusky and fused (both corresponding to the types of the same names in *D. melanogaster*) each appeared first in an individual that showed the new character in only one wing; and in each case tests showed that some of the X-bearing sperms carried the new gene, while others did not. In all cases discussed in the last two paragraphs, genetic tests have established the allelomorphism of the new mutant genes to the old ones whose names they bear.

There is thus clear evidence that mutations have arisen in the following sex-linked loci in the germ-cells of males: yellow (3 times), white (twice), lozenge (twice), dusky (twice), fused (twice), singed (once), bar (once). It should be noted that all these loci are among the more mutable ones of Drosophila.

Another class of cases to which the unequal-crossing-over hypothesis is probably not applicable is that in which mutation can be shown to have occurred at some stage other than maturation. A number of such instances are on record for Drosophila. The mosaic males described above are examples, and a longer list of cases for autosomal mutant types and for sex-linked mutations in females could easily be compiled. But since there is evidence that crossing over does very rarely occur at somatic divisions, this evidence can hardly be considered decisive. In the case of certain types of frequently recurring somatic mutations in plants, however, the mutation occurs far too often to make an appeal to somatic crossing over seem plausible. The clearest example of this sort is the variegated pericarp of maize studied by EMERSON (1917) and others, in which a given gene mutates many separate times in a single individual plant.

Mutations are known in which there appeared to be no crossing over in the region concerned,—both in females and in males where crossing over does not normally occur at all. The previously cited cases of mutations in the X chromosomes of males are examples. These may seem to furnish conclusive evidence that mutation need not be accompanied by crossing over. There is, however, one possibility that needs to be considered in this connection.

Recent results (not yet published, but soon to appear) obtained with triploid females (BRIDGES and ANDERSON) and with females having unlike attached X's (ANDERSON, L. V. MORGAN, and STURTEVANT) have shown that crossing over must normally occur when the homologous chromosomes are doubled: that is, in a "four-strand stage" (in diploid females). These results show also that crossing over may occur between only two of the strands at a given level. Now, if it be supposed that *sister* strands may cross over with each other, there will result chromosomes in which no rearrangement of mutant genes has occurred, since sister strands come from the division of one chromosome and will be identical in the genes that they carry. Yet it is conceivable that such crossing over might be unequal, and in such a case might lead to the production of a new mutation that did not appear to be due to crossing over.

The data presented in this paper show that such an event must be extremely rare in the case of bar, since no clear case was found of bar mutation (in a female) unaccompanied by evident crossing over between forked and fused. The few exceptional cases may be accounted for in this way; but, as pointed out when they were described, it seems at least equally probable that all of them are due to experimental errors. We must conclude that sister strands do not cross over with each other; or, if they do, that the crossing over is rarely, if ever, unequal.

It is therefore unlikely that apparent non-crossover mutations in other loci are to be referred to crossing over between sister strands.

"Presence and Absence" and Quantitative View of Mutation

It will be observed that the hypothesis advocated in this paper makes bar, double-bar and round by reversion (or infrabar, double-infrabar and round by reversion) represent quantitative variations of the same substance. In the case of bar and round, the hypothesis is the same as the original and most special type of quantitative view, the "presence and absence" hypothesis. But the present scheme differs from the earlier ones in that it is based on definite evidence for the occurrence of unequal crossing over. That is, the mechanism whereby the quantitative differences are brought about is an essential part of the hypothesis. In the preceding section we have seen that there is definite evidence to show that unequal crossing over is not usual in the production of new mutant types. It is especially noteworthy that this evidence was derived in part from the white locus of Drosophila and the variegated locus of maize,—two of the best-known examples of loci that have produced large series of multiple allelomorphs. It is clear, therefore, that the bar case does not furnish support to the idea that mutations in general are quantitative in nature. Even with respect to multiple allelomorphs, where the quantitative view has often been urged, it is obvious that, at least in the cases of white and variegated, the bar evidence does not in any way support that view.

Are Deficiencies Due to Unequal Crossing Over?

The "section-deficiencies" described by BRIDGES (1917, 1919) and by MOHR (1919, 1923 b) are probably to be interpreted as due to losses of definite sections of chromosomes. It will be observed that bar reversion has here been treated as due to the loss of a very short section; it may accordingly be described as a deficiency that is too short to show the lethal effect and other properties of the previously described deficiencies. When the case is stated in this way, the question at once arises: is it probable that notch and other deficiencies have also arisen through unequal crossing over? If so, the contrary crossover should be a chromosome that was double for a region corresponding to the deficient section. Such a chromosome has never been identified, but it may be doubted if it would be detected even if present. Furthermore, it might well be lethal even in heterozygous females, in which case it would not be capable of detection.

There is evidence that deficiencies may arise in other ways than by unequal crossing over. In at least one case (BRIDGES and MORGAN 1923) the section missing from a second chromosome was found to be present, but attached to a third chromosome. In this case, then, the deficiency can not have been due to unequal crossing over. The first deficiency described, that for forked and bar (BRIDGES 1917), occurred first as a single female that had obtained the deficient X from her father. Here the deficiency arose (either in a male or very early in the cleavage of a female zygote) at a time when crossing over (and bar reversion) does not normally occur. In the case of notch, also, there is evidence that the deficiency may originate at stages other than maturation. LANCEFIELD (1922) records the occurrence of a notch (probably corresponding to that of *D. melanogaster*) in *Drosophila obscura;* the mutation was first detected as two females from a pair mating that gave numerous offspring. In this case the deficiency must have originated in the gonial cells of one parent, unless the two notch females received their notch chromosomes from the father, in which case it is just possible that they came from two sperms derived from a single spermatocyte. But in this case the hypothesis of unequal crossing over remains as improbable as before. I have observed two cases in *D. melanogaster* that represent "somatic" (i.e., not occurring at the maturation divisions) occurrences of notch. In one case three notch females were produced from a single mother. The X's of the mother were attached, and the notch daughters, like all their sisters, did not carry a paternal X. These three females were all sterile, so here it was not possible to demonstrate that the new type was actually notch; but the numerous characters of notch make the identification very probable. The other case also occurred in a line in which the females all had attached X's. A female, from a line with no notch ancestry, was notch in the left wing but not in the right. The offspring showed that this female was, like her mother, heterozygous for several sex-linked genes. These included scute, 3 units to the left of notch, and crossveinless, 10 units to the right of it. Some of the eggs of the mosaic carried notch, but many of them did not. Furthermore, tests showed that scute and crossveinless were in opposite chromosomes in both types of eggs; that is, the mutation to notch occurred at a cleavage division, and was not accompanied by crossing over between scute and crossveinless. From these three instances we may conclude that the notch deficiency may arise at stages in the life cycle at which crossing over and bar reversion do not normally occur, and, in the third case, there is definite evidence that crossing over did not occur. While it may still be supposed that unequal crossing over will sometimes give rise to section deficiencies, the evidence indicates that the three best-known examples of section deficiencies in Drosophila have not arisen in that way.

Unequal Crossing Over and the Exact Nature of Synapsis

The data on crossing over have all indicated consistently that when two chromosomes cross over they do so at exactly corresponding levels. The case of bar is the first one in which any inequality of crossing-over levels has been detected; and we have seen in the preceding sections that an analysis of other possible instances of such an occurrence makes it probable that they must be explained in some other way. The case of bar is clearly quite exceptional. But it does serve to suggest that the exact correspondence of crossover levels, that is so constant, is not to be referred to a property common to all the genes. For unequal crossing over occurs in females that are homozygous for bar or for infrabar, and in such females these loci are alike in the two X chromosomes that cross over unequally. It is difficult to imagine how the chromosomes can pair so extremely exactly as they must do, unless in some way like genes come to lie side by side. But the present case indicates that this interpretation will have to be applied with some caution.

Summary

1. Sixteen different kinds of changes at the bar locus are shown to occur exclusively, or nearly so, in eggs that undergo crossing over at or near the bar locus.
2. This result can be explained if it is supposed that such crossovers are unequal, so that one daughter chromosome gets two representatives of the bar locus while the other receives none.
3. Only one mutation in this locus has been shown to have occurred in the germ track of a male. This one gave rise (from bar) to a new and less extreme allelomorph called infrabar.
4. Infrabar does not appear to represent a quantitative change in the bar gene.
5. When, by unequal crossing over, bar and infrabar come to lie in the same chromosome, they maintain their separate identities, and may be recovered again.
6. In such double forms the two elements also maintain their sequence in the same linear series as the rest of the genes. It is thus possible to obtain

bar-infrabar and also infrabar-bar. These two types look alike, but can be distinguished by their origin and by the usual tests for determining the sequence of genes.

7. Facet counts are given for all the possible combinations of the following members of the bar series: round, infrabar, bar, double-infrabar, bar-infrabar, double-bar.

8. Analysis of these data shows that two genes lying in the same chromosome are more effective on development than are the same two genes when they lie in different chromosomes.

9. A general survey makes it seem improbable that many mutations in other loci are to be explained as due to unequal crossing over.

LITERATURE CITED

BRIDGES, C. B., 1917: Deficiency. Genetics **2:** 445–465.

1919: Vermilion deficiency. Jour. Gen. Physiol. **1:** 645–656.

1921: Triploid intersexes in *Drosophila melanogaster*. Science N. S. **54:** 252–254.

BRIDGES, C. B., and MORGAN, T. H., 1923: The third chromosome group of mutant characters of *Drosophila melanogaster*. Carnegie Inst. Washington Publ. 327. 251 pp.

EMERSON, R. A., 1917: Genetical studies of variegated pericarp in maize. Genetics **2:** 1–35.

HERSH, A. H., 1924: The effect of temperature upon the heterozygotes in the bar series of Drosophila. Jour. Exper. Zoöl. **39:** 55–71.

HERSH, R. K., 1924: The effect of temperature upon the full-eyed race of Drosophila. Jour. Exper. Zoöl. **39:** 43–53.

KRAFKA, J., 1920: The effect of temperature upon facet number in the bar-eyed mutant of Drosophila. Jour. Gen. Physiol. **2:** 409–464.

LANCEFIELD, D. E., 1922: Linkage relations of the sex-linked characters of *Drosophila obscura*. Genetics **4:** 335–384.

MAY, H. G., 1917: Selection for higher and lower facet numbers in the bar-eyed race of Drosophila and the appearance of reverse mutations. Biol. Bull. **33:** 361–395.

METZ, C. W., 1916: Chromosome studies in the Diptera. II. The paired association of chromosomes in the Diptera and its significance. Jour. Exper. Zoöl. **21:** 213–279.

MOHR, O. L., 1919: Character changes caused by mutation of an entire region of a chromosome in Drosophila. Genetics **4:** 275–282.

1923 a: A somatic mutation in the singed locus of the X chromosome of *Drosophila melanogaster*. Hereditas **4:** 142–160.

1923 b: A genetic and cytological analysis of a section deficiency involving four units of the X chromosome in *Drosophila melanogaster*. Zeitschr. indukt. Abstamm. u. Vererb. **32:** 108–232.

MORGAN, L. V., 1922: Non-criss-cross inheritance in *Drosophila melanogaster*. Biol. Bull. **42:** 267–274.

MORGAN, T. H., and BRIDGES, C. B., 1916: Sex-linked inheritance in Drosophila. Carnegie Inst. Washington Publ. 237. 87 pp. 2 *pls*.

MULLER, H. J., 1920: Further changes in the white-eye series of Drosophila and their bearing on the manner of occurrence of mutation. Jour. Exper. Zoöl. **31:** 443–473.

SEYSTER, E. W., 1919: Eye facet number as influenced by temperature in the bar-eyed mutant of *Drosophila melanogaster*. Biol. Bull. **37:** 168–182.

STURTEVANT, A. H., and MORGAN, T. H., 1923: Reverse mutation of the bar gene correlated with crossing over. Science N. S. **57:** 746–747.

TICE, S. C., 1914: A new sex-linked character in Drosophila. Biol. Bull. **26:** 221–230.

WEINSTEIN, A., 1918: Coincidence of crossing over in *Drosophila melanogaster*. Genetics **3:** 135–173.

ZELENY, C., 1919: A change in the bar gene of Drosophila involving further decrease in facet number and increase in dominance. Jour. Gen. Physiol. **2:** 69–71.

1921: The direction and frequency of mutation in the bar-eye series of multiple allelomorphs of Drosophila. Jour. Exper. Zoöl. **34:** 203–233.

1922: The effect of selection for eye facet number in the white bar-eye race of *Drosophila melanogaster*. Genetics **7:** 1–115.

1923: The temperature coefficient of a heterozygote with an expression for value of a germinal difference in terms of an environmental one. Biol. Bull. **44:** 105–112.

ZELENY, C., and MATTOON, E. W., 1915: The effect of selection upon the "bar-eye" mutant of Drosophila. Jour. Exper. Zoöl. **19:** 515–529.

Author's Notes, 1961

Several of the loose ends in this paper were re-examined later, with results reported in paper no. 16.

It was shown later by Muller, Prokofyeva-Belgovskaya, and Kossikov (1936, C. R. Acad. Sci. U.S.S.R. **1:** 87–88) and by Bridges (1936, Science, **83:** 210–211) that Bar is due to the duplication of a small section of chromosome (salivary gland bands 16A1 through 16A6). Double-bar has this section present three times. This result puts the data of the present paper in a different light, but does not invalidate any of the conclusions. The position effect, for example, still stands as stated.

14

SEQUENCE OF CORRESPONDING THIRD-CHROMOSOME GENES IN *DROSOPHILA MELANOGASTER* AND *D. SIMULANS* †

It has been shown (Sturtevant, 1920; 1921*a,b*) that *Drosophila melanogaster* and *D. simulans* may be crossed, and so it is possible to test suspected corresponding genes in the two species. There is, for example, a recessive vermilion-like eye-color (known as scarlet) in each species. If scarlet of either species is crossed to the wild type (or to any other eye-color) of the other species, the resulting hybrids have the wild-type or "red" eye-color. Therefore each species carries the dominant wild-type allelomorph of the scarlet mutant of the other species. But if scarlet *melanogaster* is crossed to scarlet *simulans,* the resulting hybrids have scarlet eyes. Therefore the two scarlets must represent modifications of the same wild-type gene, or at least of wild-type genes that have like effects on development. Such genes as the two scarlets are described in this laboratory by the term "corresponding," rather than (as formerly) "allelomorphic"—since it seems desirable to reserve the latter term for genes that regularly separate at maturation. In the present case the fact that all the hybrids are sterile prevents any test of the segregation.

A study of thirteen corresponding genes in the X-chromosomes of *melanogaster* and *simulans* (Sturtevant 1921*a,* and unpublished data) shows that the sequence of the loci concerned is the same in the two species. But it was shown by Sturtevant (1921*c*) that the sequence is not identical in the third chromosomes. If the three corresponding genes scarlet (st), peach (p), delta (Δ, formerly called "deltoid" in *simulans*) are mapped, the sequence is found to be stpΔ in *melanogaster,* stΔp in *simulans.*

The present paper is a report on three additional pairs of corresponding genes in this chromosome, which make it possible to compare the maps in more detail. The three loci concerned, already well known in *melanogaster,* are sepia, hairless, and claret. In the cases of sepia and claret the mutant types of *simulans* are closely similar in appearance to the types of the same names in *melanogaster,* and it was a simple matter to obtain sepia hybrids and claret hybrids by crossing the types concerned. The evidence that these

† In collaboration with C. R. Plunkett. Reprinted from the Biological Bulletin, Vol. 50, 1926, pp. 56–60.

are corresponding genes is thus of the same nature as in the cases of scarlet, peach, and the sex-linked genes previously reported (Sturtevant, 1921*a*, *b*).

Hairless is, however, dominant in both species, so that the same difficulty was encountered here as in the case of delta—the hybrids show the mutant character even when it is introduced from only one parent. But in each species hairless, like delta, is lethal when homozygous; accordingly this lethal effect was used to test the correspondence of the hairless genes. Since the earlier tests of delta were not as clear-cut as might be desired, the new tests were carried out with flies containing both genes, and served to show that both loci correspond.

D. melanogaster females that were Δ/H—i.e., carried delta in one third chromosome and hairless in the other one—were crossed to *simulans* males that were also Δ/H. There resulted 169 hybrid offspring: 164 delta hairless, 4 delta, 1 hairless, 0 wild-type. If the delta genes do not correspond, so that Δ/Δ is viable, the cross should give half as many delta flies as delta hairless; if the hairless genes do not correspond, the hairless class should also be half as large as the delta hairless class. The few delta and hairless offspring produced are evidently due to crossing over in the mothers, 1.5 per cent. of whose eggs are expected to carry neither Δ nor H. The result was checked by crossing *melanogaster* females that were ΔH/+ by *simulans* males that were Δ/H. From this cross, all the ΔH eggs should give inviable zygotes, and equal numbers of delta and hairless offspring should result. There were obtained the following hybrids: 1 delta hairless, 225 delta, 226 hairless, 0 wild-type. The one delta hairless remains unexplained, for in this case the mothers carried C_{III}, which should have prevented all crossing over between Δ and H. This individual probably represents a new mutation, though it is possible that it was really homozygous for Δ or for H, since the somatic effects produced by these genes are in many respects on the same characters but in opposite directions, so that heterozygosis for one of them might conceivably tend to counteract the lethal effect of the other one. Since all the *melanogaster-simulans* hybrids are sterile, it was not possible to test this exceptional individual. In any case, this one specimen cannot vitiate the conclusion that both delta and hairless are corresponding genes in the two species.

Fig. 1 shows the linkage relations of the six corresponding genes in the

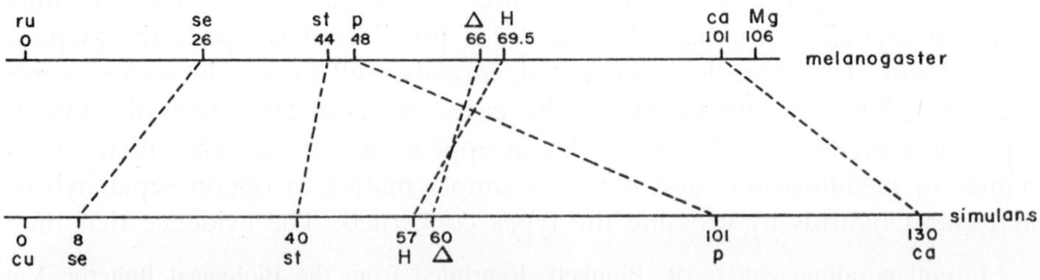

Figure 1

two species, together with the left-most and right-most known loci of each species.

The *melanogaster* map is based on the values given by Bridges and Morgan (1923). The *simulans* map is based on unpublished data, including counts from crosses involving several genes not here recorded. The intervals in this map are not yet accurately determined, because of a lack of favorable mutations; but they are more likely to be lengthened than to be shortened by further data. The sequence of the corresponding genes in *simulans* is clearly established, as the following data will show. Claret cannot be distinguished somatically from peach, so its position has been determined with respect to minute, which is very close to it.

Table I.

$\frac{+\ \mathrm{st}\ +}{\mathrm{se}\ +\ \Delta}$ ♀ × se st + ♂.

Non-Crossovers.		Single Crossovers.				Double Crossovers.		Total.
0.		1.		2.		1, 2.		
st	seΔ	Δ	sest	stΔ	se	+	sestΔ	
121	142	70	44	43	79	21	8	528

Table II.

$\frac{+\ \mathrm{H}\Delta\ +}{\mathrm{st} + + \mathrm{p}}$ ♀ × st + + p♂.

0.		1.		2.		3.		1, 2.		1, 3.		2, 3.		1, 2, 3.		T.
HΔ	stp	p	stHΔ	Hp	stΔ	HΔp	st	Δ	stHp	+	stHΔp	H	stΔp	Δp	stH	
320	330	69	75	7	11	181	181	3	0	41	29	3	1	0	1	1,252

Minute (which is dominant) sometimes overlaps wild-type in appearance; accordingly only the minute offspring have been recorded in Table III. In the experiment reported in Table IV. this overlapping seemed to be slight,

Table III.

$\frac{+ + + +}{\mathrm{st}\Delta\mathrm{pM}}$ ♀ × st + p + ♂.

0.	1.	2.	3.	1, 2.	1, 3.	2, 3.	1, 2, 3.	T.
stΔpM	ΔpM	pM	M	stpM	stM	stΔM	ΔM	
126	48	77	51	22	25	28	4	381

Table IV.

$\frac{+\ \Delta\ +\ M}{st\ +\ ca\ +}$ ♀ × st + ca + ♂.

0.		1.		2.		3.		1, 2.		T.
ΔM	stca	stΔM	ca	stM	Δca	stcaM	Δ	M	stΔca	
39	19	4	2	24	18	0	2	8	3	119

so all the offspring are recorded. It is possible, however, that the two M-ca crossovers may be due to such overlapping. More extensive counts from M/ca females, in experiments in which no other loci were studied, have not given any crossing over. There is an unusually low scarlet delta crossover value shown in Table IV., but the count is so small that no special significance is to be attached to this fact.

There can now be no doubt that one half or more of the right limb of the third chromosome (in terms of map distance) has been inverted in one or the other species—since there is no clue as to which is the original sequence. The left-hand limit of the inverted section lies between scarlet and peach of *melanogaster,* which is presumably very close to the spindle-fiber attachment, and is certainly in a region where the mutant genes are mapped very close together and where coincidence of crossing over is very high (Bridges and Morgan, 1923). It will be interesting, as more mutant genes are studied in *simulans,* to see how this same material behaves when it lies further from the middle of the chromosome.

In the earlier account (Sturtevant, 1921*c*) it was suggested that the rearrangement of parts demonstrated for the *melanogaster* and *simulans* third chromosomes might be explained as resulting either from a duplication or from an inversion of a section. The new data here reported speak strongly for the inversion interpretation.

One might surmise that this inversion of a piece of a chromosome is the fundamental factor in the specific difference between *melanogaster* and *simulans* and the sterility of their hybrids. But it has been shown (Sturtevant, 1920) that at least some of the factors concerned in the inviability of certain classes of the hybrids are in the X chromosomes, where there has certainly been no such inversion.

This case shows that identity of sequence is not a necessary condition for correspondence of genes in different species. It also suggests that we need not be surprised should such an inversion be found within a species.

BIBLIOGRAPHY

Bridges, C. B., and Morgan, T. H., 1923: The Third-chromosome Group of Mutant Characters of *Drosophila melanogaster*. Carnegie Inst. Wash., publ. 327, 251 + 10 pp., 3 plates.

STURTEVANT, A. H., 1920: Genetic Studies on *Drosophila simulans*. I. Introduction. Hybrids with *D. melanogaster*. Genetics, **5:** 488–500.

1921 a: Genetic Studies on *Drosophila simulans*. II. Sex-linked Group of Genes. Genetics, **6:** 43–64.

1921 b: Genetic Studies on *Drosophila simulans*. III. Autosomal Genes. General discussion. Genetics, **6:** 179–207.

1921 c: A Case of Rearrangement of Genes in *Drosophila*. Proc. Nat. Acad. Sci., **7:** 235–237.

Author's Note, 1961

The delta hairless hybrid from ΔH/+ *melanogaster* by Δ/H *simulans* was probably due to one of the rare double crossovers that occur in females heterozygous for inversions such as "C_{III}" (see paper no. 15).

15

A CROSSOVER REDUCER IN *DROSOPHILA MELANOGASTER* DUE TO INVERSION OF A SECTION OF THE THIRD CHROMOSOME †

A wild stock of *D. melanogaster,* collected at Prague in 1925 by Dr. A. Brozek, was found to contain a crossover reducer for the right half of the third chromosome. This reducer (called $C_{III\ B}$), like those previously described by Sturtevant, Muller, Payne and Ward, lies in the region affected. Females heterozygous for $C_{III\ B}$ have been tested for crossing over throughout the chromosome. It is evident from these data that crossing over is substantially normal to the left of stripe[1] (locus 62.0), but greatly reduced to the right of that locus. Tests of flies resulting from crossing over have shown that $C_{III\ B}$ itself lies to the right of stripe, in the region chiefly affected.

$C_{III\ B}$ thus affects crossing over in the same regions as C_{III} (Muller 1916) and $C_{III\ PR}$ (Payne 1924). The results of Muller, and results of my own that are still largely unpublished, show that C_{III} and $C_{III\ PR}$ each give free crossing over in the region concerned when they are homozygous; but C_{III} over $C_{III\ PR}$ gives the reduced crossing over characteristic of either when heterozygous. That is, C_{III} and $C_{III\ PR}$ are not allelomorphic. Similar tests show that $C_{III\ B}$ is allelomorphic to C_{III} but not to $C_{III\ PR}$. The conclusions reached here as to the nature of $C_{III\ B}$ thus apply also to C_{III}.

Most of the linkage data have been obtained from females that carried one $C_{III\ B}$ chromosome and one chromosome with the mutant genes scarlet, stripe, sooty, rough and claret [1]. Such females, mated to males carrying all the recessive genes concerned, have given the results shown in Table 1.

† Reprinted from the Biologisches Zentralblatt, Vol. 46, 1926, pp. 697–702.

[1] The mutant genes here discussed are listed and described by Bridges and Morgan (1923) and by Morgan, Bridges, and Sturtevant (1925). There follows a list, showing the name, character affected, symbol, and locus on the standard third chromosome map. Stubble and Minute-g are dominant types; all the others are recessive.

scarlet (eye color): st: 44.0
Stubble (bristle shape): Sb: 58.2
stripe (thorax shape and pattern): sr: 62.0
sooty (body color): e^s (allelomorph of ebony): 70.7
rough (eye surface): ro: 91.1
claret (eye color): ca: 100.7
Minute-g (bristle size): Mg: 106.2.

Table 1

$$♀ \frac{C_{III\,B}}{st\ sr\ e^s\ ro\ ca} \times ♂\ st\ sr\ e^s\ ro\ ca$$

Wild-type	st sr e^s ro ca	sr e^s ro ca	st	e^s ro ca	st sr	st sr e^s	e^s ro	Total
2214	2058	219	238	3	4	1	1	4738

Tests of these offspring have shown that the wild-type, scarlet, and scarlet stripe individuals carry $C_{III\,B}$; the scarlet stripe sooty rough claret, stripe sooty rough claret and sooty rough claret ones do not. That is, $C_{III\,B}$ lies to the right of stripe. The single sooty rough individual also carried $C_{III\,B}$; it was not present in the scarlet stripe sooty individual. These data are all consistent with the interpretation that $C_{III\,B}$ occupies a single locus that lies to the right of claret, though it is then necessary to suppose that the scarlet stripe sooty individual came from a double crossover egg (e^s-ro, ca-$C_{III\,B}$). That is, there were two double crossovers in which one or both of the crossovers were to the right of sooty—but there were no single crossovers in this region. This result seems improbable, and the data now to be described lead to a different and more plausible interpretation of $C_{III\,B}$.

Females homozygous for $C_{III\,B}$ were produced by crossing the lines derived from the scarlet stripe and sooty rough crossovers just discussed. When such females were tested, the results showed clearly, in several different experiments, that the sequence was now scarlet stripe rough sooty instead of the usual scarlet stripe sooty rough. Figures 1 and 2 show the intervals of the usual map (from Bridges and Morgan 1923) and of that based on homozygous $C_{III\,B}$ data. Scarlet (locus 44 on the standard map) is arbitrarily taken as the zero point for this comparison, since loci to the left of scarlet have not yet been studied in homozygous $C_{III\,B}$.

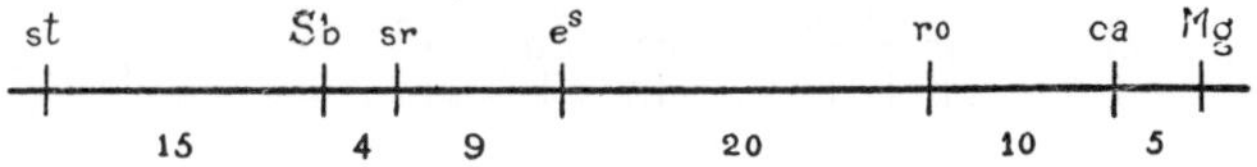

Figure 1

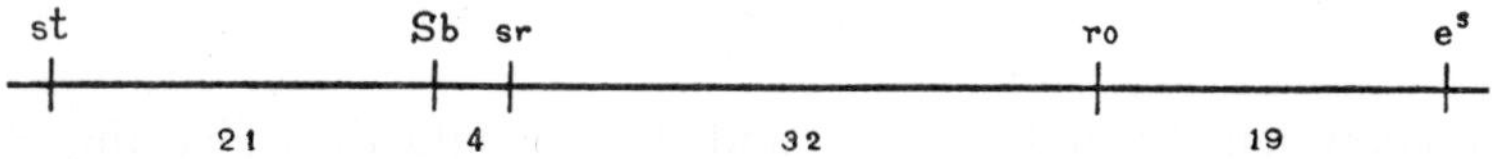

Figure 2

The natural conclusion from these data is that $C_{III\,B}$ is not a gene but is itself simply an inverted section including the loci of sooty and rough. I have suggested (Sturtevant 1921) that such an inversion should produce a crossover reducer effective only when heterozygous; and Sturtevant and Plunkett

(1926) have shown the existence of an inverted section in the third chromosome of *D. simulans* as compared to that of *D. melanogaster*.

In the present case such an interpretation may be elaborated as follows. The original $C_{III\ B}$ chromosome was $+^{st}\ +^{sr}\ +^{ca}\ +^{ro}\ +^{e}$, instead of the usual $+^{st}\ +^{sr}\ +^{e}\ +^{ro}\ +^{ca}$ (the reason for including claret in the inverted section will appear below). The females whose offspring are recorded in Table 1 thus may be represented as in Figure 3, where „$+^{1}$" and „$+^{2}$" represent the end loci of the inverted section. In these females the chromosomes failed to

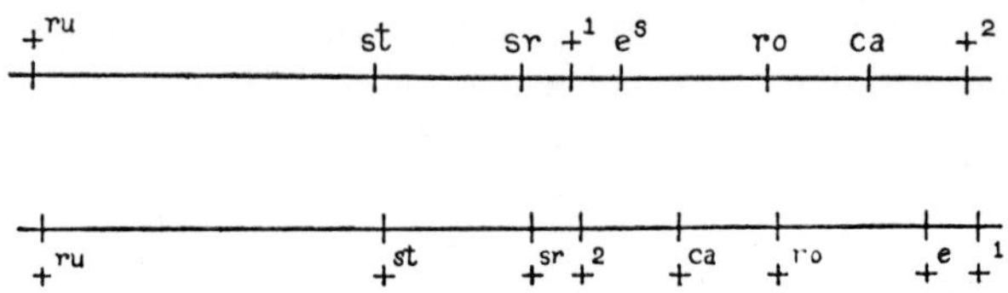

Figure 3

conjugate normally in the region between $+^{1}$ and $+^{2}$, because like parts were not opposed. Crossing over consequently failed to occur. The reduction in crossing over between stripe and sooty may mean that the inverted section actually reaches almost to stripe, or it may mean that there is a reduction in crossing over adjacent to the inverted section.

From such a female we must, on the hypothesis suggested, suppose that double crossing over gave a chromosome that was $+^{st}\ +^{sr}\ +^{ca}$ ro e^{s}. A glance at Figure 3 will show that such a result demands an additional hypothesis; for, as the formula stands, double crossing over would give $+^{st}\ +^{sr}$ e^{s} ro —i.e. the normal sequence, for these four loci. The observed result may be accounted for if we suppose that the chromosomes in question happened, in this one egg, to conjugate so that the inverted section was opposed to the corresponding genes in the normal chromosome, as in Figure 4. Double cross-

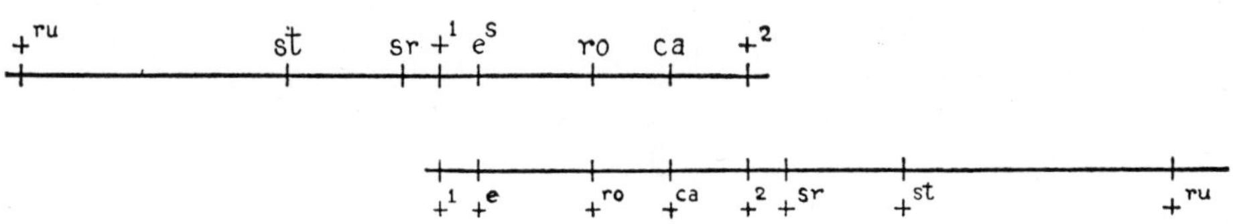

Figure 4

ing over (between $+^{1}$ and $+^{e}$, $+^{ro}$ and $+^{ca}$) would then give the observed combination. The scarlet stripe sooty individual of Table 1, that did not carry $C_{III\ B}$, was due to double crossing over in the intervals $+^{e} - +^{ro}$ and $+^{ca} - +^{2}$.

It may be noted further that single crossing over anywhere in the region where like is opposed to like would now give rise to chromosomes that would have the genes proper to one end of a normal chromosome represented twice, and those proper to the other end entirely lacking. In diploid Drosophilae such chromosomes would presumably be lethal; but in the cases observed a second crossover restored the normal complement of genes and made the new chromosome capable of forming part of a normally functioning zygote. In fact, tests show that the sooty rough $C_{III\ B}$ chromosome does not even have a recessive lethal effect, so that it must be supposed to contain one and only one complete set of genes characteristic of the third chromosome.

A comparison of Figures 1 and 2 shows that the directly comparable values st Sb, Sb sr, and e^s ro are not strikingly different. The only real discrepancy occurs in the case of st Sb, which is furthest from the inverted section and is also well known (Bridges and Morgan 1923) to be more variable in frequency of crossing over than are regions further to the right. If one supposes that the inversion has not seriously affected crossing over in the component parts of the right end, analysis of these maps leads to the conclusion that the inverted section extends to the end of the normal chromosome. Not more than 9 of the 32 units between sr and ro in Figure 2 can represent normal unchanged material, since the sr e^s value in Figure 1 is 9, and e^s is known to be in the inverted piece. At least 23 units that lie to the left of ro in Figure 2 must thus be supposed to lie to the right of it in Figure 1; but the locus furthest to the right among those known (Mg) is only 15 units beyond ro. One is tempted to predict that the standard map will ultimately be lengthened by about 8 units through the finding of new loci; but it is equally probable that some or all of these 8 units are to be accounted for by supposing slight changes in frequency of crossing over. In any case, it seems certain that claret, the most commonly used locus near the right end, lies in the inverted section; since only on this assumption can the st sr e^s crossover of Table 1 be simply explained.

The two crossovers in the presence of heterozygous C_{III} that were described by Muller (1918, pp. 444 and 449) can both be explained as coming from chromosomes that conjugated as in Figure 4. Both were then double crossovers within the inverted section. One of them was also a crossover in some region to the left of the inverted section, indicating that the free ends shown in Figure 4 may bend around and conjugate with each other accurately enough to allow crossing over. The second case reported by Muller was interpreted by him as resulting from two independent crossings over in different individuals, but the present interpretation seems to fit the facts as reported better than does that view, and is in fact necessitated by the discovery that C_{III} is an inversion.

There are indications in the literature which suggest that some of the other crossover reducers may also be due to inversion of sections of chromosomes. Payne (1924), for $C_{III\ PL}$ and $C_{III\ PR}$, and Ward (1923, for $C_{II\ CyL}$ and $C_{II\ CyR}$, report a few double crossovers in regions where single crossovers are rare or absent. These crossovers seem not to have been carefully tested; but the experiments are now being repeated in order to see if such rare double crossovers are of the same nature as those obtained from $C_{III\ B}$.

LITERATURE CITED

BRIDGES, C. B., and MORGAN, T. H., 1923: The third-chromosome group of mutant characters of *Drosophila melanogaster*. Carnegie Inst. Washington, publ. 327: 251 pp.

MORGAN, T. H., BRIDGES, C. B., and STURTEVANT, A. H., 1925: The genetics of *Drosophila*. Bibliogr. Genet. **2:** 1–262.

MULLER, H. J., 1916: The mechanism of crossing over. Amer. Nat. **50:** 193–221, 284–305, 350–366, 421–434.

1918: Genetic variability, twin hybrids and constant hybrids, in a case of balanced lethal factors. Genetics **3:** 422–499.

PAYNE, F., 1924: Crossover modifiers in the third chromosome of *Drosophila melanogaster*. Genetics **9:** 327–342.

STURTEVANT, A. H., 1919: Inherited linkage variations in the second chromosome. Carnegie Inst. Washington, publ. 278: 305–341.

STURTEVANT, A. H., 1921: A case of rearrangement of genes in *Drosophila*. Proc. Nat. Acad. Sci. **7:** 235–237.

STURTEVANT, A. H., and PLUNKETT, C. R., 1926: Sequence of corresponding third-chromosome genes in *Drosophila melanogaster* and *D. simulans*. Biol. Bull. **50:** 56–60.

WARD, L., 1923: The genetics of curly wing in *Drosophila*, another case of balanced lethal factors. Genetics **8:** 276–300.

Author's Notes, 1961

The analysis given fails to take into account the centromere and its effects on segregation. These relations were cleared up later (see paper no. 22).

The conclusions as to the length of the inversion—now known as In(3R)C—were essentially confirmed by Bridges (1937, Carnegie Year Book, **36:** 301), whose analysis of the salivary gland cells shows that it extends from section 92E (between Sb and e) to 100F2 (just 3 bands proximal to the end of the chromosome).

16

A FURTHER STUDY OF THE SO-CALLED MUTATION AT THE BAR LOCUS OF *DROSOPHILA* † [1]

Introduction

An earlier study (STURTEVANT 1925) of the "mutations" at the bar locus of *Drosophila melanogaster* led to the conclusion that these mutations are due to unequal crossing over at the bar locus. Reverted bar is no-bar, having no allelomorph at the bar locus; double-bar (the ultra-bar of ZELENY) has two bar genes lying next each other in the linear series of genes.

The present paper gives the results of a few experiments designed to clear up certain points left doubtful by the earlier study.

Time of Occurrence of Bar Mutation

ZELENY (1921) recorded five cases in which more than one reversion appeared in a single culture, and concluded from this evidence that reversion probably occurs, sometimes at least, in oogonial cells. Since the five cultures in question were not from individual females they do not prove that several reversions may be produced by a single mother, but they are at least suggestive. If, following this suggestion, one assumed that unequal crossing over leading to bar reversion occurred at some stage earlier than the maturation divisions, it would be expected that normal crossing over between forked and fused would occur after it. The result should be that about 3 percent of the "mutations" at the bar locus would appear to be unaccompanied by forked-fused crossing over—and the data on record show a small number of such exceptional cases. A further consequence, open to experimental test, should be that bar reversion would not interfere with crossing over in neighboring regions.

Accordingly an experiment has been carried out with females homozygous for infrabar (used instead of bar in order to make the separation for eye-color easier), and heterozygous for garnet-2 forked and fused (see figures 1 and 2).

† Reprinted from Genetics, Vol. 13, 1928, pp. 401–409.

[1] Contribution from the CARNEGIE INSTITUTION OF WASHINGTON.

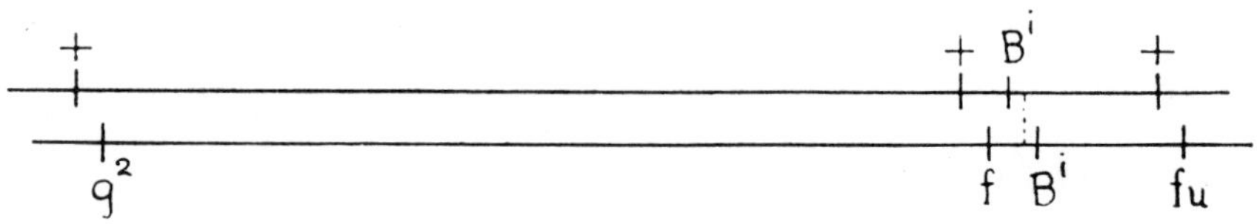

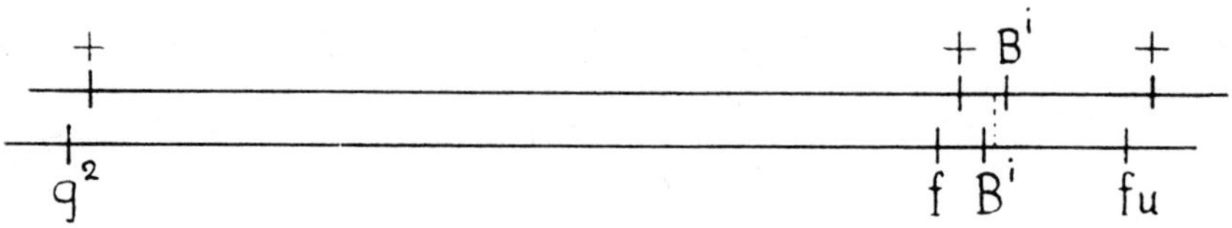

Figure 1 (above) and Figure 2 (below)

Table 1

Offspring from $\frac{B^i}{g^2fB^if_u}$ ♀ × $g^2fB^if_u$ ♂.

B^i									B^iB^i				+			
0		1		2		1,2		Total	♀*		♂		♀*		♂	
+	g^2ff_u	ff_u	g^2	f_u	g^2f	f	g^2f_u		f_u	g^2f	f_u	g^2f	f_u	g^2f	f_u	g^2f
19536	14939	2351	2678	574	647	4	7	40736	4	2	2	6	4	7	11	4

* The "mutant" females were, of course, all heterozygous for B^i, received from the father.

A small series (1647 individuals) from $\frac{f\ B^i}{g^2\ \ B^if_u}$ also gave a reversion (ff_u) that was not a g^2f crossover. In the two series there is a total of 41 "mutant" individuals, all of them crossovers between forked and fused, but none of them garnet-2 forked crossovers. There are 5040 garnet-2 forked crossovers among 40736 unmutated offspring of the table, or 12.4 percent, in good agreement with the 12.1 percent recorded by MORGAN, BRIDGES and STURTEVANT (1925).

If mutation and garnet-2 forked crossing over are unrelated, the chance that none of the 41 mutant individuals should be a garnet-2 forked crossover is therefore $(1 - .124)^{41} = .004$, or 1 in 250. One may safely conclude that the unequal crossing over that gives rise to mutation at the bar locus interferes with crossing over in the garnet-2 forked region, just as does ordinary crossing over between forked and fused, and hence that the unequal crossing over occurs at the same time as ordinary crossing over.

There must, therefore, be other explanations of the two relations that led to the experiment. The occurrence of mutations not accompanied by forked fused crossing over is open to two interpretations, as previously pointed out (STURTEVANT 1925). Crossing over between sister strands is a possible explanation, but if it occurred at all would be expected to give many more apparent non-crossover mutations than have been observed. The most probable view seems to be that all such records are due to experimental errors. It is to be noted that none occurred in the present series of experiments.

As pointed out above, ZELENY'S data do not prove that more than one mutation was produced by a single mother in any case. However, in the present series there were four separate cultures in which a single mother produced two mutant offspring, as follows:

Culture	*Mutant offspring*
21794	$B^iB^if_u$, $g^2fB^iB^i$
22254	f_u, g^2f
23611	f_u, f_u
23898	$g^2fB^iB^i$, g^2f

Examination of these data shows that in three of the four cases the two individuals cannot have been produced by one crossing over.[2] In the first case, for example, the crossing over that produced a $B^iB^if_u$ chromosome would also give a g^2f (figure 1)—but could not give $g^2fB^iB^i$, the combination actually recovered. It follows that the production of more than one mutant offspring by a single mother must be due to the occurrence of more than one unequal crossing over in that mother. Whether or not such coincidences are more frequent than would be expected with a random distribution can hardly be determined without the collection of a vast amount of accurate data.

Relative Numbers of Reversions and Double-Bars

On the hypothesis of unequal crossing over a homozygous bar stock should produce equal numbers of double-bar and wild-type "mutants," but all observers have found distinctly more wild-types than double-bars. I have suggested (STURTEVANT 1925) that this is due to two factors—lower viability of double-bar and greater likelihood of overlooking mutant individuals. The viability factor can be directly measured. Using the inbred stocks prepared for the facet counts recorded in my earlier paper, I have made counts from double-bar over round mated to round and to bar. A small series gave 60 double-bar over bar females to 82 bar over round, and 126 double-bar males to 161 round males. Equality is expected in each case, and the deviations are due to differential viability. It follows that, in the combinations met with as mutant individuals in a homozygous bar stock, double-bar has a viability about 75 percent that of round.

[2] Unless by chance one of the rare "normal" forked fused crossovers followed the mutation in one egg and not in the other in each case. The chance that this should have happened in all three instances is negligible.

ZELENY (1921) records the production of 52 reversions and 3 double-bars among 85008 specimens examined. In addition there were at least 5 double-bars that were not tested, so that we may take 52:8 = 6.5:1 as the observed ratio. ZELENY argues that there were probably very few mutant individuals overlooked—probably only a few of the double-bar females. I obtained (STURTEVANT 1925) 8 reversions and 2 double types from homozygous bar, 18 reversions and 3 double types from homozygous infrabar, and 2 reversions and 3 double types from bar over infrabar—a total of 28:8 = 3.5:1. This is somewhat closer to the expected equality than ZELENY'S 6.5:1, but the experiments here reported for homozygous infrabar gave 27:14 = 1.9:1. This is still a definite excess of reversions, over what can be accounted for by differential viability, especially as the viability difference is less for double-infrabar than for double-bar, but more carefully conducted experiments and more experience having lowered the ratio from 3.5:1 to 1.9:1, I feel confident that there is still room for technical improvement sufficient to remove the remaining discrepancy.

Position Effect

It was shown (STURTEVANT 1925) that two bar genes lying in the same chromosome are more effective in reducing facet number than are the same two genes when they lie in different chromosomes. It seemed possible that such a relation might also hold for non-allelomorphic genes, but the one test, now to be described, has given a negative result.

The two dominant genes delta and hairless lie 3 units apart in the third chromosome (BRIDGES and MORGAN 1923). Hairless causes a reduction in bristle number, and delta partially counteracts this effect. These genes were selected for study because they lie close together, and because bristle number is convenient for quantitative work. An inbred strain was established, in which repeated brother-sister matings and repeated use of crossovers between delta and hairless made it probable that modifiers were uniform. Matings were made in parallel, delta × hairless and delta hairless × wild-type. All offspring from these matings were reared at 26°C. The results are shown in table 2.

Table 2

Number of dorsocentrals.

Mating	3	4	5	Total	Mean
Δ × H	5	95	0	100	3.95
ΔH × +	10	167	1	178	3.95

While the numbers here are small, the exact agreement in mean number of dorsocentrals makes it seem certain that there is no position effect.

As pointed out in my earlier paper, the pairing of homologous chromosomes that occurs in somatic divisions in Drosophila suggests that there may be a position effect normally present of such a nature that allelomorphs reinforce each other's effects. Such an effect was offered as an explanation of the different dominance relations observed in triploids and in cases of translocation (MORGAN, BRIDGES and STURTEVANT 1925). A new set of facts makes this interpretation now seem improbable. I have shown (STURTEVANT 1926) that the C_{III} present in ebony and other stocks is due to the inversion of a large section of the third chromosome, and there is now evidence that the $C_{II\ L}$ of the curly stock and two other $C_{III\ R's}$ are due to similar inversions. In flies heterozygous for such inversions the position relations of many genes are greatly changed, but no difference in somatic appearance has yet been detected in such flies. It follows that changes in position of this order are not usually of developmental significance. One may conclude that the association of the two elements in double-bar is much more intimate than the association between identical loci in homologous chromosomes, and that the "position effect" perhaps rests on something more than mere closeness together.

Facet Counts of Derived Types

Facet counts recorded (STURTEVANT 1925) for a reversion from bar and one from infrabar suggested that these derived rounds might be different from the wild-type, since both strains gave lower facet numbers with bar-infrabar (one of them also with double-bar and in males) than did wild-type. The inference that there is a wild-type allelomorph of bar, absent in reversions, and having an effect on facet-number opposite to that of bar, was further tested by studying an infrabar from bar-infrabar over round, and a bar from double-bar over round. Both of these derived types when tested against double-bar, gave somewhat larger facet-numbers than the controls (original infrabar and original bar, respectively). It was suggested that perhaps the derived types carried wild-type allelomorphs, which partially counteracted the effects of the bar and infrabar genes. However, the differences were small in all these experiments, and it was emphasized that more tests were needed.

Several new derived types have now been studied, and have not borne out the assumption of a wild-type allelomorph.[3]

Two new infrabars, both from double-infrabar over reverted infrabar,

[3] The reverted double-infrabar and the first derived bar arose in the inbred forked stocks that were used in the earlier study. The other derived types were crossed at least five times to these stocks before counts were made. All the other stocks used in the present experiments were the same as those described in the earlier paper. The arguments for supposing modifiers to be alike are the same as in the earlier experiments. The temperature was 25° ± 1°.

would have been expected to give the same results as the old infrabar, even if there is an effective wild-type allelomorph—since such an allelomorph should not be present in either the old or the derived types. The result of a test against double-bar was: control (old infrabar) 35.7 facets, new infrabar 33.6 and 35.4 facets, respectively (individuals counted, 51, 30, and 30). The expectation is clearly correct; there are no significant differences.

A new reverted double-infrabar was tested, and at the same time the reverted infrabar of the earlier experiments was used. These and wild-type were tested against double-bar. Forty females of each type were counted, with these results: wild-type, 43.8; reverted infrabar, 44.4; reverted double-infrabar, 43.2. The reverted infrabar differs from wild-type in the opposite direction from the earlier test, and none of the differences here seem significant.

Two new bars, both from double-bar over round, were tested in three ways, as shown in table 3.

Table 3

Facet counts.

	OVER B^iB^i		HOMOZYGOUS ♀		MALE	
	n	*M*	*n*	*M*	*n*	*M*
Bar (control)	45	37.4	25	61.8	30	69.8
First derived bar	45	38.2	25	54.4	30	67.6
Second derived bar	45	35.4	25	49.6	60	61.0

Both of the derived types give lower facet numbers than the control, and in the case of the second derived type this difference is probably significant—but the difference is in the opposite direction from that found in the earlier experiments.

The conclusion seems to be justified that the observed differences are due to uneliminated modifiers or to chance environmental differences. In any case the new results do not at all support the hypothesis of a wild-type allelomorph affecting facet-number.

The three homozygous bar types recorded above averaged 55.3 facets; a double-bar over round series of 40 females reared at the same time gave a mean of 43.8. This gives a position effect of the same kind as the data in the earlier paper, but rather less in extent.

Reverted Infrabar and Crossing Over

Reverted bar or infrabar may be considered as a deficiency which is so short that it does not have the lethal effect of most deficiencies.

It has been shown by BRIDGES (1917, 1919) and MOHR (1923) that females heterozygous for the usual type deficiency give no crossing over in the section covered by the deficiency. This fact suggested another test of the relation between reverted bar and wild-type. It has been shown (STURTEVANT 1925) that any double-type over wild-type gives about 0.1 percent of single-type, the mutant individuals always being forked-fused crossovers. That is, in such heterozygous females there is crossing over between the two elements of the double-type. A test has now shown that reverted infrabar gives the same result as normal with double infrabar (table 4).

Table 4

Offspring from $\frac{B^iB^i}{fr_{ev}f_u}$ ♀ × $fr_{ev}f_u$ ♂.

0		1		2		TOTAL	MUT.
B^iB^i	ff_u	f_u	fB^iB^i	$B^iB^if_u$	f		fB^i
1474	1415	1	3	35	42	2968	2

Is a Wild-Type Allelomorph of Bar Normally Present?

The results reported in this paper indicate that reverted bar and reverted infrabar do not differ from wild-type in their effect on facet-number or in their crossing over relations in heterozygotes with double-types. No other kind of test appears to be available, so the simplest conclusion is that ordinary stocks carry no allelomorph of bar. This is consistent with the fact that the change from round to bar has been detected only once in *D. melanogaster,* and no similar mutation has occurred in any other species. One may surmise that the stock that gave rise to the original bar mutation had some sort of allelomorph already present—though other assumptions may also be made. It should be noted that, if a normal allelomorph were present in a stock and underwent unequal crossing over, as bar and infrabar do, the only stable condition would be that in which no bar allelomorph was present. While unequal crossing over does not alter the number of bar allelomorphs present in a stock, any strain without such allelomorphs will breed true, while any strain in which they are present will still give occasional individuals lacking them. If it be granted that Drosophila populations show great fluctuations in numbers, and often are reduced to very few individuals—as is very probable from their habits—then it would, in fact, be expected that unequal crossing over would have reduced most strains to the "no-bar" condition.

Summary

1. "Mutation" at the bar locus interferes with crossing over in the neighboring garnet-forked interval, indicating that unequal crossing over at the bar locus occurs at the same time and by the same mechanism as normal crossing over.

2. Homozygous bar or infrabar stocks probably give equal numbers of reversions and of double-types; but the latter are somewhat less viable and much less easily detected, so that reversions are found more often. Improved technique has, however, decreased the observed difference.

3. The dominant mutant types Delta and Hairless show no "position effect" on bristle number—that is, $\Delta H/+$ and Δ/H do not differ in bristle number. This and other evidence suggests that relative position of genes is not usually significant in development—though new data agree with old in showing that it is significant in the case of bar.

4. New data indicate that reverted bar (or reverted infrabar) is not different in facet-number from wild-type, and that derived types in general show no effect of their past history.

5. Double-infrabar over reverted infrabar gives infrabar.

6. The last two points indicate that reverted bar (or reverted infrabar) is not different from wild-type, and the conclusion is drawn that ordinary wild stocks carry no bar allelomorph.

LITERATURE CITED

BRIDGES, C. B., 1917: Deficiency. Genetics **2:** 445–465.

1919: Vermilion-Deficiency. Jour. Gen. Physiol. **1:** 645–656.

BRIDGES, C. B., and MORGAN, T. H., 1923: The third chromosome group of mutant characters of *Drosophila melanogaster*. Carnegie Inst. Washington, publ. 327, 251–x pp. *3 pl.*

MOHR, O. L., 1923: A genetic and cytological analysis of a section deficiency involving four units of the X-chromosome in *Drosophila melanogaster*. Zeitschr. indukt. Abstamm. u. Vererb. **32:** 108–232.

MORGAN, T. H., BRIDGES, C. B., and STURTEVANT, A. H., 1925: The genetics of Drosophila. Bibliographia Genetica **2:** 1–262.

STURTEVANT, A. H., 1925: The effects of unequal crossing over at the bar locus in Drosophila. Genetics **10:** 117–147.

1926: A crossover reducer in *Drosophila melanogaster* due to inversion of a section of the third chromosome. Biol. Zentralbl. **46:** 697–702.

ZELENY, C., 1921: The direction and frequency of mutation in the bar-eye series of multiple allelomorphs of Drosophila. Jour. Exp. Zool. **34:** 203–233.

Author's Note, 1961

The discussion of a possible "normal allelomorph" of bar needs revision, in the light of the salivary gland analysis that is described in the notes on paper no. 13.

17

THE CLARET MUTANT TYPE OF *DROSOPHILA SIMULANS*: A STUDY OF CHROMOSOME ELIMINATION AND OF CELL-LINEAGE †

PART I. GENETIC STUDY OF CLARET

Origin, Early History, Description, and Locus of Claret

A stock of *Drosophila simulans* from Austin, Texas, was examined in 1924 by Dr. C. R. PLUNKETT, and was found to be contaminated with *D. melanogaster*. There were also two eye-colors present among the *simulans* specimens—wild-type and a pinkish color, resembling the peach previously known in *simulans*. This mutant eye-color was shown by PLUNKETT to be a new third-chromosome recessive, corresponding to the previously known claret of *melanogaster*. Because of the contamination present in the original stock, the source of the claret gene remains uncertain, though it is most probable that it arose in the Austin stock, since the type was then unknown, and has not since been discovered in any other *simulans* stock.

The locus of claret has already been reported (STURTEVANT and PLUNKETT '26). It lies at 130 in the third chromosome, very close to the locus of Minute. These two are the furthest to the right among the known loci in this chromosome. It is not certain which of the two is further to the right, but they are less than 1 unit apart. In appearance the character is much like the claret of *melanogaster*, with which it corresponds. It also closely resembles garnet and peach *simulans* or peach *melanogaster*. Scarlet claret is a yellowish color similar to scarlet claret *melanogaster* or scarlet peach in either species. The claret gene usually produces no unexpected effects except on the eggs of homozygous claret females. Claret males, and both males and females heterozygous for claret, breed freely and rarely produce unusual results.[1] PLUNKETT soon found that claret females were often sterile, and rarely or never gave numerous offspring, so the stock has been carried on since that

† Reprinted from the Zeitschrift für wissenschaftliche Zoologie, Vol. 135, 1929, pp. 323–356

[1] Females heterozygous for claret sometimes produce mosaics of a type distinct from those discussed in this paper. These are now being studied.

time by mating heterozygous females to claret males, or by mating together flies with Minute in one chromosome and claret in the other. The latter method gives a "balanced" stock, as Minute and claret rarely cross over; the homozygous Minutes all die, and the homozygous claret females breed rarely or not at all.

Offspring of Claret Females

Preliminary tests with claret females showed among their offspring a surprisingly large number of gynandromorphs, and a number of specimens answering to the general description of "Minutes," but quite different in appearance from the previously known Minute of *simulans* that was referred to above. Later results leave no room for doubt that these specimens had but one fourth chromosome; they will be called Diminished in this paper, this being the term applied by BRIDGES ('21) to haplo-IV *melanogaster*.

In order to determine more about the nature of the gynandromorphs just mentioned, claret females were mated to yellow and to forked males, these characters both being sex-linked and identifiable in all bristlebearing parts of the body. Gynandromorphs were again found to be frequent, and it was at once noted that in practically all of them the male parts were yellow or forked (according to the character the father showed). They must, therefore, have represented eggs that started to develop as females (XX), and in which the maternal (wild-type) X was eliminated from one of the early cleavage cells. Another unexpected class also appeared in these experiments—namely, yellow or forked males—i.e., non-disjunctional exceptions. The totals obtained from such matings are shown in table 1, in which are included also a few results from matings to singed, another recessive sex-linked bristle character.

Table 1

ca ♀ × y, sn, or f ♂.

	+	Dm	Dm mos.	Total
♀	243	198	35	476
gyn	45	26	21	92
Reg ♂	165	152	9	326
Exc ♂	105	140	12	257
Total	558	516	77	1151

The male parts showed the paternal character in all but 11 of the gynandromorphs of this table. It is probable that some at least of these 11 exceptional individuals began their development as non-disjunctional exceptional females (see below). The low productivity of claret females may be judged from the fact that the 1151 flies of table 1 represent the total output of 44 fertile matings, nearly all of which were small mass cultures made up with from three to six claret females. Not one of these cultures produced as many as 100 flies. In this and all other series from claret mothers, inspection of the cultures shows that many undeveloped eggs are present in the food. In fact claret females apparently lay about the same number of eggs as their wild-type sisters.

The non-disjunctional males in these experiments were evidently accompanied by a few exceptional females, since tests of a number of the wild-type females of table 1 showed that some of them failed to carry the paternal sex-linked gene. Accordingly, a yellow stock heterozygous for claret was made up, and yellow claret females from this were mated to forked males, in order that the frequency of non-disjunctional daughters might be determined. The results are shown in table 2.

Table 2

y ca ♀ × f ♂.

	+	Dm	Dm mos.	Total
+ ♀	32	43	6	81
y ♀	9	5	1	15
gyn	7	5	2	14
y ♂	38	42	4	84
f ♂	28	55	3	86
Total	114	150	16	280

One culture of this series also gave three specimens that were probably intersexes; they will be discussed later. The next two types of matings to be described also give information on the nature and frequency of exceptional females, so discussion of table 2 may be postponed for the present.

It seemed probable that some of the exceptional females produced were equational exceptions—i.e., homozygous for some sex-linked genes for which their mothers were heterozygous. Two experiments were carried out to test this possibility. Yellow claret females, heterozygous for the sex-linked genes dusky and forked, were crossed to wild-type males in the first experiment; in the second, dusky claret females heterozygous for yellow and forked were crossed to wild-type males. The results are shown in tables 3 and 4.

In each of these experiments one of the gynandromorphs had both X's of the female parts maternal (yellow and dusky, respectively), showing that elimination of one X had occurred in an egg that started as an exceptional female. In all the other gynandromorphs of these two tables both male and female parts were wild-type (as regards sex-linked genes). Six of the exceptional females, and the female parts of one of the gynandromorphs, showed sex-linked recessive characters for which the mother had been only hetero-

Table 3

y ca ♀ (heterozygous for dy and f) x + ♂.

	+	Dm	Dm mos.	Total
+ ♀	108	170	19	297
y ♀	4	5	3	12
gyn...............	9	14	2	25
y ♂	104	137	11	252
+ ♂	54	231	11	296
Total............	279	557	46	882

Table 4

dy ca ♀ (heterozygous for y and f) × + ♂.

	+	Dm	Dm mos.	Total
+ ♀	86	160	11	257
dy ♀	9	6	0	15
gyn...............	10	11	6	27
dy ♂	71	106	6	183
+ ♂	60	137	9	206
Total............	236	420	32	688

zygous. These were as follows: table 3—2 dy, 1 f; table 4—4 y, 0 f. These give the following percentages:

yellow, $4/16 = 25\%$
dusky, $2/13 = 15\%$
forked, $1/29 = 3\%$

While these data are not extensive, it is evidently significant that they give decreasing frequencies of homozygosis as one moves to the right. This same relation has been found by ANDERSON ('24) in studying primary non-disjunctional exceptions (females) in *D. melanogaster,* and is there evidently due to the facts that crossing over occurs in the "four-strand" stage, and that primary non-disjunction is due to a suppression of the reduction division for the spindle-fiber end of the chromosome. The frequency of occurrence of equational exceptions for any locus thus is a measure of the distance of that locus from the attachment-point of the spindle-fiber (see discussions by ANDERSON '25, L. V. MORGAN '25, MORGAN, BRIDGES and STURTEVANT '25). We may conclude that crossing over occurs after both chromosomes have undergone the equational split, and that the spindle-fiber of the *simulans* X is attached at the "right" end, both conclusions agreeing with those already drawn for *melanogaster*. As will appear later, the cytological findings also suggest a close similarity between the X's of the two species.

It has been shown by BRIDGES ('16) and SAFIR ('20) that, in *melano-*

gaster, males arising from primary non-disjunction have no Y chromosome (are "XO" in constitution), and are sterile. Fifteen of the exceptional males recorded in tables 1 to 4 were tested, and all were found to be sterile. The three regular males tested were all fertile, and the sterility of the exceptional males cannot have been due to any mutant genes, since most of them were wild-type in character. It seems justifiable to conclude that the exceptional males from claret mothers are actually XO in constitution, and also that XO males are sterile in *simulans* as in *melanogaster.*

The exceptional females of tables 2 to 4 should be XXY in constitution, and should therefore give secondary non-disjunction. Thirteen of them have been tested; seven of these gave exceptions and three of the others gave less than 40 offspring, so that their failure to produce exceptions is certainly not significant. There was a total of 19 exceptions among 2108 offspring. The percentage of exceptions (0.9) is less than that previously recorded for *simulans* (2.9—see STURTEVANT '21). However, it seems certain that all the exceptional females were really XXY in constitution, since experience has shown that in *melanogaster* the frequency of secondary exceptions is extremely variable.

Crossing Over in Claret Females

The claret females in tables 3 and 4 were heterozygous for two sex-linked genes, and their regular sons were therefore available for a determination of the amount of crossing over that occurred. In some cases it was not certain that all the mothers of a culture were of the same constitution —i.e., some may have been $\frac{\text{y}}{\text{y dy f}}$ while others were $\frac{\text{y dy}}{\text{y \quad f}}$. After elimination of such doubtful cultures the counts show the following crossover proportions:

y f — 48/105 = 45.7%
dy f — 29/218 = 13.3%

The "standard" values, based in each case on over 18,000 flies, are: y f, 44.0; dy f, 21.7. The y f value is too large to be sensitive to slight variations (because of the occurrence of double crossovers), and the numbers from claret females are small; but the dy f value is probably significantly decreased. A further check on this is furnished by the tests of exceptional daughters of claret females that were themselves heterozygous for dusky and forked; such females gave 91/409 = 22.3%, or practically the standard value. It is evident that the regular sons of claret females give a significantly decreased dy f crossing over value. The interpretation of this relation is not clear, in the absence of adequate data on the frequency of crossing over in the eggs that give rise to exceptional females, for without those data it is uncertain whether claret decreases crossing over, or crossing over is normal but non-disjunction is more frequent in crossover eggs.

The Sex-ratio among Primary Exceptions

Tables 2 to 4 show, among the females and gynandromorphs, 44/743 = 5.9% of non-disjunctional exceptions; tables 1 to 4 show, among the males, 845/1690 = 50.0% of exceptions. As has been shown above, the exceptional males were sterile, and the exceptional females included numerous "equationals." For these (and other) reasons it is clear that the above percentages concern primary exceptions—i.e., the claret mothers had no Y chromosomes. It has been observed by BRIDGES ('16), SAFIR ('20), MAVOR ('24) and ANDERSON ('24) that primary exceptions in *melanogaster* are more often male than female. The summary given by MORGAN, BRIDGES and STURTEVANT ('25) shows a ratio of 4.2 males to 1 female in MAVOR'S X-ray experiments. ANDERSON obtained a ratio of 5.5 males to 1 female in his X-ray experiments. It is evident in the present series that the frequency of non-disjunction is not the same in all the tables, so for a direct comparison one may use the exceptions of tables 2 to 4. These show 588 exceptional males and 44 exceptional females and gynandromorphs, or a ratio of 13.4 to 1.

The excess of male exceptions found in *melanogaster* has already been interpreted; an interpretation of the still greater excess here may be approached by an analysis of the probable ways in which exceptions may be produced.

a. Both X's may pass to the same pole at the second maturation division of the egg. This should give equal numbers of eggs with 2 X's and with none (unless there is some sort of preferential maturation, which seems unlikely). The result would be equal numbers of XXX (superfemales, not observed and certainly rarely, if ever, viable), XXY (exceptional females), XO (exceptional males) and YO (inviable) zygotes. This is the only probable method of producing exceptional females; we may conclude that it is also responsible for one-thirteenth of the exceptional males.

b. One X (or both of them) may fail to pass to either pole at maturation. This will give rise to eggs without an X, and to exceptional males but not exceptional females. It is the same process as that which must be assumed to occur when an X is eliminated after fertilization to give rise to a gynandromorph. It has also been given as the probable cause of the excess of exceptional males in *melanogaster*. It may be pointed out that if such an elimination occurs at the first maturation division or at an oogonial division it will also be likely to give a similar end result—eggs with one X (giving regular offspring) or with none (giving exceptional males), but no eggs with two X's (giving exceptional females).

c. Both maternal X's may be eliminated at the first cleavage division of an XX (regular female) zygote. This will give rise to an XO (exceptional) male. It is not a probable event in ordinary cultures, but the high frequency of elimination in cleavage stages, shown in the present data (by the large number of gynandromorphs), makes it worth considering as a possibility.

There appears to be no available method of distinguishing between methods b and c, but one or the other must occur in nearly one-fourth of the viable eggs of claret females. It is probable that they also occur in many other eggs that do not give viable zygotes, for either would give a YO (inviable) combination if the sperm carried a Y instead of an X. Here, then, is an explanation of a part of the infertility of claret females—though there is clearly more mortality than can be accounted for on this basis alone.

Gynandromorphs

Tables 1 to 4 show a total of 1153 females, 158 gynandromorphs, and 1690 males. In the gynandromorphs of table 1 the female parts were all wild-type; the male parts were wild-type (maternal) in 11, but showed the mutant characters of the father in the remaining 81 specimens. In tables 2 to 4 there were two specimens that showed maternal sex-linked genes in both parts; all the rest were wild-type in the female parts and showed the paternal sex-linked characters (forked in the 14 from table 2, wild-type in the 50 from tables 3 and 4) in the male parts. It is probable that most or all of the 11 specimens from table 1 with wild-type male parts were wholly maternal, like the 2 exceptional specimens from tables 3 and 4.

These results are all consistent with the view that the gynandromorphs are caused by elimination of a maternal X at an early division. MORGAN and BRIDGES ('19) have shown that elimination is the commonest cause of gynandromorphs in *melanogaster,* but have discussed other possible causes. Partial fertilization or independent development of sperm nuclei are not to be considered in the present case, because neither part of a gynandromorph is claret unless the father carries claret, and because second-chromosome recessive genes present in the father never show in the gynandromorphs. It follows that both the male and the female parts carry autosomes from both parents. The occurrence of eggs with two independent nuclei, each fertilized by a separate sperm, is also not an available interpretation; for such an origin should give the male parts maternal as often as paternal (since regular sons are as frequent as are non-disjunctional exceptions).

The elimination hypothesis in the present case involves the assumption that it is nearly (absolutely?) always the maternal X that is eliminated. This is not in agreement with the results of MORGAN and BRIDGES ('19), who found, among the rare gynandromorphs that occur in most *melanogaster* stocks, 15 cases in which the maternal X had been eliminated and 15 in which the paternal had been eliminated. In a small series of miscellaneous *simulans* gynandromorphs I found (STURTEVANT '21) the maternal X eliminated once, the paternal three times. However, in Minute-n *melano-*

gaster, the first recorded strain with a high frequency of gynandromorphs, BRIDGES ('25) showed that it is again always the maternal X that is eliminated. This case differs from that of claret in that Mn, that is responsible for the elimination, is itself located in the chromosome that is eliminated; and the results of STERN ('27) with Minute-w, Minute-y, and Minute-β show that the Minute-bearing chromosome is eliminated even when it was received from the father. The elimination interpretation in the claret case is also strengthened by the facts that X's are eliminated from some eggs to produce the exceptional males, and that the fourth chromosome is also often eliminated, as will appear in the next section.

Diminished

Tables 1 to 4 show a total of 1187 normal-bristled flies, 1643 diminished, and 171 specimens mosaic for the diminished character. The diminished flies have small bristles, the last section of the fourth vein is usually imperfect, they develop slowly, and both sexes are usually sterile. They resemble the diminished of *melanogaster* (BRIDGES '21), but lack the conspicuous "trident" pattern of the thorax. This pattern is normally present, though faint, in *melanogaster;* in diminished it is much more distinct. But *simulans* normally has no trace of a trident pattern, and it is accordingly not surprising to find that the two types differ in this respect.

I have tested the fertility of many diminished flies, and have succeeded in getting offspring three times. These were all from diminished females, mated to normal males; the offspring were 15 diminished females, 5 diminished males, 47 wild-type females, and 65 wild-type males. Attempts to breed from the 20 diminished flies of these cultures were all unsuccessful. I have also obtained offspring from three diminished mosaic females, but these included no diminished offspring. In my experiments diminished males have always been sterile, but Dr. C. R. PLUNKETT informs me that he has twice gotten offspring from them. One of his cultures was a backcross for diminished and the third chromosome recessive scarlet; all four classes were produced, showing that diminished is not in the third chromosome.

The best evidence that the diminished offspring of claret females are haplo-IV is furnished by the cytological results. Three oogonial figures were found in each of two specimens, and these six figures agreed in showing only one small round chromosome. The clearest figure from each female is here reproduced (figures 1b and 1c).

Correlation between Elimination of X and of IV

A summary of tables 1 to 4 shows that rather more than half (54 percent) of the offspring of claret females are diminished, and about

5 percent are mosaics for the diminished character. Examination of the various classes (with respect to X chromosomes) shows that the same frequencies hold fairly well for each such class, with two striking exceptions: among the gynandromorphs the frequency of diminished mosaics rises to 20%, and among the exceptional males the frequency of diminished rises to 67%. That is to say: when an X is eliminated before cleavage in any specimen the chance is greater that the specimen will also lack one chromosome IV; and when an X is eliminated after fertilization the chance is greater that the specimen concerned will also lack a chromosome IV in some (but not all) cells of the body. More exact data are available for 27 of the gynandromorphs that were also diminished mosaics. Twenty-one (i. e., 78%) of these had two kinds of tissue—"normal" (i. e., not diminished) female, and diminished male; four had normal female, normal male, and diminished male tissue; the remaining two had normal female, diminished female, and normal male parts. It follows that in over three-fourths of these specimens the X and IV were eliminated at the same division, while in the other cases they were eliminated at different divisions (apparently most often the X at an earlier division than the IV). Not only were the two chromosomes usually eliminated at the same division, but in every such case they were eliminated from the same one of the two daughter cells; for no specimens were found with diminished female and normal male parts. A study of these 21 normal female-diminished male mosaics shows that they have about the same frequencies of male parts as the whole series of gynandromorphs. Of the 36 selected regions used in studying the whole series of gynandromorphs, 6 of the 21 had seven or less male, 4 had 12 or 13 male, 7 had 15 to 19 male, and 4 had 21 to 24 male (compare with frequencies for all claret gynandromorphs recorded in Part II of this paper). In one specimen the male parts included only the right third leg and part of the right wing-disc. It seems clear, therefore, that "double elimination" (of X and IV) occurs at various ones of the cleavage divisions and is not (as I first suspected) directly due to imperfect fusion of the pronuclei

Tests for Elimination of Chromosomes II and III

Since the maternal X and IV are often eliminated in eggs laid by claret females, it may be surmised that the same fate sometimes befalls chromosomes II and III. If this happens, claret females mated to males carrying second or third chromosome recessives should produce some offspring showing these recessives in part or all of their bodies. Such matings have been made—most of the fathers of the cultures summarized in tables 3 and 4 were from stocks homozygous for autosomal recessives, and a few other such matings have been made. Table 5 shows the numbers obtained from various recessives.

Table 5

Recessives in father	Offspring ♀	gyn.	♂	Total
black	297	25	481	803
spread plum	140	15	219	374
black spread plum	23	0	37	60
black polychaete spread plum	35	2	52	89
sepia	24	3	59	86
scarlet	5	1	10	16
scarlet peach	6	1	7	14
Total	530	47	865	1442

The paternal recessive character appeared in just one specimen, which belongs in the second row of the table. Yellow claret females were mated to spread plum males, producing 154 offspring, all of them of the usual classes except that one diminished (not-yellow, i. e. non-disjunctional) male had a left eye that was plum in color, smaller than the right eye in size, and roughish in texture. Attempts to measure the size of the facets were not satisfactory, because the roughening was found, on greater magnification, to involve much irregularity in the shape of the facets. It seems probable that this region was completely haploid, like the regions described by BRIDGES ('25a) in occasional Minute-n *melanogaster*. However, more critical data will be necessary to establish the occurrence of haploid tissue in the offspring of claret females.

There can be little doubt that chromosomes II and III are often eliminated in the eggs of claret females, and that such eggs are inviable. Such an assumption furnishes an explanation of the high mortality always encountered; it can, however, be checked only by cytological study of the fertilized eggs of claret females. Such a study has been undertaken by Dr. A. F. HUETTNER.

Probable Occurrence of Intersexes

Culture number 19311 was made up by mating yellow claret females to dusky forked males. A total of 67 offspring of the usual classes resulted. In addition, however, there appeared three specimens, wild-type for all the mutant characters concerned, that strongly resembled the triploid intersexes of *melanogaster* described by BRIDGES ('21a). All had sex-combs and heavily veined wings; one had no spermathecae, gonads apparently both ovo-testes, external genitalia incomplete male; another had complete female genitalia, both internal and external, but very small ovaries. Three wild-type sisters and one diminished sister of these specimens were tested; all gave normal results. Two sisters failed to produce offspring when tested.

Two other cultures (one from claret female by garnet forked male, the other from yellow claret female by black male) have each given a single specimen, wild-type for all mutant characters, that strongly suggested the triploid intersexes of *melanogaster*.

These results suggest that claret females sometimes produce functional diploid eggs; but unless triploid females can be obtained from them, or some equally convincing test be applied, this supposition must remain questionable.

Discussion and Summary

The oogonial chromosome groups have been examined in two claret females. Nothing unusual appears. Since no normal figure of *simulans* has ever been published, one of these is here reproduced (fig. 1 a) together

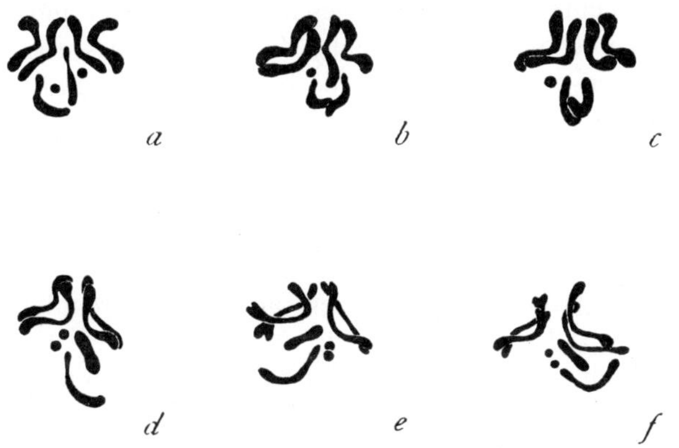

Fig. 1. Chromosomes of gonial divisions of *Drosophila simulans*: a claret female; b, c diminished females; d, e, f wild-type male.

with three spermatogonial figures from a wild-type male (fig. 1 d, 1 e, 1 f). It will be seen that these agree closely with the figures for *melanogaster*, except that the X shows a rather more distinct constriction, and the Y is shorter and lacks the hook present in *melanogaster*. Both of these peculiarities are constant in the *simulans* figures that have been studied.

Oogonial Chromosomes of Claret Females

The foregoing account has shown that the eggs laid by claret females give the following unusual results:

1. Over 3% of all the flies produced are gynandromorphs, in which a maternal X has been eliminated.

2. About 6% of the females produced arise from primary non-disjunction, and are XXY in constitution.

3. About 50% of the males produced are non-disjunctional exceptions, XO in constitution and sterile.
4. Among the eggs giving rise to regular sons, there is a marked reduction in crossing over in the right end of the X chromosome.
5. Some of the exceptional females are "equational" exceptions.
6. About 40% of all the offspring produced are haplo-IV.
7. About 4% are haplo-IV in part but not all their bodies.
8. There is a very high mortality, presumably due in large part to elimination of chromosomes II and III.

With the exception of the 4th point, all these results can be brought under one heading if we suppose that, in claret females and in the early stages of their offspring, the rate of division of the chromosomes is slowed down in relation to the rate at which the rest of the mitotic mechanism functions. This retardation occurs only at the stages near maturation and early cleavage. It results in chromosomes occasionally failing to pass to either pole at division (elimination-points 1, 3, 6, 7, 8) or more rarely, in both homologues passing to the same pole (non-disjunction-points 2, 5, and probably occasionally 3, 6, and 8).

It is not clear why the claret gene should have such an effect. The oogonial stages look normal, and have normal chromosomes. Claret males, females heterozygous for claret, and *melanogaster* females homozygous for the corresponding claret of that species, all produce regular results in no way suggestive of the behavior of claret *simulans* females. Further light will probably come from a study of the eggs by cytological methods.

PART II. DISTRIBUTION OF PARTS IN THE GYNANDROMORPHS

Introduction

Many of the claret females were mated to yellow, forked, or singed males (tables 1 and 2). Most of the gynandromorphs produced had lost a maternal X in their male parts, so that their female parts were wild-type while their male parts were yellow, forked, or singed. Under these conditions all bristle-bearing parts of the body can be identified as to the mutant characters and consequently as to sex. A series of mimeographed outlines of flies was prepared, and each part of each gynandromorph was checked off as to sex on these standard outlines. In this way 96 detailed "maps" were obtained showing the distribution of parts.

These diagrams have been used, first, to determine the stage of development at which elimination occurred, and, second, in a study of the derivation, or "cell-lineage," of the surface of the adult fly. As a preliminary to the discussion of these questions it will be well to briefly review what is known of the development of the surface portions.

Literature on Muscid Development

HUETTNER ('23, '24) has studied the fertilization and early cleavage of *Drosophila melanogaster*. The cleavage is of the usual insect type, resulting in the production of numerous nuclei, each surrounded by a protoplasmic island, in the central portion of the egg. At about the 256 nuclei stage (i. e., after the eighth cleavage division) these nuclei migrate to the surface, and there form a layer of cells—the blastoderm. Some nuclei, however, remain in the central portion of the egg, and give rise to the yolk-nuclei, which do not directly take part in the further development of the individual. Those of the cleavage nuclei that reach the surface in the posterior end of the egg give rise to the germ-cells. HUETTNER is of the opinion that these (the "pole-cells") are not constant in number, and do not necessarily come from any given one of the earlier cleavage nuclei. There is no available information as to the pattern of the cleavage in *Drosophila* or in any related insect, though it appears probable that the early divisions are not oriented in any fixed pattern, so that the cells in a given part of the blastoderm have no constant genetic relationship to those in any other given part. This, however, is only a surmise.

After the formation of the blastoderm the development has not been described for *Drosophila;* but it is altogether probable that the events are much like those in the *Muscidae,* which have been described in detail by many authors. The main points of the story were made out by WEISMANN, and elaborated by KOWALEVSKY, GANIN, METSCHNIKOFF, VAN REES, and many others. The present summary has been compiled chiefly from the account given in KORSCHELT and HEIDER ('99), from the papers of NOACK ('01) and REITH ('25) on the embryonic development, and from that of SNODGRASS ('24) on metamorphosis; the reader is referred to these papers for a full list of the earlier literature. SNODGRASS' paper is especially useful in the present case because it deals with *Rhagoletis,* a form usually considered to be somewhat closer to *Drosophila* than are *Musca* and *Calliphora,* on which nearly all the earlier work was done.

The embryo first appears as a "germ-band" lying on the ventral surface of the egg. A longitudinal furrow appears in this band, and the edges of the furrow then grow together. The parts thus brought below the surface give rise to the mesodermal and endodermal portions of the larva, and hence do not enter into the history of the surface of the adult. The amniotic folds, so conspicuous in many insects, are here almost lacking. There is a very small cephalic amniotic fold, and a larger posterior one; but both are temporary, the folds simply flattening out again instead of growing together over the embryo. There appears not to be a growth of the germ-band dorsally, so that it is probable that the ectoderm of the dorsal region of the embryo simply differentiates in place, from the blastoderm cells of the region concerned.

The larva, on hatching, is made up of 12 segments—1 cephalic, 3 thoracic, and 8 abdominal. The external part of the cephalic segment really represents little more than the "neck" of the embryo, or of the adult, as the embryonic head is largely invaginated before hatching. The surface (as well as most of the internal organs) of the adult develops from a few small islands of tissue present in the larva—the imaginal discs. As a result, most of the larval ectoderm disappears in the adult; and this is of prime importance in evaluating the evidence from the mosaics under discussion in this paper. Six imaginal discs are described in the head, and four in each of the thoracic and abdominal segments. From the anterior dorsal region of the larval pharynx (i. e., from what was the outer dorsal surface of the embryonic head) arise a pair of "frontal sacs." Each of these bears at its apex a bulb, the distal surface of which is the optic bud, or imaginal disc of the eye. Just proximal to the bulb is a swelling, the floor of which is the antennal disc. These two discs—the optic and antennal of one side—thus lie close together when first formed, and apparently they grow together very early. The right and left frontal sacs come off from the pharynx very close to each other, but each is fairly long, so that it is difficult to judge how closely related the right and left discs are. The other two head discs arise from the floor of the larval pharynx, just below the mouth of the frontal sacs; they give rise to the proboscis.

Of the four imaginal discs in the first thoracic segment, the two dorsal give rise to the humeri, the two ventral to the first pair of legs. The ventral discs of the second and third thoracic segments give rise to the second and third pair of legs. The dorsal discs of the second thoracic segment give rise to the wings, and to most of the dorsum of the thorax, i. e., the mesonotum and scutellum. The dorsal discs of the third segment give rise to the halteres and the small metanotum; these parts could not be classified in the mosaics here discussed, so that the dorsal discs of the metathorax do not further concern us.

Each of the dorsal discs of an abdominal segment gives rise to half of the dorsal sclerite (tergite) of the adult segment; each of the ventral discs gives rise to half of the ventral sclerite (sternite). The so-called first abdominal segment of the adult is presumably developed from the first and second larval segments, this conclusion being based chiefly on the occurrence of two pairs of spiracles. Accordingly what are here considered as the first five abdominal segments of the adult account for six of the larval segments; the remaining two give rise to the genital and anal portions, which are only incidentally discussed in this paper.

The gonads, dissected out in many of the mosaics, have a simple embryonic history. As stated above, they are derived from those blastoderm cells that come to lie at the posterior pole of the egg. This group of cells is early separated from the rest of the embryo, and then develops directly into the adult gonads. The history of the spermathecae and sperm-pump is not clear.

Stage at Which Elimination Occurs

From the account just given it is possible to select a number of organs that must be supposed to develop from separate imaginal discs. In many cases it is clear that an organ that must be supposed to develop from a single disc is itself mixed in one or more of the 96 gynandromorphs, so for the preliminary tabulation it was necessary to arbitrarily select some

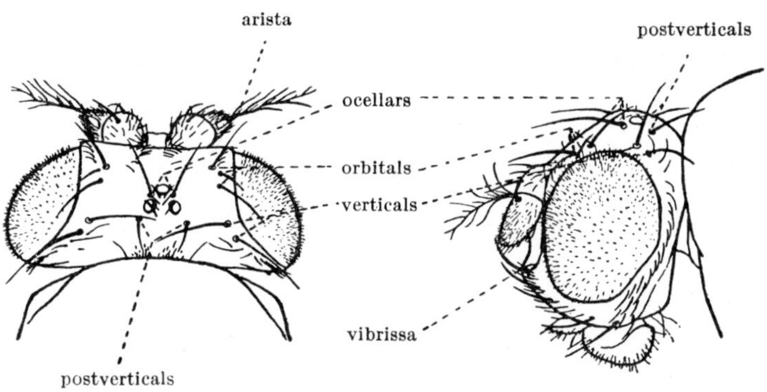

Fig. 2. Top and side views of head.

given portion of an organ to be considered. The list of regions used is as follows:[1]

1. Antenna.
2. Outer vertical bristle.
3. Upper humeral "
4. Posterior dorsocentral bristle.
5. Posterior sternopleural "
6. Apex first tibia.
7. Apex second "
8. Apex third "
9. First abdominal tergite (lower posterior corner).
10. Second " " " " "
11. Third " " " " "
12. Fourth " " " " "
13. Fifth " " " " "

[1] The positions of the bristles referred to here and in the discussion that follows may be seen by reference to figures 2 and 3.

14. First " sternite (outer " "
15. Second " " " " "
16. Third " " " " "
17. Fourth " " " " "
18. Fifth " " " " "

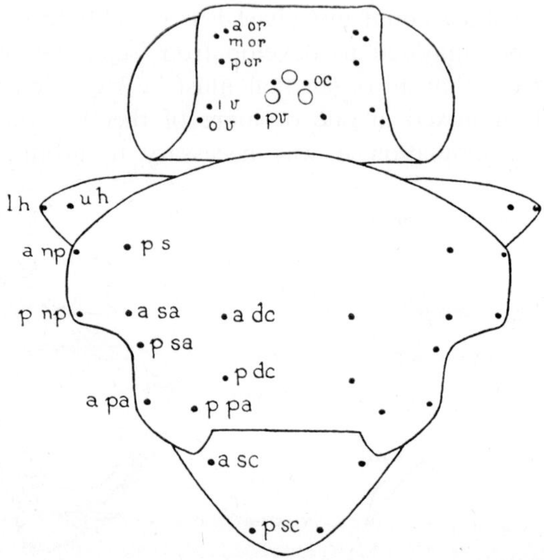

Fig. 3. Diagram of head and thorax, *adc* anterior dorsocentral, *anp* anterior notopleural, *a or* anterior orbital, *a pa* anterior postalar, *a sa* anterior supraälar, *a sc* anterior scutellar, *iv* inner vertical, *lh* lower humeral, *mor* middle orbital, *oc* ocellar, *ov* outer vertical, *pdc* posterior dorsocentral, *pnp* posterior notopleural, *por* posterior orbital, *ppa* posterior postalar, *ps* presutural, *p sa* posterior supraälar, *p sc* posterior scutellar, *pv* postvertical, *u h* upper humeral.

(After PLUNKETT, 1926.)

Number 5, the posterior sternopleural, may be questioned, as it is mesothoracic and no separate pleural disc is described here; it was included because preliminary study showed it not to be closely correlated with any other region here listed.

Each of these regions occurs on each side of each specimen, so that there are 36 regions per fly to be tabulated. The number of these that were male was counted for each separate one of the 96 diagrams; the results are shown in table 6.

There appear to be modes at 18, 9 and 5 parts male. These correspond to 50%, 25% and 13% of the possible 36 parts. Evidently they correspond also to cases in which one of the first two, first four, or first eight cleavage nuclei were male—i.e., to elimination of an X chromosome at the first, second, or third cleavage division respectively. There are also some cases in which definitely more than 18 parts were male; these presumably correspond to cases where elimination occurred more than once, so that 3 of the first 4

Table 6

Number of male parts in gynandromorphs.

Number of parts	Number of specimens	Number of parts	Number of specimens
0	2	21	3
1	4	22	2
2	4	23	2
3	4	24	0
4	2	25	1
5	6	26	1
6	2	27	2
7	4	28	0
8	1	29	0
9	10	30	0
10	2	31	1
11	2	32	3
12	4	33	1
13	2	34	2
14	2	35	0
15	3	36	1
16	5	Total	96
17	5		
18	9		
19	4		
20	0		

nuclei (or some similar combination) were male. Roughly we may divide the data thus:

1/2 male	40 ± %
1/4 male	20 ± %
less than 1/4 male . .	25 ± %
more than 1/2 male . .	15 ± %

This is only an approximation, as the data do not fall into very sharp classes. There is, furthermore, the possibility of a relatively high frequency of repeated elimination, even in those cases that are not significantly more than 50% male. Added to this uncertainty is the fact that the 36 selected regions give descendants of only a rather small sample of the whole blastoderm of the egg; it is also probable that this is not a random sample, since the abdominal regions seem to be more closely related than are the thoracic ones, as will appear below. The most that can be concluded is that the elimination occurs chiefly, or perhaps exclusively, at early cleavage divisions. It probably always occurs before the segregation of the imaginal discs, since only one specimen has been found that showed both male and female parts in a region that must be supposed to come from one disc, unless male parts were also present in neighboring regions. It may be noted that such specimens (with male parts present only as patches within single discs) are by far the commonest type of mosaics in the Minute-n *melanogaster* described by BRIDGES ('25).

Cell Lineage of Gynandromorphs

The gynandromorphs come from eggs that, since early cleavage, have contained both male and female cells. The distribution of male and female parts of those parts of the body surface that carry hairs or bristles is known for the 96 gynandromorphs for which diagrams were prepared. It is evident that a systematic study of these diagrams can give considerable information as to the cell-lineage of *Drosophila,* for parts that are genetically closely related will more often be alike as to sex than will parts more remotely related. The present section gives the results of such a study.

Most parts of the body are themselves occasionally mosaic. Table 7 shows the frequency with which this was observed for several parts of the body.

Table 7

One side of head	43	3rd leg	0
Mesonotum	24	1st abdominal tergite	1
Wing	6	2nd abdominal tergite	1
Sternopleura	2	3rd abdominal tergite	0
1st leg	2	4th abdominal tergite	2
2nd leg	2	5th abdominal tergite	1

One may conclude from these data that one side of the head includes descendants of a larger sample of blastoderm nuclei than do the other parts recorded; which was to have been expected, since each side of the head probably comes from three separate imaginal discs, whereas each of the other parts (except the sternopleura) must be supposed to develop from a single disc. The other parts include all the regions (outside of the head) that have been observed to be mixed in any case, among those that must be supposed to come from single discs. Among these, it is evident that the wing-disc (which gives rise not only to the wing but also to the mesonotum) is most often mixed. This presumably means that it comes from a larger number of blastoderm cells than do the other discs.

The accompanying tables (8, 9, 10, 11, 12) show the available statistics for the relationships between various regions of the 96 gynandromorphs.[2] The numbers given show numbers of times the two regions in question were different (one male and the other female), except in the table (8) for right-left relations, where it has been necessary to double the observed numbers

[2] The data concerning relationships within the mesonotal area are not included, but will be discussed later in this paper.

to make them comparable to the other data (since the table deals with 96 whole flies, whereas the others deal with 192 half flies, both the right and left halves being taken into account).

Table 8

Numbers of times homologous organs on opposite sides of the body were different in the 96 gynandromorphs. The observed numbers are here doubled, to make them comparable with those of tables 9 to 13.

palpus	56	third leg	82
vibrissa	60	first tergite	84
antenna	54	first sternite	72
ocellar	52	second tergite	86
postvertical	44	second sternite	70
outer vertical	48	third tergite	80
humerus	72	third sternite	76
first leg	66	fourth tergite	76
anterior dorsocentral	118	fourth sternite	84
posterior notopleural	108	fifth tergite	78
scutellum	100	fifth sternite	92
second leg	84		

The numbers in these tables furnish rough indexes of the remoteness of relationship of the parts concerned. For clearly those parts that are more often different must be parts that more often come from separate cleavage nuclei. Conversely stated, the smaller the number of times separated, the more closely related are the parts.

Examination of the original diagrams (or of the numerous published figures of *Drosophila* gynandromorphs) indicates that in the thorax and abdomen the right and left halves of the body develop separately after the formation of the imaginal discs. When one half of the mesonotum has hairs of a different type from those of the other half, the dividing line regularly comes between

Table 9

Numbers of times various portions of the head were different in the 96 gynandromorphs.

palpus and vibrissa	6	lower orbital and upper orbital	5
palpus and antenna	7	middle orbital and upper orbital	3
vibrissa and antenna	11	middle orbital and inner vertical	9
vibrissa and lower orbital	10	middle orbital and outer vertical	14
vibrissa and middle orbital	14	upper orbital and ocellar	8
vibrissa and outer vertical	23	upper orbital and inner vertical	6
antenna and lower orbital	6	ocellar and postvertical	6
antenna and middle orbital	6	ocellar and inner vertical	12
antenna and upper orbital	9	postvertical and inner vertical	10
antenna and ocellar	9	postvertical and outer vertical	16
lower orbital and middle orbital	4	inner vertical and outer vertical	7

Table 10

Numbers of times various portions of the thorax were different in the 96 gynandromorphs.

humerus and first leg	43
humerus and presutural	31
humerus and scutellum	26
humerus and anterior sternopleural	39
first leg and anterior sternopleural	34
first leg and second leg	25
anterior notopleural and anterior sternopleural	18
posterior notopleural and second leg	37
scutellum and third leg	37
anterior sternopleural and posterior sternopleural	2
anterior sternopleural and second leg	15
posterior sternopleural and second leg	15
posterior sternopleural and third leg	29
second leg and third leg	30

Table 11

Numbers of times various portions of the abdomen were different in the 96 gynandromorphs.

first tergite and first sternite	19
first tergite and second tergite	13
first sternite and second sternite	10
second tergite and second sternite	15
second tergite and third tergite	8
second sternite and third sternite	12
third tergite and third sternite	19
third tergite and fourth tergite	15
third sternite and fourth sternite	10
fourth tergite and fourth sternite	23
fourth tergite and fifth tergite	18
fourth sternite and fifth sternite	14
fifth tergite and fifth sternite	28

Table 12

Numbers of times various portions in separate body regions were different in the 96 gynandromorphs.

vibrissa and first leg	56
outer vertical and humerus	67
outer vertical and first leg	58
postvertical and humerus	67
postvertical and presutural	69
vibrissa and fifth tergite	82
palpus and fifth sternite	82
scutellum and first tergite	46
third leg and first sternite	39

the two middle acrostichal rows of hairs. In the few cases where this rule does not hold there is clear evidence (in other regions) that one half is itself mosaic, making it probable that even in these rare cases each disc has contributed exactly one-half of the mesonotum. In the case of the abdomen there is no clear "landmark" for the median dorsal line, but apparently the same rules hold. Only one exceptional case has been seen, where one half of the tergite had forked hairs and bristles, and these extended as a tongue well into the not-forked half. It is quite possible that in this case the supposedly not-forked half really came from a mixed disc; in any case it is clear that in nearly all cases each dorsal abdominal disc gives rise to just one half of a tergite. The same is true also of the relation of the ventral abdominal discs to the sternites, and of the ventral thoracic discs to the legs.

In the head the relations are not so clear. Heads do occur that are male on one side and female on the other, with the dividing line strictly median; but they are not common. Often the line is a little "off center," so that both ocellars or both postverticals lie on one side of it—but usually when the head is mosaic the dividing line is far from median. One might expect to find six regions in the head, corresponding to the six imaginal discs described (paired eye, antenna and proboscis discs); but such regions are not at all obvious from the data.

Table 8 shows that the highest (doubled) number of times homologous regions on opposite sides of the head were different was 60 (for the vibrissa); the lowest number for the rest of the body was 66 (for the first leg). These relations agree with the general impression just stated. As for the distinctness of the regions on one side of the head that presumably come from separate imaginal discs, table 9 shows that there are no numbers as high as the right-left difference, but does not indicate that there are any distinct regions on a side. If such regions were present the parts should fall into closely related groups, each group being much less closely related to the members of any other group. A study of the data in the table, plotted on a diagram of the head, indicates that no such relations occur, but rather that the closeness of relationship as judged by these data is correlated with nearness together on the surface of the head and with nothing else. The conclusions to be drawn from this result will be discussed after the description of the relations found in the thorax.

The values for the thorax given in table 10 are consistently higher than those for the head, in agreement with the greater distances apart of the beginnings of the thoracic imaginal discs of the larva. It will be observed that most of the lower values in the table concern the sternopleural bristles, and it will be recalled that these bristles are mesothoracic, lying between the mesonotum (wing-disc) and the middle leg. There are only two imaginal discs described for the mesothorax: from which of these are the sternopleural bristles derived? The present data clearly indicate—from either one. The lowest value of the table is that for the relation of anterior to posterior sternopleural; the three next lowest are for sternopleurals to middle leg and to anterior notopleural (wing-disc). To study the relations in another way, those cases were collected in which the mesonotum and middle leg of one side were different, and neither was itself mosaic. There were 25 such cases; in 13 the sternopleurals were like the leg, in 12 they were like the mesonotum. Among all the 96 gynandromorphs there was no case in which the sternopleurals were different both from the mesonotum and from the middle leg. These results mean that the sternopleurals are developed from the tissue that comes to lie in the appropriate region. It makes no difference where that tissue comes from. One may surmise that the result depends simply on which of the two imaginal discs concerned happens to grow faster and first comes to occupy the sternopleural region. This interpretation finds some support in the data; for it would be expected that female tissue might grow somewhat faster than male (since females develop faster than males), and among the 25 cases just mentioned the sternopleurals were female in 17, male in 8. This difference is of doubtful statistical significance; but it should be pointed out that the same disproportion occurs when the sternopleurals are like the leg (9 female, 4 male) and when they are like the mesonotum (8 female, 4 male). Since the difference is also in the direction that might have been surmised beforehand, it seems probable that it is significant.[3]

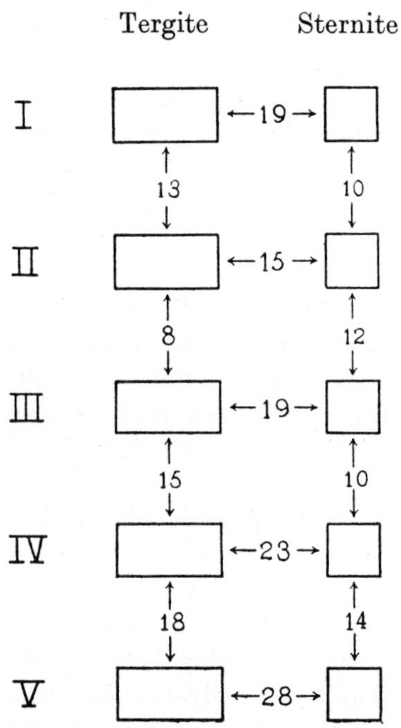

Fig. 4. Relation of abdominal sclerites in 96 gynandromorphs. The numbers indicate times the two sclerites indicated were different.

It is probably to such relations as these that one should refer the failure to distinguish separate imaginal disc regions in the head, rather than to a developmental difference between *Drosophila* and the forms whose imaginal discs have been described. It may be supposed that there is no predetermined boundary between the tissues that come from the optic and antennal discs, for example; the parts contributed depend merely on which disc happens first to occupy the region that is to develop into a given part. It is probable, judging by the data at hand (appearance of individual specimens as well as the statistics of table 9), that in the head this indeterminate growth of the discs extends also across the median line, which is evidently rarely, if ever, the case in the thorax and abdomen. Evidence will be presented below, in connection with the data on the relations between the parts of the mesonotum, to show that there is a similar indeterminate growth within an imaginal disc.

[3] As was shown above, the male portions of gynandromorphs are more often haplo-IV than are the female portions, and haplo-IV individuals develop relatively much more slowly than do males. It seemed possible, therefore, that the relations just described might be due to the effects of the fourth chromosome on growth-rates, rather than to the sex of the parts concerned. Examination of the data leaves this possibility still open. Ten of the 25 gynandromorphs concerned had the male parts in question haplo-IV, the female parts not so. In these 10 specimens the sternopleurals were female in 8, male and haplo-IV in 2. More data will be necessary to make clear the relative effects of the X and fourth chromosomes.

The data of table 11, showing the relations of the abdominal discs, are shown graphically in fig. 4. The values are on the whole intermediate between those found for cephalic and for thoracic regions. They are consistently very much lower than the right-left values for the abdomen, as was the case also for the thoracic region. On the whole, the values average lower for discs adjacent along the longitudinal axis than for the relations between tergite and sternite of the same body segment. This result probably means that adjacent dorsal (or ventral) imaginal discs come from regions of the blastoderm that lie closer together than do those that give rise to corresponding dorsal and ventral discs.

Table 12 shows a few of the many possible pairs of regions lying in different main body regions. These values are, as might have been expected, uniformly high. They indicate that the thorax and the first abdominal segment are about as closely related as are different discs in the thorax, while the head and thorax are scarcely more closely related than are homologous regions on opposite sides of the body. Both these results are consistent with the positions in which the imaginal discs arise in the larva.

Cell Lineage of the Mesonotum

In the preceding discussion no account has been given of the relations found within the wing-disc region—which in practice means the relations between the mesonotal bristles. This region has been reserved for separate treatment because there are available two additional sets of data for comparison. PLUNKETT ('26) has published a figure of this region (here reproduced as fig. 3), in which the distances between the various bristles are proportional to the actual distances as measured on the surface of the fly.[4] BRIDGES ('25) has reported the frequent occurrence of "piebalds" or mosaics with small male regions, in Minute-n females. I have collected data on the distribution of the male parts in 41 mixed wing-discs of Minute-n females, and Dr. C. STERN has allowed me to incorporate with these his observations on 71 mixed wing-discs from the same source.

It is thus possible to draw up three separate tables showing the following relationships between each possible pair of mesonotal bristles:

(a) Number of times different in the 24 mixed wing-discs found among the 96 claret gynandromorphs (table 13).

(b) Relative distance apart, as measured on PLUNKETT's diagram (table 14).

[4] This figure, and also the Minute-n data next described, concern *Drosophila melanogaster*, while the claret data are from *D. simulans*. This difference is certainly of little, if any, significance, since the thoracic region appears to be identical in form in the two species.

(c) Number of times both male in the 112 mixed wing-discs of Minute-n females (table 15).

In the case of the claret gynandromorphs the two scutellars were alike in every case. If the posterior of these is left out of account, the remaining values of table 13 are found to be very closely correlated with the distances shown in table 14. Even if the values concerning the posterior scutellar be

Table 13

Numbers of times stated pairs of mesonotal bristles were different in the 24 mixed wing-discs of the 96 gynandromorphs.

	anp	*pnp*	*pst*	*asa*	*psa*	*apa*	*ppa*	*adc*	*pdc*	*sc*
anp	—	5	2	12	13	17	21	14	16	22
pnp	5	—	7	5	8	14	17	11	13	17
pst	2	7	—	10	11	17	21	14	16	22
asa	12	5	10	—	3	9	12	8	7	12
psa	13	8	11	3	—	6	9	7	6	11
apa	17	14	17	9	6	—	3	5	3	5
ppa	21	17	21	12	9	3	—	8	4	2
adc	14	11	14	8	7	5	8	—	4	9
pdc	16	13	16	7	6	3	4	4	—	6
sc	22	17	22	12	11	5	2	9	6	—

Table 14

Distances apart (in arbitrary units) of stated pairs of bristles in figure 3.

	anp	*pnp*	*pst*	*asa*	*psa*	*apa*	*ppa*	*adc*	*pdc*	*asc*	*psc*
anp	—	5	4	6	9	12	15	12	15	19	25
pnp	5	—	6	4	5	8	11	11	12	15	21
pst	4	6	—	5	7	12	13	9	12	17	23
asa	6	4	5	—	2	7	9	7	9	13	19
psa	9	5	7	2	—	4	6	7	7	10	16
apa	12	8	12	7	4	—	4	9	6	7	13
ppa	15	11	13	9	6	4	—	7	3	4	10
adc	12	11	9	7	7	9	7	—	5	11	16
pdc	15	12	12	9	7	6	3	5	—	6	12
asc	19	15	17	13	10	7	4	11	7	—	6
psc	25	21	23	19	16	13	10	16	12	6	—

included a correlation table makes it clear that nearness together on the surface and closeness of relationship are highly correlated. Since the claret data include only 24 mixed discs, while 55 pairs of bristles are concerned, the correlation coefficient would have a spuriously low probable error; its absolute value is of so little statistical significance that the coefficient has not been calculated.

The Minute-n data cannot be easily related to the other two series, since table 15 makes it clear that the different mesonotal bristles are not equally likely to be male. Among the 112 mixed discs the posterior postalar was male in 50, the posterior scutellar in 13, the other bristles in numbers between

Table 15

Numbers of times stated pairs of bristles were both male in the 112 mixed wing-discs observed in Minute-n piebalds.

	anp	*pnp*	*pst*	*asa*	*psa*	*apa*	*ppa*	*adc*	*pdc*	*asc*	*psc*
anp	—	18	18	16	14	8	11	8	5	3	0
pnp	18	—	19	21	20	12	16	12	7	5	1
p st	18	19	—	20	17	9	12	16	7	4	1
a sa	16	21	20	—	30	18	23	15	9	6	2
p sa	14	20	17	30	—	25	30	18	13	8	3
a pa	8	12	9	18	25	—	36	14	14	11	5
ppa	11	16	12	23	30	36	—	19	20	12	7
a dc	8	12	16	15	18	14	19	—	23	8	4
pdc	5	7	7	9	13	14	20	23	—	9	4
a sc	3	5	4	6	8	11	12	8	9	—	9
p sc	0	1	1	2	3	5	7	4	4	9	—
alone	5	0	6	2	1	2	2	8	2	5	4
Total	24	25	32	37	41	38	50	42	30	21	13

Table 16

Coefficients of association for Minute-n piebalds.

	anp	*pnp*	*pst*	*asa*	*psa*	*apa*	*ppa*	*adc*	*pdc*	*asc*	*psc*
anp	—	+0,94	+0,83	+0,73	+0,52	+0,10	+0,03	−0,11	−0,20	−0,29	−1,00
pnp	+0,94	—	+0,89	+0,92	+0,85	+0,37	+0,47	+0,27	+0,04	+0,05	−0,59
p st	+0,83	+0,89	—	+0,72	+0,45	−0,18	−0,20	+0,35	−0,18	−0,29	−0,69
a sa	+0,73	+0,92	+0,72	—	+0,92	+0,45	+0,49	+0,10	−0,10	−0,13	−0,50
p sa	+0,52	+0,85	+0,45	+0,92	—	+0,74	+0,75	+0,21	+0,19	+0,04	−0,35
a pa	+0,10	+0,37	−0,18	+0,45	+0,74	—	+0,97	−0,02	+0,36	+0,44	+0,11
ppa	+0,03	+0,47	−0,20	+0,49	+0,75	+0,97	—	−0,13	+0,55	+0,28	+0,04
a dc	−0,11	+0,27	+0,35	+0,10	+0,21	−0,02	−0,13	—	+0,78	+0,02	−0,17
pdc	−0,20	+0,04	−0,18	−0,10	+0,19	+0,36	+0,55	+0,78	—	+0,43	+0,11
a sc	−0,29	+0,05	−0,29	−0,13	+0,04	+0,44	+0,28	−0,02	+0,43	—	+0,92
p sc	−1,00	−0,59	−0,69	−0,50	−0,35	+0,11	+0,04	−0,17	+0,11	+0,92	—

these limits. The simplest method of treating these data has seemed to be by means of YULE's coefficient of association $\left(\frac{(AB \times ab) - (Ab \times aB)}{(AB \times ab) + (Ab \times aB)}\right)$, the total number of individuals in the population being taken as 112, the number of mixed discs. This method of treating the data has certain statistical drawbacks, but is probably adequate for present purposes. The coefficients are shown in table 16.

Correlation tables indicate that these coefficients are also closely correlated (negatively) with the distances on PLUNKETT's diagram. They are likewise negatively correlated, but apparently less strongly so, with the values for claret gynandromorphs. Here again the nature of the data does not seem to warrant calculation of the coefficients, or publication of the correlation tables (the necessary data for these are of course contained in tables 13, 14, and 16).

These results indicate that there is no determinate cell-lineage within the wing-disc. This conclusion is clearly substantiated by examination of the individual mosaics, either in the claret or in the Minute-n series. For example: both dorsocentrals may be alike but different from both postalars, or the posterior dorsocentral and posterior postalar may be alike but different from the corresponding anteriors, or any one of the four may be different from the other three. Such relations occur for any group of mesonotal bristles one examines. However, it is clear that related cells remain adjacent, for one finds definite patches that are male—in almost every case a single such patch on one wing-disc. In the Minute-n series (and also some of the claret series) the hairs also could be identified, since the mutant differences concerned were yellow, singed, or forked-5, all of which produce distinguishable effects on the hairs as well as the bristles. Study of such specimens shows that the male tissue in every case forms a single definite clearly-marked patch, with no outlying spots, and with no female tissue in the patch. Evidently the surface of the mesonotum is formed from cells freely proliferating, without any subsequent intermingling of cells from different origins. Furthermore, the development is indeterminate, in that the genetic relationship of two trichogenic cells is, on the average, inversely proportional to their distance apart, and not dependent on whether the line joining them is longitudinal or makes any given angle with the long axis of the fly. Otherwise stated, cell proliferation in the anlagen of the mesonotum is random, and the differentiation of parts that occurs later is independent of the origin of the cells but is somehow brought about by their position.

Genital Organs

Most of the claret gynandromorphs were dissected, but in many cases it was not possible to be certain of the internal conditions—in some cases apparently because the gonads were very poorly developed, but probably also in some cases because of inefficient dissecting.

In no case were spermathecae (female organs) and chitinized sperm-pump (male organ) found in the same specimen. In every case where the external genitalia were wholly male or wholly female, these internal chitinized parts (if found) agreed with them. The gonads, however, might be of either sex with either spermathecae or sperm-pump. However, if the abdomen was externally wholly of one sex the gonads were always found to agree with it.

Four cases were found in which both ovarian and testicular tissue was present. In no case were sections of such gonads studied, but the identification from external appearance is probably adequate. The relations in these four specimens were briefly as follows:

A. One testis with paragonium, one small ovary. Sperm-pump present. Externally, one egg-guide and one half a "genital tergite" (male) with a clasper.

B. One small ovary, other gonad mixed. No spermathecae. Externally, one egg-guide, one clasper.

C. One testis, with normal paragonium, one mixed gonad with small paragonium. One large and one small spermatheca; penis present. Externally, one egg-guide, one half a genital tergite with clasper.

D. Two small testes, one small median ovary attached to the vas. Two paragonia, sperm-pump. External genitalia normal male.

The three gonads present in specimen D are anomalous; otherwise these specimens show the relations to be expected from the results of HUETTNER ('23) on the origin of the germ-cells in *Drosophila* and from the general results of the present study of gynandromorphs.

The Mechanism of Differentiation in Drosophila

The gynandromorphs described in this paper, and the earlier ones described for *Drosophila* by MORGAN and BRIDGES ('19) and others, make it quite clear that the cleavage of the egg is indeterminate. There is no definite pattern among the mosaics, which can only mean that the cleavage nuclei are distributed to the blastoderm differently in different embryos. It follows that there are no potential differences among the nuclei before they migrate to the surface of the egg.

The work of REITH ('25) on the egg of *Musca* shows that the different regions of the peripheral protoplasm of the egg are already determined (as to their future development) before the outward migration of the cleavage nuclei. REITH cauterized various portions of the surface of the egg. He found that the uninjured portions developed just as they would have done in an uninjured egg, regardless of the position of the cauterized region. He was able to show that any given region of the embryo may be injured without inhibiting the development of the other regions. The work of PAULI ('27) on constricted eggs and centrifuged eggs of *Calliphora* and *Musca* confirms REITH's conclusions.

It seems probable that every part of the blastoderm is "self-differentiating," though the technique used is capable of demonstrating this only in a general way and for large unit areas. There remain two ways of interpreting the results. There may be a cytoplasmic differentiation, or the different potentialities of different portions of the blastoderm may be functions of their position. There is at present no evidence to show which is true for the embryo.

Casual inspection of *Drosophila* gynandromorphs usually gives the impression that they are mostly male on one side of the body, female on the other. I have never seen a specimen that was strictly so, though approximations to this pattern are not rare. The casual impression is due to the fact that the median dorsal line of the fly (especially in the thorax and abdomen) forms a sharp line of demarcation between regions derived from corresponding imaginal discs. That is to say: the history of the dorsal thoracic and abdominal discs is determinate, in that they rarely or never produce parts that belong, structurally, to the opposite side of the body. Examination of teratological specimens of *Drosophila* confirms this conclusion. The so called "half-thorax" specimens evidently represent cases in which one wing-disc has been more or less completely suppressed; in such cases the wing-disc of the other side forms its own half of the mesonotum, but does not form any of the region proper to the missing disc (see fig. 33, p. 77, MORGAN, BRIDGES and STURTEVANT '25).

Some of the adjacent discs on one side of the body appear also to be wholly distinct. The dorsal prothoracic (humerus) and dorsal mesothoracic (wing and mesonotum), for example, appear never to encroach on each other. But, as shown above, the various cephalic discs do not seem to give rise to sharply marked-off portions of the surface of the head, and the sternopleural bristles come equally often from the dorsal and from the ventral mesothoracic discs. In the case of the sternopleurals a study of the characters of the parts concerned was also shown above to support the view that differential growth rates for the two discs are responsible for this variability. Either disc is capable of producing sternopleurals, and the one that grows into the region first does produce them. That is, the production of sternopleural bristles depends, not on the nature of the cells concerned (i.e., on which imaginal disc produced them, or what portion of the blastoderm their ancestors occupied), but on the position those cells occupy in relation to some (or all?) of the surrounding parts.

MORGAN and BRIDGES ('19) concluded that each separate part of a *Drosophila* gynandromorph developed the sexual characters and mutant characters proper to its own genetic constitution—i.e., that such characters were independent of the constitution of the rest of the individual. This conclusion has been adopted in more recent papers on *Drosophila* mosaics (with the exceptions now to be discussed), including the present account. It is certainly the usual relation, and only because it is so has it been possible to use the gynandromorphs in any study of developmental problems. For the present paper such a relation has so far been taken for granted, but it is so important an assumption for the arguments used that it must be examined. The facts are that when the parts of gynandromorphs differ in more than one respect, the different characters are completely correlated. If, from the pedigree, yellow and forked are known to lie in the same chromosome of the gynandromorph, then all yellow parts are found to be forked, and all forked parts are yellow. It turns out that such correlation holds invariably for yellow, forked, singed, white, prune, ruby, garnet, dusky, and sex (the only characters checked in the present series except diminished, which is unsatisfactory for reasons given above, but almost certainly does agree with the other characters named). A larger series of characters has been shown to be so correlated in *melanogaster*. These relations can only mean that the characters in question are dependent only on the constitution of the tissue of the part concerned.

Certain exceptional cases have, however, been described for *melanogaster*. Gynandromorphs in which one eye or part of an eye is genetically vermilion often fail to show the vermilion character (STURTEVANT '20, BRIDGES '25). Presumably the vermilion character depends, for its development, on the nature of some other part of the body, but there is no evidence to show what other part is concerned. The bar gene may be regarded as an inhibitor of facet development. A study of its action in Minute-n piebalds (STURTEVANT '27) has shown that it has different effects on different parts of the eye. Posterior to the middle of the eye is a region where facets are never inhibited. Anterior to this is a region where facets are inhibited, unless not-bar tissue is near by, in which case the inhibition is not effective. Near the anterior edge of the eye facets are inhibited even in small isolated patches of not-bar tissue. Here, then, is a second case in which the nature of the development of certain cells is influenced by the constitution of other (in this case neighboring) cells.

There is another problem which may be soluble when more data are available from mosaics—viz., at what developmental stage are the characters of a part determined? In ordinary gynandromorphs the two kinds of tissue have been distinct since early cleavage, but in Minute piebalds elimination occurs usually after the imaginal discs are laid down, so that the tissue which is ultimately male went through the first part of its development with the female constitution. When the male patches of Minute-n piebalds are yellow the yellow color seems usually to be less marked than in wholly yellow flies or in ordinary gynandromorphs with yellow parts. It is, however, not clear whether this is due to the small size of the yellow patches or, as just suggested, to the fact that the normal allelomorph of the yellow gene was present in the ancestors of these cells throughout their early history. STERN ('27) reports that, in Minutes that are mosaics for ebony, the ebony character is visible in regions where it is strongly developed in flies that are wholly ebony, but is not visible in regions where it is weakly developed in such flies. He also suggests that the time factor may be of importance. PLUNKETT ('26) has likewise suggested the possible influence of the time factor in discussing a single mosaic that indicated the scute gene as an exception to the principle of independent differentiation.[5]

LITERATURE CITED

ANDERSON, E. G., 1924: X-rays and the frequency of non-disjunction in Drosophila. Papers Mich. Acad. Sci. Arts Lett. **4:** 523–526.

——— 1925: Crossing over in a case of attached X chromosomes in Drosophila melanogaster. Genetics **10:** 403–417.

BRIDGES, C. B., 1916: Non-disjunction as proof of the chromosome theory of heredity. Genetics **1:** 1–52, 107–163.

——— 1921: Genetical and cytological proof of non-disjunction of the fourth chromosome of Drosophila melanogaster. Proc. Nat. Acad. Sci. **7:** 186 to 192.

——— 1921a: Triploid intersexes in Drosophila melanogaster. Science, n. s. **54:** 252–254.

——— 1925: Elimination of chromosomes due to a mutant (Minute-n) in Drosophila melanogaster. Proc. Nat. Acad. Sci. **1:** 701–706.

——— and MORGAN, T. H., 1923: The third-chromosome group of mutant characters of Drosophila melanogaster. Carnegie Inst. Wash., publ. 327, 251 pp.

HUETTNER, A. F., 1923: The origin of the germ cells in Drosophila melanogaster. Journ. Morph. **37:** 385–423.

——— 1924: Maturation and fertilization in Drosophila melanogaster. Journ. Morph. Physiol. **39:** 249–265.

KORSCHELT, E. and HEIDER, K., 1899: Text-book of the embryology of invertebrates. Vol. 3, 441 pp. London.

MAVOR, J. W., 1924: The production of non-disjunction by X-rays. Journ. Exper. Zool. **39:** 381–432.

MORGAN, L. V., 1925: Polyploidy in Drosophila melanogaster with two attached X chromosomes. Genetics **10:** 148–178.

——— T. H. and BRIDGES, C. B., 1919: The origin of gynandromorphs. Carnegic Inst. Wash., publ. 278, 1–122.

——— and STURTEVANT, A. H., 1925: The genetics of Drosophila. Bibliogr. Genet. **2:** 1–262.

MULLER, H. J. and DIPPEL, A. L., 1926: Chromosome breakage by X-rays and the production of eggs from genetically male tissue in Drosophila. British Journ. Exper. Biol. **3:** 85–122.

NOACK, W., 1901: Beiträge zur Entwicklungsgeschichte der Musciden. Zeitschr. f. wiss. Zool. **70:** 1–57.

PAULI, M. E., 1927: Die Entwicklung geschnürter und centrifugierter Eier von

[5] MULLER and DIPPEL ('26) have reported two cases, in each of which they suppose that a cell in a young ovary lost one X chromosome. The descendants of these cells continued to develop as ovarian tissue, though their new constitution was that characteristic of a normal testis. While this conclusion seems a priori not improbable, the data given are inadequate. A simple genetic test of the exceptional offspring recorded was available but seems not to have been applied. In the absence of such a test it seems to me improbable that they had the constitution the authors assign to them, and if that conclusion is adopted the data have no bearing on the mechanism of ovarian determination.

Calliphora erythrocephala und Musca domestica. Zeitschr. f. wiss. Zool. **129:** 483–540.

PLUNKETT, C. R., 1926: The interaction of genetic and environmental factors in development. Journ. Exper. Zool. **46:** 181–244.

REITH, F., 1925: Die Entwicklung des Musca-Eies nach Ausschaltung verschiedener Eibereiche. Zeitschr. f. wiss. Zool. **126:** 181–238.

SAFIR, S. R., 1920: Genetic and cytological examination of the phenomena of primary non-disjunction in Drosophila melanogaster. Genetics **5:** 459–487.

SNODGRASS, R. E., 1924: Anatomy and metamorphosis of the apple maggot, Rhagoletis pomonella. Journ. Agric. Research **28:** 1–36.

STERN, C., 1927: Über Chromosomenelimination bei der Taufliege. Die Naturwiss. **15:** 740–746.

STURTEVANT, A. H., 1920: The vermilion gene and gynandromorphism. Proc. Soc. Exper. Biol. Med. **17:** 70–71.

——— 1921: Genetic studies on Drosophila simulans. II. Sex-linked group of genes. Genetics **6:** 43–64.

——— 1927: The effects of the bar gene of Drosophila in mosaic eyes. Journ. Exper. Zool. **46:** 493–498.

——— and PLUNKETT, C. R., 1926: Sequence of corresponding third-chromosome genes in Drosophila melanogaster and D. simulans. Biol. Bull. **50:** 56–60.

Author's Notes, 1961

Stern (1940, Journ. Morphol. **67:** 107–122) has come to different conclusions about the origin of the sternopleural bristles. He studied specimens in which one of the imaginal discs had failed to evert, giving individuals which lacked one wing, or one middle leg, and the associated surface structures. He found that when the wing and associated mesopleural surface were absent the sternopleurals were present, whereas absence of the middle leg was always associated with the absence (or, occasionally, presence but definite abnormality) of the sternopleurals.

He concludes that the sternopleurals arise from cells of the leg disc, although he recognizes as a possible alternative "an interpretation which attributes the potency of forming a sternopleural to either disc but makes it dependent on some influence related to the ventral disc." I should conclude, taking all the data together, that this alternative is more probable—i.e., that there is an inducer of sternopleurals in the leg disc, but that cells from the wing disc may respond to it if they come to lie in the appropriate position.

The most essential point is that, in the 13 cases in which the sternopleurals were different in composition from the second leg, they were always of the same composition as the structures derived from the wing disc.

18

GENETIC AND CYTOLOGICAL STUDIES ON *OENOTHERA*

III. THE TRANSLOCATION INTERPRETATION †

Introduction

After CLELAND's discovery of the regular zig-zag method of separation of the ring chromosomes (CLELAND, 1924), it might have occurred to geneticists familiar with the oenotheras that any chromosome of the ring was separating from the two adjacent chromosomes because these chromosomes contained the homologs of different parts of the first chromosome. But it was not until BELLING (1927) suggested segmental interchange (reciprocal translocations) between non-homologous chromosomes as an interpretation of ring formation in Datura that the idea was applied to the oenotheras.

The first printed account of the application of BELLING's hypothesis to the oenotheras was by HÅKANSSON in 1928. This paper was followed shortly by a very able discussion of the general problem by DARLINGTON (1929), and by a more thorough application of the hypothesis to certain crosses by CLELAND and BLAKESLEE (1930).[1] In the last named paper the subject was treated as strictly hypothetical and no clear distinction was made between those complexes which must have one particular arrangement of chromosome ends and those which might have more than one arrangement and still satisfy all the known conditions.

An increase in the available knowledge of chromosome configurations in *Oenothera* hybrids permits a more thorough analysis of certain complexes at this time. We have attempted to follow the numbering of chromosome

† In collaboration with Sterling Emerson. Reprinted from the Zeitschrift für induktive Abstammungs- und Vererbungslehre, Vol. 59, 1931, pp. 395–419.

[1] MEURMANN (1929) and HÅKANSSON (1930) have referred to an account of the application of the translocation hypothesis to certain *Oenothera* hybrids in an unpublished part of the doctor's thesis of the first author of the present paper. The idea had actually been suggested to him by the first author after BLAKESLEE's discussion of ring formation in Datura at the Philadelphia (1926) meeting of the A.A.A.S., the application to *Oenothera* having in fact been originally suggested by BELLING in conversation. The interpretation of *Oenothera* cytology and genetics based on the translocation hypothesis follows naturally from BELLING's original postulation for Datura, and was in fact implied by BELLING himself (1927).

ends as listed by CLELAND and BLAKESLEE whenever possible, but have been careful not to name chromosomes for which the checks were incomplete. It has been necessary to disregard some of the chromosomes named by CLELAND and BLAKESLEE, but this method has the advantage of showing the limits to which chromosome regions can be identified in different complexes with the available information.

Assumptions

In accounting for the formation of rings in *Oenothera* as a probable result of reciprocal translocations, three unproved premises are essential. There are certain indications, however, that each of these premises is probably true. The same points have been raised by previous writers (see BELLING, CLELAND, and DARLINGTON) but will be presented here in different form.

1. Chromosomes observed to pair in the configuration stage of meiosis are considered to be strictly homologous, and those which form parts of large rings are understood to be at most only partially homologous. This assumption seems reasonable since both members of a pair (which can sometimes be distinguished by genetic factors) behave alike in all known hybrids (i.e., they give the same configuration in combination with any given complex—compare the known configurations of *excellens* and ^{h}*franciscana* in various hybrids), while chromosomes that do not pair may give different configurations in their respective hybrids.

2. Homologous regions of chromosomes are expected to be attracted to each other whether or not they lie in similar chromosomes. In other words, the right region of a chromosome in a ring will have its homolog in the left end of the chromosome on its right, and so on around the ring. This assumption depends in part upon our general ideas concerning the attraction between homologous chromosomes (see especially DOBZHANSKY, 1931) and is supported by the evidence supporting the first assumption. The chief support for the second premise comes from the great degree of accuracy with which we can predict chromosome configurations by means of this assumption.

3. A full double complement of chromosome "ends" is considered necessary for the formation of a viable zygote. A zygote is not expected to be viable if it contains three "ends" of one type and one of some other (instead of two of each) even though it contains fourteen whole chromosomes. Experimental evidence shows that extra chromosomes may be present, but the addition of extra ends prevents the formation of closed rings (see discussion of trisomic mutations and triploids by HÅKANSSON, 1930, and DARLINGTON, 1930). In one case a plant was found which lacked half of one chromosome and this again prevented the formation of a closed ring (STURTEVANT, in the first paper of this series). It follows that any irregularities of the sort postulated above should be detectable cytologically. So far there has been no

known complex regularly giving this behavior. This may be taken as supporting the above contention.

Cytological data. The chromosome configurations of the various species and hybrids used in the following analysis together with reference to the authority for each are listed in table 1. In this table species names are not used but species and hybrids alike are listed by the complexes of which they consist. The scheme used to indicate the number of chromosomes in a group (ring or pair) is the same as was employed in the first two studies of this series. The configurations produced by the various combinations of certain complexes are shown graphically in table 2 (pp. 150 and 151). Predictions of certain combinations are also included in this table; they are represented by the figures in parentheses.

Table 1

A list of chromosome configurations of the species and hybrids to be used in the following analysis:

accelerans·flavens: 6, 6, 2; STURTEVANT, first study of this series.
accelerans·gaudens: 6, 4, 2, 2; ibid.
accelerans·jugens: 8, 4, 2; ibid.
accelerans·maculans: 10, 2, 2; ibid.
accelerans·velans: 14; ibid.
acuens·gaudens: 14; CLELAND and OEHLKERS, 1929.
acuens·hHookeri: 4, 4, 2, 2, 2; ibid.
acuens·stringens: 4, 4, 2, 2, 2; CLELAND and OEHLKERS, 1930.
acuens·truncans: 14; CLELAND and OEHLKERS, 1929.
acuens·velans: 6, 4, 2, 2; ibid.
albicans·elongans: 12, 2; ibid.
albicans·flavens: 12, 2; CLELAND, 1928.
albicans·gaudens: 8, 6; ibid.
albicans·hHookeri: CLELAND and BLAKESLEE, 1930.
albicans·N♂: 12, 2; STURTEVANT, first study of this series.
albicans·punctulans: 10, 2, 2; CLELAND and BLAKESLEE, 1930.
albicans·stringens: 12, 2; CLELAND and OEHLKERS, 1929.
albicans·velans: 14; ibid.
deprimens·flavens: 12, 2; ibid.
deprimens·gaudens: 10, 2, 2; ibid.
deprimens·stringens: 14; OEHLKERS, 1926; CLELAND and OEHLKERS, 1929.
deprimens·velans: 10, 2, 2; CLELAND and OEHLKERS, 1929.
elongans·flavens: 8, 2, 2, 2; ibid.
elongans·hfranciscana: 4, 4, 4, 2; EMERSON, second study of this series.
excellens·flavens: 4, 4, 2, 2, 2; CLELAND and BLAKESLEE, 1930.
excellens·hfranciscana: 2, 2, 2, 2, 2, 2, 2; EMERSON, second study of this series.
excellens·gaudens: 10, 2, 2; CLELAND and BLAKESLEE, 1930.
excellens·hHookeri: 4, 2, 2, 2, 2, 2; ibid.
excellens·punctulans: 12, 2; ibid.
excellens·velans: 6, 2, 2, 2, 2; ibid.
fascians·gaudens: 8, 2, 2, 2; STURTEVANT, first study of this series.
fascians·jugens: 14; ibid.

fascians·maculans: 6, 4, 2, 2; ibid.
fascians·N♂: 8, 2, 2, 2; ibid.
fascians·velans: 8, 2, 2, 2; ibid.
flavens·^{h}franciscana: 4, 4, 2, 2, 2; Emerson, second study of this series.
flavens·gaudens: 12, 2; Cleland and Oehlkers, 1929.
flavens·^{h}Hookeri: 4, 2, 2, 2, 2, 2; Cleland and Blakeslee, 1930.
flavens·N♂: 4, 4, 2, 2, 2; Sturtevant, first study of this series.
flavens·punctulans: 8, 2, 2, 2; Cleland and Blakeslee, 1930.
flavens·stringens: 4, 2, 2, 2, 2, 2; Cleland and Oehlkers, 1929.
flavens·sulfurens: 8, 4, 2; Emerson, second study of this series.
flavens·velans: 4, 4, 2, 2, 2; Cleland and Oehlkers, 1929.
^{h}franciscana·sd ^{h}franciscana: 2, 2, 2, 2, 2, 2, 2; Emerson, second study of this series.
^{h}franciscana·gaudens: 10, 2, 2; ibid; Sturtevant, first study of this series.
^{h}franciscana·^{h}Hookeri: 4, 2, 2, 2, 2, 2; ibid.
^{h}franciscana·jugens: 10, 2, 2; Sturtevant, first study of this series.
^{h}franciscana·^{h}latifrons: 4, 2, 2, 2, 2, 2; Emerson, second study of this series.
^{h}franciscana·N♂: 6, 2, 2, 2, 2; ibid; Sturtevant, first study of this series.
^{h}franciscana·sulfurens: 10, 2, 2; Emerson, 1931.
^{h}franciscana·velans: 6, 2, 2, 2, 2; Sturtevant in first and Emerson in second studies of this series.
gaudens·^{h}Hookeri: 10, 2, 2; Cleland and Blakeslee, 1930.
gaudens·jugens: 14; Sturtevant, first study of this series.
gaudens·^{h}latifrons: 8, 2, 2, 2 (F_2); Emerson, 1930.
gaudens·N♂: 10, 2, 2; Sturtevant, first study of this series.
gaudens·punctulans: 8, 6; Cleland and Blakeslee, 1930.
gaudens·stringens: 14; Cleland and Oehlkers, 1929.
gaudens·sulfurens: 8, 6; Emerson, second study of this series.
gaudens·truncans: 10, 4; Cleland and Oehlkers, 1929.
gaudens·velans: 12, 2; Cleland, 1925.
^{h}Hookeri·punctulans: 10, 2, 2; Cleland and Blakeslee, 1930.
^{h}Hookeri·pubens: 12, 2; Sturtevant, first study of this series.
^{h}Hookeri·sulfurens: 10, 4; Emerson, second study of this series.
^{h}Hookeri·truncans: 14; Cleland and Oehlkers, 1929.
^{h}Hookeri·velans: 4, 2, 2, 2, 2, 2; Cleland and Blakeslee, 1930.
jugens·maculans: 14; Sturtevant, first study of this series.
jugens·N♂: 10, 2, 2; ibid.
jugens·pubens: 14; ibid.
jugens·velans: 12, 2; ibid.
^{h}latifrons·sulfurens: 12, 2; Emerson, second study of this series.
^{h}latifrons·velans: 8, 2, 2, 2; Emerson, 1930.
maculans·N♂: 12, 2; Sturtevant, first study of this series.
maculans·velans: 14; ibid.
N♂·velans: 6, 2, 2, 2, 2; ibid.
pubens·velans: 14 ? (at least no pair); ibid.
punctulans·velans: 12, 2; Cleland and Blakeslee, 1930.
stringens·truncans: 10, 4; Cleland and Oehlkers, 1930.
stringens·velans: 6, 4, 2, 2; Cleland and Oehlkers, 1929.
sulfurens·velans: 14; Emerson, second study of this series.
truncans·velans: 10, 4; Cleland and Oehlkers, 1929.

Cytological analysis of the complexes

Let *ʰHookeri* be arbitrarily represented:[2]

ʰHookeri: 1·2, 3·4, 5·6, 7·8, 9·10, 11·12, 13·14.

Flavens gives $\odot$ 4 and five pairs with *ʰHookeri;* therefore two *ʰHookeri* pairs have exchanged to give the *flavens* arrangement. Arbitrarily selecting 1·2 and 3·4 for this exchange,

Table

Chromosome configurations produced in the

The figures in parentheses represent predictions for which there are at present no direct nation of complexes due either to a zygote lethal carried

	acuens	*albicans*	*flavens*	*franciscana*	*gaudens*
accelerans......			6 6 2		6 4 2 2
acuens........	7 pairs				14
albicans.......		—	12 2		8 6
deprimens.....		—	12 2		10 2 2
elongans.......		12 2	8 2 2 2	4 4 4 2	
fascians.......		—			8 2 2 2
flavens........		12 2	7 pairs	4 4 2 2 2	12 2
franciscana....			4 4 2 2 2	7 pairs	10 2 2
gaudens.......	14	8 6	12 2	10 2 2	—
Hookeri.......	4 4 2 2 2	1 4	4 2 2 2 2 2	4 2 2 2 2 2	10 2 2
jugens..				10 2 2	14
latifrons.......			(6 4 2 2)	4 2 2 2 2 2	8 2 2 2
N♂..........		12 2	4 4 2 2 2	6 2 2 2 2	10 2 2
pubens........		—			
punctulans.....		10 2 2	8 2 2 2	12 2	8 6
stringens.......	4 4 2 2 2	12 2	4 2 2 2 2 2		14
sulfurens......		—	8 4 2	10 2 2	8 6
truncans.......	14				10 4
velans.........	6 4 2 2	14	4 4 2 2 2	6 2 2 2 2	12 2

[2] Since *ʰHookeri* is used as the starting point in the following deductions, and since chromosomes of other complexes are sometimes referred to as "exchanged *ʰHookeri* chromosomes," it might be inferred that *ʰHookeri* is to be considered the primitive or ancestral type. No such inference was intended. The *ʰHookeri* complex has been used in this way purely as a matter of convenience.

flavens: 1·4, 2·3, 5·6, 7·8, 9·10, 11·12, 13·14.

Velans gives ⊙ 4 and five pairs with *ʰHookeri,* two ⊙ 4 and three pairs with *flavens.* It therefore has the two chromosomes (1·2 and 3·4) that are present in *ʰHookeri* and not in *flavens,* and two others of *ʰHookeri* have exchanged—say 5·6 and 7·8. That is,

velans: 1·2, 3·4, 5·8, 6·7, 9·10, 11·12, 13·14.

These assumptions leave 1·2 equivalent to 3·4 and 5·6 equivalent to 7·8. These may be differentiated by considering *gaudens,* which gives ⊙ 10, two

2

various combinations of selected complexes.

cytological observations. The diagonal lines in some squares indicate an impossible combi-
in common or to the presence of a pollen lethal in each

Hookeri	*jugens*	*latifrons*	*maculans*	*N*♂	*stringens*	*velans*
(12 2)	8 4 2		10 2 2	12 2	(8 6)	14
4 4 2 2 2					4 4 2 2 2	6 4 2 2
14				12 2	12 2	14
					14	10 2 2
	14		6 4 2 2	8 2 2 2		8 2 2 2
4 2 2 2 2 2		(6 4 2 2)		4 4 2 2 2	4 2 2 2 2 2	4 4 2 2 2
4 2 2 2 2 2	10 2 2	4 2 2 2 2 2		6 2 2 2 2		6 2 2 2 2
10 2 2	14	8 2 2 2		10 2 2	14	12 2
7 pairs	(12 2)	(6 2 2 2 2)	(12 2)	(4 2 2 2 2 2)	(4 4 2 2 2)	4 2 2 2 2 2
(12 2)			14	10 2 2		12 2
(6 2 2 2 2)		7 pairs				8 2 2 2
(4 2 2 2 2 2)	10 2 2		12 2	7 pairs		6 2 2 2 2
12 2	14					14 ?
10 2 2						12 2
(4 4 2 2 2)						6 4 2 2
10 4		12 2				14
14					10 4	10 4
4 2 2 2 2 2	12 2	8 2 2 2	14	6 2 2 2 2	6 4 2 2	

pairs, with *ʰHookeri,* and ⊙ 12, one pair, with *velans* and also with *flavens.* That is, it has one chromosome that is present in *ʰHookeri* but not in *flavens* (1·2 or 3·4), and one that is present in *ʰHookeri* but not in *velans* (5·6 or 7·8). Selecting 1·2 and 5·6, one may write *gaudens:* 1·2, 5·6, (not 3·4, or 7·8).

^{h}*franciscana* gives ⊙ 4, 5 pairs, with ^{h}*Hookeri;* 2 ⊙ 4, 3 pairs, with *flavens;* ⊙ 6, 4 pairs, with *velans;* ⊙ 10, 2 pairs, with *gaudens*. It has two chromosomes like ^{h}*Hookeri* but not like *flavens* (i.e., 1·2 and 3·4), and one like ^{h}*Hookeri* but not like *velans* (5·6 or 7·8). The other one of the latter two must have exchanged with another ^{h}*Hookeri* chromosome to give ⊙ 4 with ^{h}*Hookeri* and ⊙ 6 with *velans,* for any ^{h}*Hookeri* exchange not involving 5·6 or 7·8 would give 2 ⊙ 4 with *velans*. If 5·6 had exchanged, the two pairs with *gaudens* would be impossible; for, *gaudens*·^{h}*Hookeri* has only two pairs, one of which is 5·6, and therefore no combination of 5 or 6 with anything else can give a pair with *gaudens*. Therefore 5·6 is present, and 7·8 has exchanged—say, with 9·10. Therefore,

^{h}*franciscana:* 1·2, 3·4, 5·6, 7·10, 8·9, 11·12, 13·14.

N gives 2 ⊙ 4, 3 pairs, with *flavens;* ⊙ 6, 4 pairs, with *velans* and with ^{h}*franciscana;* ⊙ 10, 2 pairs with *gaudens*. It has at least one chromosome like *velans* but not like *flavens* (1·2, 3·4, 7·10, 8·9). Since it gives no ring larger than 4 with *flavens,* it must have both 1·2 and 3·4 or both 5·8 and 6·7 like *velans,* and it must have both 1·2 and 3·4 or both 7·10 and 8·9 like ^{h}*franciscana;* but 5·8 and 8·9 are mutually exclusive, also 6·7 and 7·10. Hence 1·2 and 3·4 must be present. Of the other 5 chromosomes, 3 are present in *flavens,* 2 in *velans,* 2 in ^{h}*franciscana,* and 1 in *gaudens*. Since three of these five chromosomes are present in *flavens,* and the other two form a ring of four with *flavens,* there has been a simple exchange between some two of the last five *flavens* chromosomes. The *gaudens* result shows that 5·6 is present, for none of the other *flavens* chromosomes is common to *gaudens;* and 5·6 exchanges cannot be present in *gaudens,* since 5·6 itself is present. Therefore 7·8 is absent, for if it were present with 5·6 there would be ⊙ 4 in *velans*·*N*. If 7·8 had exchanged with 9·10 there would be either seven pairs or ⊙ 4 and 5 pairs in ^{h}*franciscana*·*N;* hence 9·10 is present. That is, 7·8 has exchanged with either 11·12 or 13·14, which are still equivalent. Selecting 13·14,

N: 1·2, 3·4, 5·6, 7·14, 8·13, 9·10, 11·12.

It will be seen that this result indicates that ^{h}*Hookeri*·*N* must give ⊙ 4, 5 pairs. The hybrids are now being grown for cytological study of this test case.

Jugens. There are two pairs with ^{h}*franciscana,* two with *N,* and only one with *velans*. Therefore one of 5·6, 7·10, and 8·9 (the three ^{h}*franciscana,* not-*velans,* chromosomes) is present and one of 5·6, 7·14, and 8·13 (the three *N,* not-*velans,* chromosomes). There is no pair with *gaudens,* which means that 5·6 is absent. 7·14 and 7·10 are mutually exclusive, as are 8·9 and 8·13. 7·10 and 8·9 are still equivalent, so either may be selected as the one present in *jugens*. Selecting 7·10, it follows that 8·13 is also present. It may be noted that this selection serves to fix 5, 6, 7, 8, 9, 10, 13, and 14, no two of which are now equivalent.

The single pair with *velans* cannot be 1·2 (since there is no pair with *gaudens*), nor can it be 5·8, 6·7, 9·10, or 13·14, since 7·10, and 8·13 are present. There remain only 3·4 and 11·12 as possibilities. These cannot be distinguished on the basis of the existing cytological data alone; but the genetic data show that **P** (see discussion below) is in the ring in *jugens·N,* and **P** is known to lie in the 3·4 chromosome of *N*. Therefore *jugens* does not have 3·4, and, by elimination, 11·12 must be present.

This gives three chromosomes of *jugens;* the other four remain unknown.

jugens: 7·10, 8·13, 11·12.

Albicans. There is a pair with *flavens,* none with ^{h}*Hookeri*. There must be either 1·4 or 2·3 (the only two *flavens,* not-^{h}*Hookeri,* chromosomes) present in *albicans*. These two are still equivalent; so we may select 1·4 as the one present. This selection serves to fix 1, 2, 3, and 4, no two of which are now equivalent.

There is a pair in *albicans·N,* none in *albicans·*^{h}*Hookeri*. It follows that *albicans* has 7·14 or 8·13 (the two *N,* not-^{h}*Hookeri,* chromosomes); but these are not equivalent, 8·13 being identified as the one present in *jugens*. There are no data to show which is the one present in *albicans*. We may therefore write

albicans: 1·4 (7·14 or 8·13).

The selections made in the cases of *jugens* and *albicans* serve to define every chromosome end, with the exception that 11 and 12 still remain equivalent. Before proceeding to discuss the various complexes alphabetically we may summarize the definitions, as follows.

Definitions

1·2: present in ^{h}*Hookeri* and *gaudens* but not in *flavens*.
- 1—in the pair of *flavens·albicans*.
- 2—in the ring of *flavens·albicans*.

3·4: present in ^{h}*Hookeri* but not in *gaudens* or *flavens*.
- 3—in the ring of *flavens·albicans*.
- 4—in the pair of *flavens·albicans*.

5·6: present in ^{h}*Hookeri* and *gaudens* but not in *velans*.
- 5—with 8 in *velans*.
- 6—with 7 in *velans*.

7·8: present in ^{h}*Hookeri* but not in *gaudens* or *velans*.
- 7—in a pair in *jugens·*^{h}*franciscana*.
- 8—in the ring in *jugens·*^{h}*franciscana*.

9·10: exchanged with 7·8 in ^{h}*franciscana*.
- 9—in the ring in *jugens·*^{h}*franciscana*.
- 10—in a pair in *jugens·*^{h}*franciscana*.

11·12: identified as the remaining chromosome of ^{h}*Hookeri* when the other six have been identified.
 11—not yet differentiated from 12.
 12—not yet differentiated from 11.
13·14: exchanged with 7·8 in *N*.
 13—with 8 in *N*.
 14—with 7 in *N*.

Accelerans. There is one pair with *flavens,* one with *N,* and none with *velans*. Therefore *accelerans* has one of 1·4, 2·3, 5·6, 7·8, and also one of 5·6, 7·13, 8·14. If 7·8 is present 7·13 and 8·14 are impossible, and 5·6 must be present—which gives two pairs with *flavens*. The possibilities therefore reduce to (a) one of 1·4 and 2·3, and one of 7·14 and 8·13, or (b) 5·6. The (a) relation may be shown to be unworkable if we compare the *flavens* and *N* combinations. Suppose *accelerans* to be 1·4, 7·14:

flavens . . . . 1·4, 2·3, 5·6, 7·8, 9·10, 11·12, 13·14

accelerans . . . 1·4 7·14

N . . . . . . 1·2, 3·4, 5·6, 7·14, 8·13, 9·10, 11·12

Examination shows that in order to give the observed ring of 12 with *N,* 2·3, 5·6, 8·13, 9·10, 11·12 must be linked together into a single ring by the *accelerans* complex. But in that case *accelerans·flavens* should give a ring of 12; the actual observation was two rings of 6. Any other supposition under (a) above leads to the same inconsistency. A similar argument shows that 1·3 and 2·4 are likewise impossible. Therefore 1, 2, 3, and 4 are all in separate chromosomes in *accelerans;* and (a) being excluded, only (b) remains possible, and 5·6 is present.

accelerans: 1· 2· 3· 4· 5·6.

Acuens. There are three pairs with ^{h}*Hookeri,* two with *velans;* hence one of the two ^{h}*Hookeri,* not-*velans,* chromosomes (5·6 and 7·8) is present. There is no pair with *gaudens,* meaning that 5·6 is absent; hence 7·8 is present. There are two rings of four with ^{h}*Hookeri;* these evidently result from two interchanges, one involving 1·2 (absent in *acuens* because there is no pair with *gaudens*), the other involving 5·6. Since there is no ring of four with *gaudens,* 1·2 and 5·6 have not exchanged with each other. There are two rings of four and three pairs in *acuens·stringens;* this means that 3·4 is one of the ^{h}*Hookeri* chromosomes that has undergone exchange in *acuens,* (see analysis of *stringens* below), and that the other is the one that has exchanged with 5·6 in *stringens*. The three pairs are the same in *acuens·* ^{h}*Hookeri* and *acuens·stringens*.

acuens: 1 of $\left\{\begin{matrix} 1\cdot3 & 2\cdot4 & 5\cdot & 6\cdot \\ 1\cdot & 2\cdot & 3\cdot5 & 4\cdot6 \\ 1\cdot & 2\cdot & 3\cdot6 & 4\cdot5 \end{matrix}\right\}$ 7·8, 2 of (9·10, 11·12, 13·14).

Albicans. 1·4 (7·10 or 8·9). See analysis above.

Deprimens. The ^{h}*Hookeri* chromosome that has exchanged with 5·6 in *stringens* (see below) is present here in its ^{h}*Hookeri* form, but further study of *stringens* will be necessary before this chromosome can be identified. Otherwise only negative conclusions can be drawn for *deprimens.*

Elongans. There are three pairs with *flavens* and one with ^{h}*franciscana,* hence *elongans* has two chromosomes of *flavens* that are not in ^{h}*franciscana;* that is two of 1·4, 2·3, 7·8, 9·10. Since there is no ring larger than 4 with ^{h}*franciscana, elongans* must have both 1·4 and 2·3 or both 7·8 and 9·10. The two remaining *flavens,* not-^{h}*franciscana,* chromosomes cannot be simply exchanged as they would then give ⊙ 4 with *flavens,* but each must be exchanged with a chromosome common to *flavens* and ^{h}*franciscana.* The pair with ^{h}*franciscana* must be the same as one of the pairs with *flavens.*

elongans: 1 of (5·6, 11·12, 13·14) and either 1·4, 2·3, 7· 8· 9· 10·
or 1· 2· 3· 4· 7·8, 9· 10.

Excellens. Gives 7 pairs with ^{h}*franciscana,* which see.

Fascians. There are three pairs with *velans,* three with *gaudens.* This is possible only if the one chromosome common to *velans* and *gaudens* is present. This will be clear if we represent *velans·gaudens* by a hypothetical scheme (known not to be correct, but serving the present purpose):

velans . . . . 1·2, 3·4, 5·8, 6·7, 9·10, 11·12, 13·14

gaudens . . . 1·2, 3·14, 4·5, 8·6, 7·9, 10·11, 12·13

If now any three *velans,* not-*gaudens,* chromosomes are selected, such as 3·4, 5·8, 6·7, it will be found impossible to select three *gaudens,* not-*velans,* chromosomes that will result in one and only one full complement. But if 1·2 is selected, the problem is simple e. g., 1·2, 3·4, 5·8, 7·9, 10·11, 6·13, 12·14, or several other arrangements will satisfy the requirements.

There are three pairs and a ring of eight in *fascians·N,* and P and V are independent. It follows that at least one of 3·4 or 5·6 is present, or perhaps both of them.

fascians: 1·2, (3·4 or 5·6 or both).

flavens: 1·4, 2·3, 5·6, 7·8, 9·10, 11·12, 13·14 (see discussion above).

^{h}*franciscana:* 1·2, 3·4, 5·6, 7·10, 8·9, 11·12, 13·14 (see discussion above).

gaudens: 1·2, 5·6 (see discussion above).

ʰHookeri: 1·2, 3·4, 5·6, 7·8, 9·10, 11·12, 13·14 (see discussion above).

jugens: 7·10, 8·13, 11·12 (see discussion above).

ʰlatifrons. Gives five pairs with *ʰfranciscana,* three with *velans,* hence it has two of 5·6, 7·10, 8·9. There is only one pair with *sulfurens;* therefore 7·10 and 8·9 are not both present, and therefore 5·6 is present. P is known to be in 3·4 of *ʰfranciscana* and independent of the ⊙ 4 in *ʰfranciscana·ʰlatifrons;* therefore *ʰlatifrons* has 3·4. *Gaudens·ʰlatifrons* F_2 has ⊙ 8, 3 pairs; F_1 could have had ⊙ 8, ⊙ 4, 1 pair, or ⊙ 8, 3 pairs (known to have one pair, 5·6). If *ʰlatifrons* does not have 1·2 it must have 1·2 exchanged with either 7·10 or 8·9 to give ⊙ 4 with *ʰfranciscana,* and *ʰlatifrons* then could not give ⊙ 4 with *gaudens* (since *gaudens* has neither 7·10 nor 8·9 and does not give ⊙ 4 with *ʰfranciscana* chromosomes). *ʰlatifrons* must therefore have 1·2, and three pairs with *gaudens* in the F_1.

ʰlatifrons: 1·2, 3·4, 5·6, (7·10, or 8·9) (11·12, or 13·14).

It may be predicted that *ʰlatifrons·ʰHookeri* will give the configuration 6, 2, 2, 2, 2, and *ʰlatifrons·flavens,* 6, 4, 2, 2.

Maculans. No definite conclusion is possible from the cytological evidence alone. The genetic results show that V is independent of the ring in *maculans·N,* and *V* is shown below to lie in 5·6. It may be observed that *maculans* gives configurations similar to *gaudens* in all five cases where they have been studied with the same opposing complex. No definite prediction is possible (except that at least one pair and not more than five will occur), but it may be surmised that *gaudens·maculans* will have several pairs.

maculans: 5·6.

N: 1·2, 3·4, 5·6, 7·14, 8·13, 9·10, 11·12 (see discussion above).

Pubens. There is one pair with *ʰHookeri,* none with *velans;* hence 5·6, or 7·8 is present. The genetic data show that N and V are linked in *pubens·N;* hence 5·6 is absent, so 7·8 must be present.

pubens: 7·8.

Punctulans. There are two pairs with *ʰHookeri,* one with *velans;* hence 5·6 or 7·8 is present. There is no pair with *gaudens;* hence 5·6 is absent, and 7·8 must be present. There is one more pair with *flavens* than with *ʰHookeri;* hence 1·4 or 2·3 is present. The pair with *velans* must be the same as the one with *excellens* and as one of those with *ʰHookeri* and *flavens* (i. e., 9·10, 11·12, or 13·14). Since *excellens* is the same as *ʰfranciscana,* 9·10 is absent in it and the pair concerned is either 11·12 or 13·14.

punctulans: 7·8, (1·4, or 2·3) (11·12, or 13·14).

Rigens. The cytological data do not suffice for analysis. RENNER's genetic data show a close parallelism between *rigens* and *velans* in the behavior of R and P which are also independent in *rigens·velans.* One may conclude with a high degree of probability that 1·2 and 3·4 are present.

Stringens. There are five pairs with *flavens,* two with *velans;* hence three of the *flavens,* not-*velans,* chromosomes (1·4, 2·3, 5·6, 7·8) are present. The absence of a pair with *gaudens* shows that 5·6 is not present, and *stringens* may therefore be written down as 1·4, 2·3, 7·8. The ring of 4 with *flavens* must be due to exchange of 5·6 with one of the three other chromosomes (9·10, 11·12, or 13·14), leaving two of these still present.

stringens: 1·4, 2·3, 7·8, 2 of (9·10, 11·12, 13·14).

Sulfurens. Two pairs with *^{h}franciscana,* none with *velans.* Therefore two of 5·6, 7·10, 8·9, must be present. There are no pairs with *gaudens;* therefore 5·6 is not present and both 7·10 and 8·9 must be present. There is one pair with *flavens* which must be 1·4 or 2·3 (the *flavens,* not-*^{h}franciscana* chromosomes).

sulfures: (1·4 or 2·3) 7·10, 8·9.

It may be predicted that *sulfurens·^{h}Hookeri* will have a ring of 10 and a ring of 4. Since either 1·4 or 2·3 is in the ring of 8 in *sulfurens·flavens,* both 1·2 and 3·4 must be in the ring in *sulfurens·^{h}Hookeri,* connected by the one *flavens* chromosome of *sulfurens. Sulfurens·^{h}Hookeri* was known to have this configuration, but the information has not been used in the analysis given.

Velans: 1·2, 3·4, 5·8, 6·7, 9·10, 11·12, 13·14 (see discussion above).

Cytological Predictions

The following combinations have not been obtained, but may be predicted:

accelerans·stringens: 8, 6. *Accelerans·flavens* gives 6, 6, 2, with 5·6 as the pair. Stringens differs from *flavens* by a single exchange, which involves 5·6. Therefore *accelerans·stringens* will have 5·6 incorporated in one of the *accelerans·flavens* rings of 6, and will have the configuration 8, 6.

accelerans·^{h}Hookeri: 12, 2. *Accelerans* has only one pair like *flavens* or *N*—namely 5·6. It has, therefore, only this one pair like *^{h}Hookeri.* Since *accelerans·velans* gives a ring of 14, *accelerans·^{h}Hookeri* will necessarily give a ring of 12 in addition to the 5·6 pair, rather than two rings, as may be seen by writing down the known chromosomes of the three complexes:

velans . . . . . 1·2, 3·4, 8·5, 6·7, 9·10, 11·12, 13·14

accelerans . . . ·8 5·6 7·

accelerans . . . 5·6, 7· 8·

^{h}*Hookeri* . . . . 1·2, 3·4, 5·6, 7·8, 9·10, 11·12, 13·14

Clearly if the first two give a ring of 14, the second two must give a ring of 12 and a pair.

^{h}*Hookeri·jugens:* 12, 2. Here the argument is, in a sense, the reverse of that applied to *accelerans·Hookeri.* If we write out the known chromosomes for ^{h}*franciscana, jugens* and ^{h}*Hookeri,* we get:

^{h}*franciscana* . . 1·2, 3·4, 5·6, 7·10, 8·9, 13·14, 11·12

jugens . . . . . 7·10, 8·13 11·12

jugens . . . . . 10·7, 8·13 11·12

^{h}*Hookeri* . . . . 1·2, 3·4, 5·6, 10·9, 7·8, 13·14, 11·12

It is clear that the ^{h}*Hookeri* configuration gives a continuous portion of a ring, ending in 9 and 14, which are also the ends of the portion of the ^{h}*franciscana* ring that can be followed. Since the four unknown chromosomes of *jugens* fall into this same ring against ^{h}*franciscana,* they must also do so against ^{h}*Hookeri,* and the 12, 2 configuration results.

^{h}*Hookeri·maculans:* 12, 2. The argument here is the same as in the case of *accelerans·*^{h}*Hookeri. Maculans* has 5·6 and gives a ring of 14 with *velans;* as shown above there must result a ring of 12 and a pair in ^{h}*Hookeri·maculans.*

^{h}*Hookeri·N:* 4, 2, 2, 2, 2, 2. This prediction is obvious, since the complement of each complex is fully known.

^{h}*Hookeri·stringens:* 4, 4, 2, 2, 2. This prediction has already been derived by CLELAND and BLAKESLEE (1930). *Stringens* has 1·4 and 2·3, (which will give one ring of 4 with ^{h}*Hookeri*), and gives a ring of 4 with *flavens;* this same ring must be present also in ^{h}*Hookeri·stringens,* since *flavens* and ^{h}*Hookeri* are alike for the last five chromosomes.

^{h}*Hookeri·*^{h}*latifrons:* 6, 2, 2, 2, 2. ^{h}*latifrons* is known to have four pairs with ^{h}*Hookeri,* i. e., 1·2, 3·4, 5·6, and either 11·12 or 13·14. ^{h}*latifrons* has either 7·10 or 8·9, and whichever is absent is interchanged with either 11·12 or 13·14, giving, in all, three chromosomes different from ^{h}*Hookeri.* ^{h}*Hookeri·*^{h}*latifrons* will therefore have a ring of six and four pairs, there being no other possible configuration for two complexes with four chromosomes in common and three different.

Flavens·ʰlatifrons: 6, 4, 2, 2. Chromosomes 1·2 and 3·4 of *ʰlatifrons* will form a ring of 4 with 1·4 and 2·3 of *flavens*. Since the remaining *flavens* chromosomes are the same as *ʰHookeri* there will be a ring of 6 and two pairs.

There are many partial predictions that may be made (such as that *stringens·ʰfranciscana* or *stringens·N* must give either 6, 4, 2, 2 or 4, 4, 4, 2) but it does not seem necessary to record them here. Some of the plants concerned, as well as three or four of the combinations just discussed, are now being grown.

Already several complete predictions have been verified—i. e., *ʰHookeri·velans* (see CLELAND and BLAKESLEE, 1930, for the argument), and *sulfurens·ʰHookeri*. Further, no inconsistencies in the reported configurations have yet appeared. That this point is significant becomes evident when one makes an error in transcribing data, or in working out conclusions: for it is very clear that an incorrect premise will almost invariably lead to conclusions at variance with observations. It is this self-consistency of all the recorded configurations that furnishes perhaps the most convincing evidence in favor of the translocation interpretation.

Localization of the Genes

R is independent of *velans* and *gaudens* in *Lamarckiana,* and may therefore be supposed to lie in the single pair that is present—i. e., in 1·2. It is uncertain which end of this chromosome carries it, so R may be said to lie in 1 or 2.

P is independent of R in *velans·gaudens* and in other combinations; therefore P is not in 1 or 2. In *flavens·velans* P and R are completely linked; this can only mean that P lies in 3 or 4. In *flavens·stringens* P is independent of Fr, which will be shown below to lie in 1·4; therefore P is in 3.

S is linked to P in every case studied, and this includes cases where 3·4 is present as a pair (Study II of this series). S is therefore in 3 or 4. In *albicans·flavens* (*suaveolens*) S is heterozygous and is linked to the zygote and pollen lethals. It is therefore in the ring of 12 and not in the pair (1·4). It follows that it is in 3.

N is linked to P and to S in every combination studied, and this again includes combinations that have 3·4 as a pair (EMERSON, study II of this series). Therefore N is in 3 or 4.

V is independent of the ring of 6 in *ʰfranciscana·N,* and is independent of P in several combinations in which 3·4 is present. These facts mean that it lies in 1·2, 5·6, or 11·12, the other possibilities being excluded.

In *maculans·N* there is a single pair, and V is in this pair, since V and N are independent and a plant from *maculans·N* × *N* that had received N and V from its mother, had 7 pairs of chromosomes. Since *maculans·velans* has no pair, *maculans* does not have 1·2 or 11·12. The only remaining possibility is that V is in 5 or 6. This argument is quite indirect, and depends on many more or less independent relations. Much more direct tests are now possible (such as that from *accelerans·N,* that has a single pair, which is 5·6), and are now being carried out.[3]

The zygote lethal of *flavens* segregates freely from the rest of the *flavens* complex in *rubens·flavens,* which has a single pair, 5·6. One may conclude that the lethal is in 5 or 6.

The recessive growth-habit of *jugens* is independent of P and the *jugens* lethal in *jugens·N* and in *jugens·^{h}franciscana.* These hybrids each have two pairs, but only 11·12 occurs in all three complexes concerned. The growth-habit gene therefore lies in 11·12. It may accordingly be predicted that homozygous recessives will appear on selfing *jugens·velans* (where it did not appear in a family of 8 plants) or *jugens·flavens;* since 11·12 exists as a pair in both cases. These tests are now being carried out.

Fr was shown by Oehlkers (1926) to segregate independently of the pollen lethal of *albicans* in *albicans·stringens.* The locus must, therefore, be in 1 or 4.

Br appears from de Vries' results to be independent of *velans* and *gaudens* in *Lamarckiana,* and may therefore be supposed to be in 1 or 2. One uncomfortable result is that *albicans·br velans,* on selfing, gave br br plants (de Vries, 1913). This combination commonly has a ring of 14, and no segregation is to be expected. The result is perhaps to be compared with Renner's (1925) exceptional family in which R segregated freely in *albicans ·velans.* One may suspect that 1·2 chromosomes occur in the *albicans* of some strains of *biennis.* However, since Br has not been found to be linked to anything else, it may be that it behaves like flower-size, discussed below. Direct tests of the linkage of R and Br are now under way.[4]

It follows from this analysis that 3·4 corresponds to the chromosome I of Shull or of Renner, so far as the genes concerned can be definitely assigned to chromosomes. Renner's chromosome II, and perhaps Shull's II, correspond to our 1·2; Renner's IV and probably Shull's III to our 5·6. In the absence of data on the configurations of the various *muricata* hybrids concerned, we are unable to determine the nature of Renner's III, though 7·8 is perhaps the most likely guess for it.

[3] Note added to proof. New data indicate that V is linked to R in *velans ·N* and to P in *accelerans ·N.* The above conclusions, and several others in this paper, are accordingly incorrect. It is probable that there is an error in Study I of this series, and that *velans · maculans* has a pair that was over looked. Re-examination of slides concerned has not given a clear result.

[4] Note added to proof. Preliminary counts indicate that R and Br are not linked in pure *Lamarckiana* known to have the 12, 2 configuration.

Genetic Predictions

It was shown in the first study of this series that P (or N) is independent of V in ^{h}*franciscana·N, fascians·N,* and *maculans·N;* and is linked to V in *jugens·N* and *pubens·N.* It may be predicted that, when *N* is combined with any complex having either 3·4 (i. e., P and N in a pair) or 5·6 (V in a pair), independence will obtain. Linkage is to be expected in case the 3·4 and 5·6 of *N* are in the same ring of the hybrid. That is, independence may be predicted for ^{h}*Hookeri·N, flavens·N, velans·N, gaudens ·N, accelerans·N,* ^{h}*latifrons·N;* linkage in *albicans·N.*

Similarly, it may be predicted that R and V will be independent in *velans·N, gaudens·N,* and other combinations where the two loci cannot be studied with the material we have available.

P (and N or S) will be independent of R in *velans·N, gaudens·N, fascians·velans, fascians·gaudens,* ^{h}*franciscana·velans,* ^{h}*franciscana·gaudens,* ^{h}*latifrons·velans,* ^{h}*latifrons·gaudens, punctulans·velans* etc. P and R will be linked in *accelerans·velans, jugens·velans, sulfurens·velans,* and *gaudens·jugens.*

The zygote lethal of *flavens* being in 5·6, more or less typical *"lutescens"* may be expected to appear on selfing any hybrid of *flavens* with any other complex containing 5·6, i. e., *N, accelerans,* ^{h}*latifrons, maculans,* ^{h}*franciscana.*

The recessive growth habit of *jugens* may be expected to appear on selfing any hybrid of *jugens* with another complex that carries 11·12—i. e., *flavens, velans,* ^{h}*Hookeri.*

Specific Objections

The serious objections that have been raised against the BELLING interpretation of chromosome rings are two in number.

1. *Oe. biennis* has the 8, 6 configuration, and independent segregation of the components of the two rings should result in four kinds of gametes. The genetic results of DE VRIES, RENNER, and others, show clearly that only two types (*albicans* and *rubens*) are produced. CLELAND (1926) recognized the difficulty, and suggested that the two rings do not segregate independently, but that the maternal components of each go to one pole at the first meiotic division, the paternal elements to the other. SHULL (1928) has justly pointed out that this assumption begs the whole question. There is, further, direct evidence (RUDLOFF 1930) for independent segregation of two rings in *flavens·subcurvans.*

The simplest solution of this difficulty seems to be in the assumption that the recombination products, carrying mixtures of *albicans* and *rubens* chromosomes, are not viable. These should constitute half of the spores; they are probably to be recognized (in the anthers) in some of the empty pollen grains, which (RENNER 1919) constitute more than half of all the grains. This proportion seems rather high to be accounted for by irregularities of segregation within the rings, since such irregularities are much less frequent in *biennis* than in such forms as *Lamarckiana* or *suaveolens* (CLELAND and OEHLKERS 1930).

It may be suspected that *albicans* is responsible for this relation, since RENNER's results show that it gives scarcely any recombination in any hybrid studied, including *albicans·curvans* with the configuration 6, 4, 2, 2. It seems likely that these facts are related; and we are inclined to suppose that they indicate that *albicans* includes one or more cases of translocations that are not reciprocal. Under such conditions it is possible that many of the recombination products do not have one and only one full set of genes and are inviable for this reason.

2. In some cases there are more independent genes than are expected from the configurations of the parents. With one exception (see below), the cases cited all include, as one of the genes concerned, one affecting flower-size. It is the usual experience (see especially papers of DE VRIES and RENNER) that flower-size shows segregation in any hybrid between species that differ in flower-size, and further that flower-size is independent of any other characters that may be segregating in the same plant. The most complete account is that of LANGENDORF (1930) on *albicans·^{h}Hookeri*. This has a ring of 14, and gives only pure *Hookeri* offspring—except that flower-size gives a 3:1 ratio, which LANGENDORF'S F_3 and backcross tests show to be a true Mendelian result. In the case of *albicans·velans,* also with a ring of 14, the genetic data are not so satisfactory, but HOEPPENER and RENNER (1929) have reported that F_2 *albicans·velans* plants with large and with small flowers both still have the ring of 14.

RENNER (1925) concluded that there was a series of genes (Co, etc.) concerned in this independently inherited flower-size, located in a chromosome in which no other differences have been detected. We see no alternative, on the view here proposed, to the supposition that the locus concerned shows free crossing over with the attachment-point in cases where it lies in a ring. This interpretation is in agreement with the data from *Drosophila* translocations, but is here adopted reluctantly, since the recognition of such a possibility makes far more difficult the experimental analysis of the relation of the chromosome rings to linkage in *Oenothera.* The difficulty here is the same as the familiar one encountered in determining the linkage groups of forms in which crossing over occurs in both sexes. In any given case it is not possible to distinguish between independent segregation of separate chromosomes and free crossing over in one chromosome pair. Positive cases of linkage are significant; cases in which linkage is not observable must be kept open to either interpretation. In *Oenothera* we have proceeded on the assumption that two loci that are known to be linked in some cases and independent in others lie in different chromosomes but (when they are linked) in the same ring. This depends on the further assumption that there are no very great differences in crossing over frequency in different combinations. That some such variations do occur is probable: but this procedure has not yet led to inconsistencies in the interpretation, and may be followed as long as such inconsistencies are absent. A more certain conclusion can probably be reached for these cases through the use of $2n + 1$ forms, but such experiments will involve much time and labor.

In the case, not involving flower-size, reported as including too many linkage groups (RENNER 1928), there exist specific reasons for desiring a more detailed account. This is the case of *flavens·curvans,* stated to have a 12, 2 configuration and to give independent segregation of three genes. In this case the origin of the tested plants is not stated, but is of importance. This combination is rarely obtained in F_1, but is usually produced by crossing *rigens·flavens* × *muricata* (see RENNER 1925). If, as seems likely, there are two rings in *rigens·flavens,* there exists the possibility that the resulting *flavens·curvans* plants may not all have the same configuration. Accordingly this case cannot be considered conclusive on the basis of the data now available.

The Nature of Oenothera "Mutations"

The "mutations" (in the sense of DE VRIES) found in *Oenothera* may be grouped under four main types: changes in chromosome number, forms resulting from new translocations, recessives segregating as a result of crossing over, and gene mutations.

The changes in chromosome number include haploidy (DAVIS and KULKARNI 1930, EMERSON 1929, STOMPS 1930, GATES 1929), triploidy (GATES 1909, LUTZ 1909, STOMPS 1912), tetraploidy (GATES 1909, LUTZ 1907, STOMPS 1912), extra chromosomes (GATES 1907, LUTZ 1909), extra fragments (LUTZ 1916, HÅKANSSON 1930), and the missing fragment referred to in part I of these Studies.

The forms resulting from new translocations have been discussed by DARLINGTON (1929). They differ from the parent strains not only in somatic characters, but also in having different chromosome configurations. The best-known example is *rubrinervis,* with a ring of 6 and four pairs (CLELAND 1925), derived from *Lamarckiana* with a ring of 12 and one pair. The data presented in part II of these Studies and analyzed in this paper make it probable that the production of *"franciscana sulfurea"* from *albicans·^{h}franciscana* was also due to the occurrence of a new translocation.

The new types due to crossing over fall into two classes; those with the same configuration as the parent strain, and those with seven pairs of chromosomes. If crossing over occurs in a ring it will result in shifting genes from one complex to the other, without disturbing the homologies of the ends of the chromosomes, as shown in the diagram. It is supposed that one

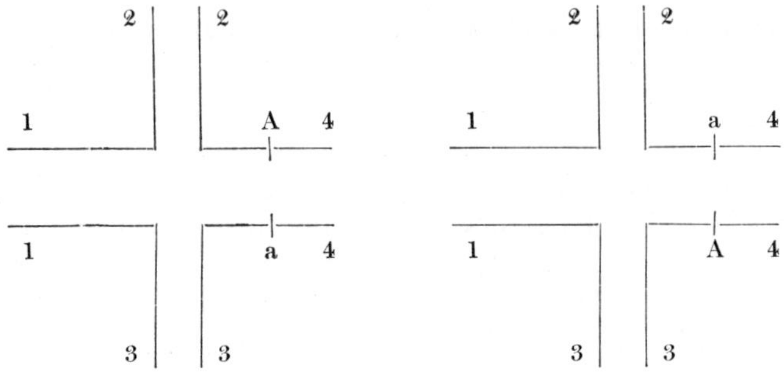

complex, made up of 1·2 and 3·4, carries gene a; the other, 1·3, 2·4, carries the allelomorph A. If crossing over occurs between the locus of A and the points of union of 4 to 2 and 3, respectively, the two complexes have exchanged their A allelomorphs, but will still be represented as 1·2, 3·4 and 1·3, 2·4, as they were before crossing over. In this way it is possible for a recessive normally present in one complex to enter the opposing complex, and thus to emerge as a viable homozygote (if the associated lethals were not included in the segment that underwent crossing over), that will still be heterozygous for most of the genes differentiating the two complexes, and that will still have the chromosome configuration of the parent strain. This is evidently the interpretation to be applied to *biennis sulfurea* (in which s, normally present in *albicans,* has entered *rubens* also—see CLELAND 1926 for chromosome configuration), *suaveolens sulfurea* (in which the s of *albicans* has entered *flavens*—see CLELAND and OEHLKERS 1930 for chromosome configuration), and *Lamarckiana nanella* (in which the n characteristic of *gaudens* has entered *velans*—see studies I and II of this series for chromosome configuration). In the case of *biennis sulfurea* the converse crossover appears to be represented by RENNER'S (1925) strain of *biennis* in which both complexes carry S. The *formosa* of *pratincola* (BLANCHARD 1929) is probably also to be referred back to a crossover.

If the lethal of a complex is lost by crossing over, the new lethal-free complex may exist in the same plant with the unmodified complex, and this plant will be homozygous for most of its genes and will have seven pairs of chromosomes. Such an explanation fits the case of *lutescens,* an almost completely homozygous *flavens* occasionally produced by *suaveolens* (DE VRIES 1918a, RENNER 1925, OEHLKERS 1923), shown in Study II of this series to have seven pairs.

Gene mutations are more or less certainly to be appealed to in accounting for the origins of rubricalyx (GATES 1915), old-gold (SHULL 1921), funifolia (Shull 1921), supplena (Shull 1927), bullata (Shull 1928a), stenophylla (de Vries 1929), and angustifolia (de Vries 1929)—all from *Lamarckiana*.

Summary

By the application of the translocation hypothesis it is possible to identify all the chromosomes of six complexes (*excellens, flavens, ^{h}franciscana, ^{h}Hookeri, N♂*, and *velans*) while partial identification can be made for thirteen others (*accelerans, acuens, albicans, elongans, fascians, gaudens, jugens, ^{h}latifrons, maculans, pubens, punctulans, stringens,* and *sulfurens*). The definitions of each chromosome are listed on p. 153.

Predictions are made for the configuration of chromosomes in certain hybrids not yet studied, p. 157.

A correlation of the genetic and cytological data shows that *R* is in 1·2, *P, S,* and *N* in 3·4 (which is identified with Shull's linkage group I), *V* and the *flavens* lethal in 5·6 (which is identified with Shull's linkage group III), and the growth-habit gene in *jugens* is in 11·12.

Predictions are made for the genetic behavior of the above genes in certain crosses of known chromosome configurations, p. 161.

"Mutations" in *Oenothera* may be due to changes in chromosome number, to occurrences of new translocations, to crossing over of a gene from one complex to another, or to gene mutations (see p. 163).

LITERATURE CITED

Belling, J., 1927. The attachment of chromosomes at the reduction division in flowering plants. Journ. Genet. **18:** 177–205.

Belling, J. and A. F. Blakeslee, 1925. On the attachment of nonhomologous chromosomes in certain 25-chromosome Daturas. Proc. Nat. Acad. Sci. **12:** 7–11.

Blanchard, F. C., 1929. The genetical constitutions of *Oenothera pratincola* and its revolute-leaved mutations. Journ. Wash. Acad. Sci. **19:** 116–125.

Cleland, R. E., 1924. Meiosis in pollen mother cells of *Oenothera franciscana sulfurea*. Bot. Gaz. **77:** 149–170.

——— 1925. Chromosome behavior during meiosis in the pollen mother cells of certain *Oenotheras*. Amer. Nat. **59:** 475–479.

——— 1926. Meiosis in the pollen mother cells of *Oenothera biennis* and *Oenothera biennis sulfurea*. Genetics. **11:** 127–162.

——— 1928. The genetics of *Oenothera* in relation to chromosome behavior, with special reference to certain hybrids. Verh. V. int. Kongr. Vererb. (Zeitschr. f. ind. Abst.-u. Vererbungsl., suppl. 2), 554–567.

Cleland, R. E. and A. F. Blakeslee, 1930. Interaction between complexes as evidence for segmental interchange in *Oenothera*. Proc. Nat. Acad. Sci. **16:** 183–189.

CLELAND, R. E. and F. OEHLKERS, 1929. New evidence bearing upon the problem of the cytological basis for genetical peculiarities of the *Oenotheras*. Amer. Nat. **63:** 497–510.

——— 1930. Erblichkeit und Zytologie verschiedener Oenotheren und ihrer Kreuzungen. Jahrb. f. wiss. Bot. **73:** 1–124.

DARLINGTON, C. D., 1929. Ring-formation in *Oenothera* and other genera. Journ. Genet. **20:** 345–363.

——— 1930. Telosynapsis or structural hybridity in *Oenothera?* Nature, May 17, 1930. 2 pp.

DAVIS, B. M., 1916. Hybrids of *Oenothera biennis* and *Oenothera franciscana* in the first and second generations. Genetics. **1:** 197–251.

——— 1926. The segregation of *Oenothera nanella-brevistylis* from crosses with *nanella* and with *Lamarckiana*. Genetics. **11:** 57–72.

DAVIS, B. M. and C. G. KULKARNI, 1930. The cytology and genetics of a haploid sport from *Oenothera franciscana*. Genetics. **15:** 55–80.

DOBZHANSKY, TH., 1931: The decrease in crossing over observed in translocations and its probable explanation. Amer. Nat. **65:** 214–232.

EMERSON, S. H., 1928a: Chromosome configuration in a dwarf segregate from *Oenothera "franciscana sulfurea."* Papers Mich. Acad. Sci. Arts Letters. **9:** 117–120.

——— 1928b: Multiple-factor inheritance in crosses between *Oenothera grandiflora* and *Oenothera franciscana*. Papers Mich. Acad. Sci. Arts Letters. **9:** 121–138.

——— 1929: The reduction division in a haploid *Oenothera*. La Cellule. **39:** 159–165.

——— 1930: The inheritance of rubricalyx bud color in crosses with *Oenothera Lamarckiana*. Proc. Nat. Acad. Sci. **16:** 796–800.

——— 1931: The inheritance of certain characters in *Oenothera* hybrids of different chromosome configurations. Genetics. **16:** 325–348.

GATES, R. R., 1907: Preliminary note on pollen development in *Oenothera lata* and its hybrids. Science. **25:** 259–260.

——— 1909: The behavior of the chromosomes in *Oenothera lata* × *Oenothera gigas*. Bot. Gaz. **48:** 179–199.

——— 1915: On the origin and behavior of *Oenothera rubricalyx*. Journ. Genet. **4:** 353–360.

——— 1929: A haploid *Oenothera*. Nature. **124:** 948.

HÅKANSSON, A., 1928: Die Reduktionsteilung in den Samenanlagen einiger *Oenotheren*. Hereditas. **11:** 129–181.

——— 1930: Zur Zytologie trisomischer Mutanten aus *Oenothera Lamarckiana*. Hereditas. **14:** 1–32.

HERIBERT-NILSSON, N., 1912: Die Variabilität der *Oenothera Lamarckiana* und das Problem der Mutation. Zeitschr. f. ind. Abst.- u. Vererbungsl. **8:** 89–231.

HOEPPENER, E. and O. RENNER, 1929: Genetische und zytologische Oenotherenstudien. II. Zur Kenntnis von *Oenothera rubrinervis, deserens, Lamarckiana-gigas, biennis-gigas, franciscana, Hookeri, suaveolens, lutescens*. Botan. Abhandl. **15:** 1–86.

LANGENDORF, J., 1930: Zur Kenntnis der Genetik und Entwicklungsgeschichte von *Oenothera fallax, rubirigida* und *Hookeri-albata*. Botan. Arch. **29:** 474–530.

LUTZ, A. M., 1907: A preliminary note on the chromosomes of *Oenothera Lamarckiana* and one of its mutants, *Oenothera gigas*. Science. **26:** 151–152.

——— 1909: Notes on the first generation hybrid of *Oenothera lata* ♀ × *Oenothera gigas* ♂. Science. **29:** 263–267.

——— 1916. *Oenothera* mutants with diminutive chromosomes. Amer. Journ. Bot. **3:** 502–526.

MEURMAN, O., 1929: Association and types of chromosomes in *Aucuba japonica.* Hereditas. **12:** 179–209.

OEHLKERS, F., 1921: Vererbungsversuche an Oenotheren. I. *Oenothera Cockerelli* und ihre Kreuzungen. Zeitschr. f. ind. Abst.- u. Vererbungsl. **26:** 1–31.

——— 1923: Vererbungsversuche an Oenotheren. II. Zeitschr. f. ind. Abst.- u. Vererbungsl. **31:** 201–260.

——— 1926: Erblichkeit und Zytologie einiger Kreuzungen mit *Oenothera strigosa.* (Vererbungsversuche an Oenotheren. IV.) Jahrb. f. wiss. Bot. **65:** 401–446.

RENNER, O., 1919: Zur Biologie und Morphologie der männlichen Haplonten einiger Oenotheren. Zeitschr. f. Bot. **11:** 305–380.

——— 1924: Die Scheckung der Oenotherenbastarde. Biol. Zentralbl. **44:** 309–336.

——— 1925: Untersuchungen über die faktorielle Konstitution einiger Komplexheterozygotischer Oenotheren. Biblioth. Genet. **9:** 168 S.

——— 1928. Über Koppelungswechsel bei *Oenothera.* Verh. V. int. Kongr. Vererb. (Zeitschr. f. ind. Abst.- u. Vererbungsl., suppl. 2). 1216–1220.

RUDLOFF, C. F., 1930: *Oenothera pachycarpa* RENNER. Genetische und cytologische Untersuchungen. Die Gartenbauwissenschaft. **3:** 499–526.

SHULL, G. H., 1921: Three new mutations in *Oenothera Lamarckiana.* Journ. Hered. **12:** 354–363.

——— 1923a: Linkage with lethal factors the solution of the *Oenothera* problem. Eugenics, Genetics, and the Family. **1:** 86–99.

——— 1923b: Further evidence of linkage with crossing over in *Oenothera.* Genetics. **8:** 154–167.

——— 1926: "Old-gold" flower color, the second case of independent inheritance in *Oenothera.* Genetics. **11:** 201–234.

——— 1927: Crossing over in the third linkage group of *Oenothera.* Proc. Nat. Acad. Sci. **13:** 21–24.

——— 1928a: A new gene mutation (mut. bullata) in *Oenothera Lamarckiana* and its linkage relations. Verh. V. int. Kongr. Vererb. (Zeitschr. f. ind. Abst.- u. Vererbungsl., suppl. 2). 1322–1342.

——— 1928b: *Oenothera* cytology in relation to genetics. Amer. Nat. **62:** 97–114.

STOMPS, T. J., 1912: Die Entstehung von *Oenothera gigas.* Ber. Deutsch. Bot. Ges. **30:** 406–416.

——— 1930: Über parthenogenetische Oenotheren. Ber. Deutsch. Bot. Ges. **48:** 119–126.

STURTEVANT, A. H. and T. DOBZHANSKY, 1930: Reciprocal translocations in *Drosophila* and their bearing on *Oenothera* cytology and genetics. Proc. Nat. Acad. Sci. **16:** 533–536.

DE VRIES, H., 1901–1903: Die Mutationstheorie. Vol. I, xii + 648 S., 8 Pl. Vol. II, xiv + 752 S., 4 Pl. Leipzig.

——— 1913: Gruppenweise Artbildung unter spezieller Berücksichtigung der Gattung *Oenothera.* Berlin. 365 S.

——— 1918a: Mutations of *Oenothera suaveolens.* Genetics. **3:** 1–26.

——— 1918b: Twin hybrids of *Oenothera Hookeri* T. and G. Genetics. **3:** 397–421.

——— 1929: Über das Auftreten von Mutanten aus *Oenothera Lamarckiana.* Zeitschr. f. ind. Abst.- u. Vererbungsl. **52:** 121–190.

19

THE USE OF MOSAICS IN THE STUDY OF THE DEVELOPMENTAL EFFECTS OF GENES†

One of the central problems of biology is that of differentiation —how does an egg develop into a complex many-celled organism? This is, of course, the traditional major problem of embryology; but it also appears in genetics in the form of the question, "How do genes produce their effects?" One of the most productive methods of studying the question in its embryological aspects is that of grafting, as elaborated by LEWIS, HARRISON, SPEMANN and others associated with them. The use of the grafting technique is also possible in studying the question from the geneticist's point of view, and much valuable information is available. The method has, however, certain limitations, some of which can be avoided by the use of spontaneously occurring mosaics, such as gynandromorphs and somatic mutants.

There are two questions that one may hope to study through the use of such mosaics, namely in what parts of the body and at what times of development are specific genes effective? The latter (the question of time of action) may also be studied by methods involving spontaneous changes in the course of development, as in GOLDSCHMIDT's experiments, or induced changes such as those described by LILLIE and JUHN; the former question (as to the place at which genes act) may also be studied by means of grafting, constriction, excision, centrifuging, or other treatment.

If a Drosophila egg starts development as a female, with two X chromosomes, it was shown by MORGAN and BRIDGES that one of these X's is occasionally eliminated from a cleavage nucleus. The cells descended from such a nucleus are male in character, and there results a gynandromorph. It often happens in experimental cultures that the male parts of such a gynandromorph have sex-linked recessive genes that are suppressed by their dominant allelomorphs in the female parts of the specimen. MORGAN and BRIDGES showed that, in general, the sex and the sex-linked characters in such gynandromorphs are completely autonomous in development. That is to say, each part develops according to its own constitution, producing the same end-results as though it had formed part of an entire individual of that same constitution.

† Reprinted from the Proceedings of the Sixth International Congress of Genetics, Vol. 1, 1932, pp. 304–307.

This remains the rule in Drosophila gynandromorphs; but a few exceptions have appeared, and these may be described briefly.

I have shown that the sex-linked dominant gene for Bar eye is an exception to the general rule. Bar acts to reduce the area of the eye, and the number of facets removed is the most convenient index of its action. Study of mosaics for Bar shows that the area occupied by facets in the normal eye may be divided into three regions with respect to the activity of the Bar gene. In the periphery of this area B removes facets not only from tissue that is genetically Bar but also from small not-Bar regions; in the central area Bar never removes facets under any condition; in the intermediate zone Bar removes facets unless not-Bar tissue is present, in which case facets develop in regions that are Bar in constitution and that lie close to not-Bar regions. I have interpreted these results as demonstrating a mutual effect of Bar and not-Bar areas on each other. It must be admitted, however, that it remains possible that we are dealing with a time relation, and that the zones differ only in the relative time at which it is determined in development that facets are or are not to be developed. Bar mosaics have been studied by the use of the Minute-n of BRIDGES, which causes eliminations late in development; if a series of early cleavage mosaics for Bar could be studied the question could be settled. Unfortunately there is no convenient method of obtaining such cleavage mosaics in quantity in *Drosophila melanogaster,* and Bar is known only in this species.

I have recently studied several of the allelomorphs of the sex-linked recessive scute in mosaics, and have found that scute is also an exception to the general rule of autonomous development. Here again most of the mosaics studied are due to eliminations occurring late in development. The results show that scute, which has as its effect the removal of specific bristles, does not remove these bristles as frequently when they lie in small patches that are scute in constitution as it does in flies that are wholly scute. Here also there is a possibility that we are dealing with a time relation, though my own interpretation of the scute series of allelomorphs leads me to suspect that the smallness of the areas scute in constitution is responsible. In this case cleavage mosaics are again needed; a few have been found, and these strongly indicate that scute shows autonomous development in large areas that result from eliminations occurring early in development.

It was observed several years ago by several investigators that yellow body-color present in the form of small spots on wild-type flies is usually not as distinctly yellow as the same area in wholly yellow flies. It has been uncertain whether this incompleteness of the yellow effect was a real one or was due to the presence of underlying not-yellow tissues partly visible through the yellow area, or possibly to some reaction in the eye or nervous system of the observer. Both of these possibilities are rendered unlikely by the recent observation that such small spots of silver body color are regularly perfectly distinct and possess sharper outlines than do yellow spots—though silver itself is a color similar to yellow but less different from wild-type. Yellow, like scute, must be classed as a gene whose effects are weakened in small spots but not in large areas of mosaics.

The first exception found to the rule of autonomous development in Drosophila was the sex-linked recessive eye color, vermilion. It has long been known that gynandromorphs with one or both eyes genetically vermilion may have eyes wild-type, vermilion, or intermediate in color. The results from mosaics due to elimination late in development were not particularly helpful here; but vermilion has recently appeared in *Drosophila simulans,* where there exists a method of obtaining early cleavage mosaics in large numbers. A series of such specimens has now been studied, and definite conclusions are available. In such mosaics, as the experiment was carried out, ovaries are always not-vermilion in constitution and testes are always vermilion in constitution. Of the specimens having eye-tissue vermilion in constitution and large enough in area to classify certainly for color, 51 of the 53 flies that had two ovaries were wild-type in eye color; all 20 of those with two well-developed testes were vermilion or intermediate in eye color; those with poorly developed testes, or with one ovary and one testis, fell into any one of the three color classes.

These results mean that the vermilion color in such mosaics is suppressed by something produced by the wild-type ovary. Three separate elements in the system may be distinguished: (1) constitution of the facet area itself; (2) constitution of the gonad; (3) constitution of some other region, or perhaps merely presence or absence of not-vermilion tissue in the fly. The third element is necessary to account for the differences between specimens with like gonads. Tabulation of such specimens indicates that this third element does not lie in the abdomen; but no particular region of the head or thorax appears to be more effective than another. One may surmise that the effective element is some internal organ not very closely related (in terms of cell lineage) to the surface areas.

These experiments were carried out by mating claret ♀ × vermilion garnet forked ♂. The garnet and forked served to identify the male regions in the 180 gynandromorphs observed. Two controls were also made: claret ♀ × garnet forked ♂ made it certain that garnet is completely autonomous in development; vermilion claret ♀ × vermilion garnet forked ♂ gave 86 wholly vermilion gynandromorphs with vermilion garnet areas, from which it is clear that poorly developed gonads alone do not interfere with the development of vermilion eye color.

The relation between eye pigment and gonad constitution described above is especially interesting in connection with the results of DOBZHANSKY, who studied *D. simulans* gynandromorphs in which the male parts were genetically white. The eye color in such mosaics showed autonomous development, as has long been known. White-eyed males also have transparent vasa efferentia and testicular envelopes, those of wild-type males being bright yellow at emergence; DOBZHANSKY showed that this yellow color does not show autonomous development. A testis or a vas efferens attached to a female (that is, under the conditions of the experiment, wild-type) duct or ovary will be yellow at emergence even though it be white in constitution. Furthermore, even if the entire genital apparatus be white and male in constitution, color may still develop, though much more slowly, if wild-type tissue is present elsewhere in the fly.

DOBZHANSKY and I have recently studied this slower development of color in somewhat more detail. It appears that, under the conditions stated, if both eyes of the mosaic are red (in some cases with very small white areas) the testes develop some color about 4 days after emergence; if both eyes are white or with a small spot of red, color does not appear in the testes until 9 to 10 days after emergence. If the eyes have both white and red present in considerable areas, the result is more variable, but color usually develops in the testes 5 to 8 days after emergence. One may conclude that there is a reciprocal relation here; in vermilion the gonads affect the eyes, in white the eyes affect the gonads.

It is clear that in most cases there is a chain of reactions between the direct activity of a gene and the end-product that the geneticist deals with as a character. One may surmise that any valid generalizations about these reactions are more likely to concern the initial links than the terminal ones. However, it is the terminal ones that are usually more open to experimental attack, since the only index to the effectiveness of a given experimental technique is the condition of the end-product. Looked at from this point of view, the type of experiment that I have described may be considered as a beginning in the analysis of certain chains of reactions into their individual links.

Author's Note, 1961

The transplantation results of Ephrussi and Beadle (1935, Bull. Biol. France Belg. 69:492–502) indicate that wild-type ovaries do not produce a substance that changes vermilion eyes toward wild-type. As suggested by Beadle (1937, Genetics, 22: 587–611) it seems probable that the results in the present paper are due to a relation, in terms of cell-lineage, between the ovaries and some other part of the body (perhaps the fat-body) that does produce such a substance. This relation probably concerns the non-germinal part of the ovary (follicle cells, etc.) rather than the germ-cells themselves, since the latter are probably not closely related (again in terms of cell-lineage) to other internal organs.

20

INVERSIONS IN THE THIRD CHROMOSOME OF WILD RACES OF *DROSOPHILA PSEUDOÖBSCURA*, AND THEIR USE IN THE STUDY OF THE HISTORY OF THE SPECIES †

Genetic studies showed several years ago that the third chromosomes of wild strains of *Drosophila pseudoöbscura* often carry suppressors of crossing-over. The salivary gland chromosome technique has made it possible to demonstrate not only that these are inverted sections, but also that there are many different inversions present in wild populations inhabiting different geographical regions. So far we have found at least fourteen different gene-sequences in wild stocks, and have found that in most geographical regions several sequences are present, though no single sequence appears to occur throughout the range of the species. There are a number of problems raised by these facts, most of which need further study; the present account is to be regarded only as a preliminary note.

One sequence has been arbitrarily designated as "standard," since it is the one in which most mutant genes are known and is also the only sequence that has been found in both of the cross-sterile races, *A* and *B*. This sequence is commonest in race *A* on the Pacific coast, from British Columbia to Lower California, but is found sporadically also in western Montana, western Nebraska, central Texas, central New Mexico (Carizozo) and southern Arizona (Santa Rita Mts.). In race *B* the standard sequence is predominant in the Sierra Nevada and the Coast Range south of the San Francisco Bay, California, but it has been found also in a single strain from the Olympic peninsula, Washington. Several other sequences represent single inversions with respect to the standard one; the "Arrowhead" sequence, found throughout the range of race *A* in British Columbia and the United States; the "Pikes Peak" sequence, found in race *A* from western Nebraska south to central Texas and southeastern Arizona; the "Klamath" sequence, found in race *B* from British Columbia south to Mono county, California; and the "Sequoia" sequence, known only in race *B* from Sequoia National Forest in California.

† In collaboration with Th. Dobzhansky. Reprinted from the Proceedings of the National Academy of Sciences, Vol. 22, 1936, pp. 448–450.

All the other known sequences differ from the standard by more than a single inversion; one of them, "Cowichan" in race *B* from Vancouver Island, differs from Klamath by a single inversion; "Chiricahua II," in race *A* from southeastern Arizona, differs from Arrowhead by a single inversion. All the others differ from all those so far named by more than a single inversion.

These facts give some information as to the historical relationships of the sequences concerned. If the sequence of the loci in the standard be arbitrarily designated *ABCDEFGH,* we may represent Klamath as *ADCBEFGH,* Arrowhead as *ABEDCFGH,* and Pikes Peak as *ABCGFEDH.* It is clear that the standard sequence could give rise to any of the three others by a single inversion, or that any of the three could give rise to standard by a single inversion. But it is equally clear that Arrowhead cannot have given rise directly (i.e., by a single inversion) either to Pikes Peak or to Klamath. If Arrowhead was the original, ancestral sequence then Pikes Peak and Klamath are related to it (and to each other) only through the standard, unless one assumes the occurrence of a complex series of coincidences (repeated breakages of the chromosome in exactly the same places). This method enables us to construct a phylogeny of the types concerned, though it does not give us a method of determining the direction in which that phylogeny should be read. For the types named the relations are shown in figure 1.

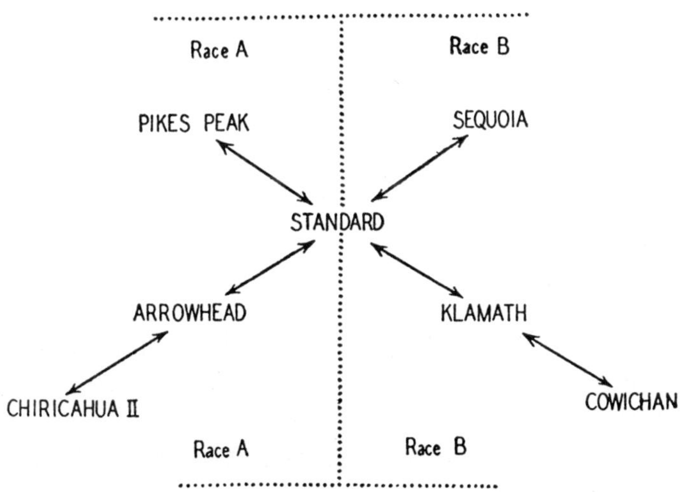

Figure 1

There is no direct method of determining which is the original sequence here; but if any one be selected, then the whole course of the evolution is thereby fixed. However, it seems likely that the standard sequence is, in fact, the ancestral one, for the following reasons among others: (1) it is centrally located in the diagram, being the only type to which several others are directly related; (2) it is the only sequence found in both races (*A* and *B*); (3) it has a wide geographical range, being equalled in this respect only by the Arrowhead sequence.

None of the other chromosomes of *Drosophila pseudoöbscura* shows a diversity of sequence at all comparable to that of the third chromosome. Tan[1] has shown that races *A* and *B* differ by having three inversions with respect to each other—one in each arm of the *X* chromosome, and one in the second chromosome. We have shown[2] that the right limb of the *X* often has an inversion associated with the "sex-ratio" gene, but the total number of sequences known for this limb is three. In the course of the present study three additional inversions have been found in wild strains: one in the second chromosome of a few strains from western New Mexico, one in the second chromosome of an *A* strain from Monterey county, California, and one in a fourth chromosome from the Mexican plateau. There are thus known from wild strains the following numbers of sequences: the left limb of *X*, two; the right limb of *X*, three; the second chromosome, four; the third, fourteen; the fourth, two. The reason for these relations remains unknown; it should, however, be pointed out that the third chromosome which shows maximum variation within the species *pseudoöbscura* is also the chromosome which shows maximum differences from the closely related species, *Drosophila miranda.*[3]

As stated, many localities yield several third chromosome sequences from a single collection. This is perhaps related to the fact that from 15 to 20 per cent of the third chromosomes tested from wild stocks have been found to carry lethal genes. This raises the question as to whether such stocks are in a condition approaching that of balanced lethals. The problem is now being studied; but two facts make us hesitate to conclude that diversity of sequence and frequency of lethals are causally related. (1) In northern and western New Mexico and in the parts of Colorado, Utah and Arizona near the point where these four states come together, extensive tests have failed to show any sequence other than Arrowhead; yet lethals were recovered here in a proportion not noticeably less than that found elsewhere. The number of tests for lethals was, however, small. (2) Preliminary tests indicate that the frequency of lethals is about the same in the second chromosome as in the third, despite the much greater uniformity in sequence in the former (complete uniformity in race *A* stocks from the regions studied).

[1] Tan, C. C., "Salivary Gland Chromosomes in the Two Races of Drosophila pseudoöbscura," *Genetics,* **20** (1935).

[2] Sturtevant, A. H., and Dobzhansky, Th., "Geographical Distribution and Cytology of 'Sex-Ratio' in Drosophila pseudoöbscura and Related Species," *Genetics,* **21** (1936).

[3] Dobzhansky, Th., and Tan, C. C., "Studies on Hybrid Sterility. III. A Comparison of the Gene Arrangement in Two Species, Drosophila pseudoöbscura and Drosophila miranda," *Zeits. ind. Abst.—Vererb.* **72,** 88–114 (1936).

21

PREFERENTIAL SEGREGATION IN TRIPLO-IV FEMALES OF *DROSOPHILA MELANOGASTER* †

Introduction

The occurrence of non-random segregation in triplo-IV females of *Drosophila melanogaster* was reported in a preliminary note (STURTEVANT 1934) of which this is the full account. As there shown, and elaborated below, all the tested fourth chromosomes may be arranged in a series, of such a nature that each chromosome will (in a triplo-IV female) pass to the haplo-IV pole at meiosis more often than will any chromosome that lies above it in the seriation. Expressed in another way, if two chromosomes (A and B) are tested against any third chromosome (C), then C will "prefer" that one of the other two that lies higher in the series.

These relations are first described in the case where one of the IV chromosomes concerned carries genes from the X chromosome. The data on other triplo-IV females are then presented. An algebraic analysis that leads to quantitative predictions is then developed.

It should be remembered throughout that the relations discussed are between whole chromosomes, not individual loci—a circumstance made experimentally feasible by the fact that crossing over in this chromosome is negligible in frequency. The chromosomes concerned are designated by the names of the mutant genes they carry, but these genes are to be regarded only as convenient markers, making it possible to follow the behavior of the chromosomes concerned.

Duplication Experiments

Males from the stock of scute-10 (= achaete − 2) were X-rayed and mated to yellow scute-D1 females. There resulted numerous wild-type daughters, and two that showed some of the characteristics of scute. These proved to have two separate modifications of the scute locus added to the achaete of the rayed males; they are designated scute-10 − 1 and scute-10 − 2. The latter (scute-10 − 2) was found to be associated with a reciprocal

† Reprinted from Genetics, Vol. 21, 1936, pp. 444–466.

translocation between the X and IV chromosomes. The break in X lies between the loci of silver and broad (to the right of the so-called "viability gene" of PATTERSON); that in IV lies between the known loci and the spindle attachment. Dr. J. SCHULTZ has examined the salivary glands of this translocation, and informs me that it is more complex than this, there being a second break in IV beyond the loci of the known genes (the piece of X being intercalated between the two ends of IV). The two new chromosomes (X with left end replaced by IV, left end of X on spindle-attachment of IV) may be obtained separately; the present study is based on flies carrying the shorter of the two as a duplication: that is, diploids plus left end of X on an extra IV spindle attachment. The "normal" X's of such hyperploids carried yellow in the experimental cultures, so that wild type body-color served as an index of the presence of the duplicating fragment.

When such flies (females) were made heterozygous for the "normal" IV chromosomes it was found (STURTEVANT 1934) that the duplication did not segregate at random with respect to these IVs. The two IVs themselves pass to opposite poles in 95 percent or more of the eggs (see below); but the duplication regularly shows a *preference* for one of them. To use the same example as previously cited, in the case of females of the constitution $\frac{y}{y}$dupl $\frac{ey^D}{gvl}$ crossed with y (not ey^D) ♂, there resulted 673 $+^y\ ey^D$, 372 $+^y\ +^{ey}$, 354 $y\ ey^D$, 806 $y\ +^{ey}$; that is, the duplication passed to the same pole as the ey^D chromosome in 67.0 percent of the cases. The result was the same whether the duplication was received from the same parent as ey^D (918/1398 = 65.7 percent) or from the same parent as *gvl* (561/807 = 69.5 percent), though in the latter case there resulted more than 50 percent recombination. The result was also the same in attached-X females carrying a Y chromosome (1118/1674 = 65.8 percent) and in separate-X females with no Y (361/531 = 67.9 percent).

Tables 1 and 2 show the results of the tests of this kind that have been carried out. Examination of these tables will show that the 26 kinds of IV chromosomes that have been tested fall into a series, approximately in the order in which they are listed in the rows of table 1. The nature of this series is such that the duplication "prefers" any chromosome to those that lie below it in the series, and this preference is greater the further apart the two chromosomes concerned lie in the series. This is shown by the fact that the values entered in each column gradually increase from the top to the bottom of the column. There are irregularities in this increase, often great enough to be statistically significant; but the general trend is unmistakable. It is probable that the discrepancies, so far as they lie beyond those to be expected from errors of random sampling, are due to differential viability or (in a few

Table 1

Percentage of cases in which the scute-10-2 duplication went to the same pole as the IV chromosome entered in each column heading, when that chromosome was tested against the ones indicated in the first column. The values entered as probable errors are the probable deviations from 50.0 percent—i.e., $67.45\sqrt{1/4n}$. The significance of the various symbols, where these are not standard designations for mutant genes, is indicated in the text and in the heading of table 2.

OPPOSED CHROMOSOME	PREFERRED CHROMOSOME								
	$+^{ci}\ ey^R$	*sv*	$ci\ ey^R$	ey^D	*ci*	X*ci*	sv^{na}	*bt*	ci^D
$+^{ci}\ ey^R$	50.0		48.5±1.8		48.5±1.4			41.4±1.8	
sv		50.0	47.2±1.3	47.8±1.6	52.2±2.5				43.9±2.4
$ci\ ey^R$	51.5±1.8	52.8±1.3	50.0	49.1±1.8	51.1±1.0	47.4±1.2	43.2±0.7	40.4±2.0	
ey^D		52.2±1.6	50.9±1.8	50.0			45.0±1.2	39.2±1.6	39.5±1.5
ci	51.5±1.4	47.8±2.5	48.9±1.0		50.0		46.9±1.0		
X *ci*			52.6±1.2			50.0	48.2±1.3		
+, $y\ sc\ w^a$			58.7±1.4	49.2±1.9			47.5±1.1		34.7±1.9
sv^{na}			56.8±0.7	55.0±1.2	53.1±1.0	51.8±1.3	50.0		43.0±1.4
+, San G.			51.9±1.3	61.1±1.8			52.7±2.2		46.3±2.1
+, Bham			60.6±1.6	52.6±1.5			57.6±1.7		51.3±1.3
bt	58.6±1.8		59.6±2.0	60.8±1.6				50.0	44.0±1.2
ci^D		56.1±2.4		60.5±1.5			57.0±1.4	56.0±1.2	50.0
+, Tonto			63.0±1.7	59.0±1.7	53.9±2.0				
bt^D			58.7±1.3	61.7±2.1			56.6±1.7		50.4±1.8
+, Red. M.			68.0±3.1	60.4±1.6			50.6±1.9		51.5±2.6
+, Arr. S.			61.6±2.9	59.0±2.0			63.1±1.4		50.3±1.9
+, Fla.			62.1±1.9	58.4±1.8					
ey		62.4±2.7	59.2±1.2				56.7±1.3	43.7±2.5	60.4±2.7
+, Seattle			61.4±1.8	62.9±1.7			54.9±1.3		
+, Sch. X			64.3±1.3	61.9±1.8					
$gvl\ ey^R$			63.8±1.6	65.2±1.4		62.5±1.4	56.0±1.7	55.0±1.4	56.5±1.7
ar	69.1±1.5		70.0±1.3	62.6±2.2			58.0±1.9		59.2±1.8
gvl		65.2±1.7	67.8±0.9	67.0±0.7	60.7±2.6		57.8±1.2	57.0±1.3	55.3±1.4
ey^4		66.4±1.9	66.4±2.2	64.0±2.9			61.3±3.5	57.2±1.7	
M^4							69.4±1.5	60.6±2.7	
ey^2			64.5±1.6	65.8±2.4	65.8±1.7		64.9±1.5	65.8±1.8	62.2±2.5

OPPOSED CHROMOSOME	PREFERRED CHROMOSOME								
	bt^D	*ey*	+, Sch. X	$gvl\ ey^R$	*ar*	*gvl*	ey^4	M^4	ey^2
$+^{ci}\ ey^R$					30.9±1.5				
sv		37.6±2.7				34.8±1.7	33.6±1.9		
$ci\ ey^R$	41.3±1.3	40.8±1.2	35.7±1.3	36.2±1.6	30.0±1.3	32.2±0.9	33.6±2.2		35.5±1.6
ey^D	38.3±2.1		38 1±1.8	34.8±1.4	37.4±2.2	33.0±0.7	36.0±2.9		34.2±2.4
ci						39.3±2.6			34.2±1.7
X *ci*				37.5±1.4					
+, $y\ sc\ w^a$				43.6±1.8		45.0±1.2			
sv^{na}	43.4±1.7	43.3±1.3		44.0±1.7	42.0±1.9	42.2±1.2	38.7±3.5	30.6±1.5	35.1±1.5
+, San G.		45.0±1.1		42.3±1.4		42.7±1.5			31.4±2.7
+, Bham		44.0±2.5		43.7±1.4		42.7±1.3			39.8±2.1
bt		56.3±2.5		45.0±1.4		43.0±1.3	42.8±1.7	39.4±2.7	34.2±1.8
ci^D	49.6±1.8	39.6±2.7		43.5±1.7	40.8±1.8	44.7±1.4			37.8±2.5
+, Tonto		49.4±2.1		47.2±1.7		42.7±1.5			
bt^D	50.0	52.1±1.7		43.2±1.5		39.0±2.1			
+, Red. M.		48.9±2.3		47.8±1.9		40.7±1.4			
+, Arr. S.		50.1±1.6		44.7±1.7		47.6±2.2			35.1±1.9
+, Fla.				42.9±1.3		50.2±1.3			
ey	47.9±1.7	50.0	47.5±1.5	44.9±1.1		44.7±1.1			
+, Seattle				51.1±2.1		50.6±2.1			
+, Sch. X.		52.5±1.5	50.0	53.1±1.4		41.4±4.4	50.0±3.7		
$gvl\ ey^R$	56.8±1.5	55.1±1.1	46.9±1.4	50.0	46.9±1.3	47.3±0.9	41.7±2.4		43.2±2.3
ar				53.1±1.3	50.0	49.5±1.1	44.7±1.8		46.5±1.7
gvl	61.0±2.1	55.3±1.1	58.6±4.4	52.7±0.9	50.5±1.1	50.0	50.8±1.1		51.3±1.1
ey^4			50.0±3.7	58.3±2.4	55.3±1.8	49.2±1.1	50.0	53.0±3.8	
M^4							47.0±3.8	50.0	47.8±3.5
ey^2				56.8±2.3	53.5±1.7	48.7±1.1		52.2±3.5	50.0

cases) to uncertainties of classification for some of the IV chromosome characters concerned.[1]

The exact seriation shown in the tables is not to be taken as accurate in

[1] The position of ey^D is different from that reported earlier (STURTEVANT 1934). This is because one of the earlier experiments was carried out with another chromosome ("Scutenick"), not further studied here, that sometimes gives a slight eye defect. These data have been discarded.

Table 2

Preference ratios for scute-10-2 duplication in tests concerning a series of wild type IV chromosomes. These came from a mutant stock (yellow⁶ scute¹² apricot) and from the following wild stocks: San Gabriel Canyon, California; Birmingham, England; Tonto, Arizona; Redwood Meadow, California; Arroyo Seco, Pasadena, California; St. Augustine, Florida; Seattle, Washington.

OPPOSED CHROMOSOME	PREFERRED CHROMOSOME							
	+, *y sc* w^a	+, San G.	+, Bham.	+, Tonto	+, Red. M.	+, Arr. S.	+, Fla.	+, Seattle
ci ey^R	41.3±1.4	48.1±1.3	39.4±1.6	37.0±1.7	32.0±3.1	38.4±2.9	37.9±1.9	38.6±1.8
ey^D	50.8±1.9	38.9±1.8	47.4±1.5	41.0±1.7	39.6±1.6	41.0±2.0	41.6±1.8	37.1±1.7
ci				46.1±2.0				
sv^{na}	52.5±1.1	47.3±2.2	42.4±1.7		49.4±1.9	36.9±1.4		45.1±1.3
ci^D	65.3±1.9	53.7±2.1	48.7±1.3		48.5±2.6	49.7±1.9		
ey		55.0±1.1	56.0±2.5	50.6±2.1	51.1±2.3	49.9±1.6		
gvl ey^R	56.4±1.8	57.7±1.4	56.3±1.4	52.8±1.7	52.2±1.9	55.3±1.7	57.1±1.3	48.9±2.1
gvl	55.0±1.2	57.3±1.5	57.3±1.3	57.3±1.5	59.3±1.4	52.4±2.2	49.8±1.3	49.4±2.1
ey^2		68 6±2.7	60.2±2.1			64.9±1.9		

detail, though the general neighborhood in which each chromosome is placed is clearly correct. The method used in obtaining this sequence, together with a numerical scale for judging the relative degrees of difference concerned, will be described below.

It will be observed that there is no evident relation between the position of a chromosome in the seriation and the mutant genes it carries. The top and the bottom chromosomes are both characterized by carrying recessive eyeless mutants, as is also *ey*, that lies near the middle of the series. Seventh from the top is a wild-type chromosome, seventh from the bottom is another; and between these lie several other wild-type chromosomes, at least some of which certainly differ from each other.

Several of the tested chromosomes are derived ones. The first, $+^{ci}ey^R$, (from Dr. C. B. BRIDGES), arose by reverse mutation from the third, *ci* ey^R. It is probable that these two are really identical in their preference properties. The *gvl* ey^R chromosome (also from Dr. BRIDGES) arose by crossing over between *ci* ey^R and *gvl;* it is intermediate, definitely different from both, and much more like *gvl*, from which its spindle-attachment is derived. The "+, Sch. X" chromosome was obtained (by Dr. J. SCHULTZ) by crossing over between ey^D and *ci*. Here the crossover is much lower in the seriation than either of the chromosomes from which it was derived. Its spindle-attachment comes from ey^D, which BRIDGES (1935) has shown to be a compound chromosome, with a diploid laterally attached duplication. This duplication may be important in determining the properties of the ey^D chromosome; it is not present in the "+, Sch. X" chromosome.

The "X *ci*" chromosome arose as a crossover in one of my triplo-IV experiments. The *ci* (and the spindle-attachment) came from *ci* ey^R, the $+^{ey}$ either from ci^D or from *gvl*. In either case the crossover is intermediate but more like the chromosome from which its spindle-attachment came. Evidently there is no one locus responsible for the preference properties of a chromosome, but it is possible that the spindle-attachment end is most important.

The relation of the duplication to the IV chromosomes is evidently due to the fact that it contains IV chromosome materials. A similar (undescribed) duplication, found in the "shaven" stock, carries the loci from *y* to silver, inclusive, but presumably has an X spindle-attachment. Experiments with it, comparable to those in table 1, gave the following values (the numbers given represent percentage preference for the first-named chromosome, which is the higher one in the seriation): *sv* – *ci* ey^R, 54.0 ± 3.2; *sv* – *ey*, 46.6 ± 1.9; ci^D – *ey*, 48.7 ± 2.7; *sv* – ci^D, 47.5 ±4.3. These values are clearly not significantly different from 50.0. Similar results were obtained with duplications 107 and 118 (DOBZHANSKY 1934), though here one of the IV chromosomes concerned was a wild type of unknown position in the series.

On the other hand, another duplication derived from an X-IV translocation (an undescribed mottled notch received from DEMEREC) gave clear indications of preference, although the fragment of X concerned is considerably longer than that of scute-10 – 2. The values obtained here were: *ci* ey^R – +*y sc* w^a, 60.0 ± 2.0; +*y sc* w^a – *gvl ey*, 57.4 ± 1.9; +*y sc* w^a – *gvl*, 66.9 ± 1.8. These are clearly significant deviations from randomness, in the same direction as the corresponding sc^{10-2} values.

Triplo-IV Females

When three IV chromosomes are present in a single female, there are three possible types of segregation that give two chromosomes to one pole and one to the other (AB/C, AC/B, BC/A). With the existing material it has not been possible to determine the relative frequencies of all three types of segregation simultaneously in experiments in which contrary classes gave a viability control (except by the use of progeny tests, which have not been carried out). In some of the data the three frequencies were determined (without the viability control), while in others the viability control was obtained but only one of the three frequencies was determined.

Two samples of the raw data are shown in tables 3 and 4. In table 3 appear the results of testing a series of different chromosomes against ey^D and ci^D.[2] Here the type of segregation determined is ey^D ci^D/x (x indicates any other chromosome listed in table 3); the viability control is similar to that in a linkage backcross. Contrary classes should be equal; inspection of the table shows that in every case the ey^D character reduced the viability materially. Other experiments indicate that this is not due to the fact that the gene occurs most often in triplo-IV flies, for this condition evidently does not interfere with the viability appreciably—in some cases at least it appears to improve it slightly. The table shows a general tendency for the percentage of tested segregation type to increase as one passes from the top to the bottom of the

[2] Both ey^D and ci^D are fully dominant when present in single dose in triplo-IV flies. It has not been found possible to distinguish such specimens from the corresponding diplo-IVs.

Table 3

Offspring of triplo-IV females of the constitution $ey^D/ci^D/x$, mated to diplo-IV males not carrying ey^D or ci^D.

X	$ey^D ci^D$	+	ey^D	ci^D	TOTAL	PERCENT $ey^D ci^D$/x
sv	181	242	250	347	1020	41.5
ci ey^R	67	82	163	168	480	31.1
ci	41	57	97	128	323	30.3
+, *y sc* w^a	44	36	49	97	226	35.4
sv^{na}	95	110	170	195	570	36.0
+, SanG.	64	125	111	157	457	41.4
+, Bham	66	84	60	140	350	42.8
bt	127	187	188	270	772	40.7
ey	71	89	77	115	352	45.5
+, Seattle	187	204	232	263	886	44.1
gvl ey^R	129	158	131	179	597	48.1
gvl	144	252	148	238	782	50.6
ey^4	101	137	56	98	392	60.7
ey^2	110	180	77	144	511	56.8

Table 4

Offspring of triplo-IV females of the type A/A/B, mated to BB males.

A	B	NOT B	B	TOTAL	PERCENT B
sv	*gvl* ey^R	181	74	255	29.0
sv	*gvl*	577	144	721	20.0
sv	ey^4	174	42	216	19.4
sv	ey^2	202	49	251	19.5
ci ey^R	sv^{na}	633	163	796	20.5
sv^{na}	*ci* ey^R	1227	183	1410	13.0
ci ey^R	*gvl*	791	243	1034	23.5
gvl	*ci* ey^R	771	92	863	10.7
gvl ey^R	*ci* ey^R	185	26	211	12.3
sv^{na}	*gvl* ey^R	976	213	1189	19.9
sv^{na}	*gvl*	1494	415	1909	21.8
sv^{na}	ey^4	1160	389	1549	25.1
ey^4	sv^{na}	659	58	717	8.1
sv^{na}	ey^2	440	155	595	26.0

series of table 1. As in that table, there are exceptions; here the value for *sv* is most aberrant. The explanation of this aberrant value is unknown, though it occurred in several independent tests and appears to be real. The general rule suggested by these results and by those of table 1 is: chromosomes high in the series tend to pass to the diplo-IV pole, those low tend to pass to the haplo-IV pole.

Table 4 shows the results from experiments of the type AAB, where the frequency of B eggs is determined. The frequency of this type of segregation is twice this value; the other two types of segregation are equivalent (AB/A), and their frequency may therefore be calculated $\left(=\frac{1-2B}{2}\right)$. Here there is no viability control, though in every case it may be surmised that the "B" class is slightly decreased from the true value by this factor.

Table 5

Frequencies of types of segregation from triplo-IV females.

FEMALE AND TYPE OF SEGREGATION	NUMBER OFFSPRING	PERCENT	FEMALE AND TYPE OF SEGREGATION	NUMBER OFFSPRING	PERCENT
$sv\ sv/gvl\ ey^{R}$	255	58.0	$ey^{D}\ ci^{D}/bt$	772	40.7
$sv\ sv/gvl$	721	40.0	$ey^{D}\ ci^{D}/ey$	352	45.5
$sv\ sv/ey^{4}$	216	38.9	$ey^{D}\ ci^{D}/+$Seattle	886	44.1
$sv\ sv/ey^{2}$	251	39.0	$ey^{D}\ ci^{D}/gvl\ ey^{R}$	597	48.1
$sv\ ey^{2}/ci\ ey^{R}$	770	21.6	$ey^{D}\ ci^{D}/gvl$	782	50.6
$ey^{D}\ ci^{D}/sv$	1020	41.5	$ey^{D}\ ci^{D}/ey^{4}$	392	60.7
$sv\ gvl/gvl\ ey^{R}$	276	26.8	$ey^{D}\ ci^{D}/ey^{2}$	511	56.8
$sv\ ey^{2}/gvl\ ey^{R}$	493	23.5	$ci\ ci^{D}/ey^{4}$	249	48.2
$sv\ gvl/ey^{2}$	278	45.0	$sv^{na}\ sv^{na}/gvl\ ey^{R}$	1189	35.8
$sv\ ey^{2}/gvl$	278	35.2	$sv^{na}\ sv^{na}/gvl$	1909	43.5
$gvl\ ey^{2}/sv$	—	19.8	$sv^{na}\ sv^{na}/ey^{4}$	1594	50.2
$ci\ ey^{R}\ ci\ ey^{R}/sv^{na}$	796	41.0	$sv^{na}\ sv^{na}/ey^{2}$	595	52.1
$ci\ ey^{R}\ ci\ ey^{R}/gvl$	1034	47.0	$gvl\ ey^{R}\ gvl/sv^{na}$	573	25.4
$ey^{D}\ sv^{na}/ci\ ey^{R}$	341	28.2	$sv^{na}\ gvl/gvl\ ey^{R}$	1022	29.5
$ey^{D}\ ci^{D}/ci\ ey^{R}$	480	31.1	$sv^{na}\ gvl\ ey^{R}/gvl$	—	45.1
$sv^{na}\ sv^{na}/ci\ ey^{R}$	1410	26.0	$gvl\ ey^{R}\ ey^{4}/sv^{na}$	338	22.4
$ci\ ey^{R}\ bt/sv^{na}$	320	35.6	$sv^{na}\ ey^{4}/gvl\ ey^{R}$	361	30.4
$sv^{na}\ ci^{D}/ci\ ey^{R}$	820	23.9	$sv^{na}\ gvl\ ey^{R}/ey^{4}$	—	47.2
$sv^{na}\ bt^{D}/ci\ ey^{R}$	280	22.6	$sv^{na}\ gvl/ey^{4}$	398	35.2
$ci\ ey^{R}\ sv^{na}/gvl\ ey^{R}$	217	36.8	$sv^{na}\ ey^{4}/gvl$	398	32.6
$sv^{na}\ ey^{4}/ci\ ey^{R}$	1074	17.4	$gvl\ ey^{4}/sv^{na}$	—	32.2
$ci\ ey^{R}\ ci^{D}/gvl$	831	54.2	$ey^{4}\ ey^{4}/sv^{na}$	717	16.2
$ci^{D}\ gvl/ci\ ey^{R}$	831	19.1	$ci^{D}\ ey/gvl$	247	38.0
$ci\ ey^{R}\ gvl/ci^{D}$	—	26.7	$ci^{D}\ gvl/ey$	247	33.2
$bt^{D}\ ey^{4}/ci\ ey^{R}$	484	27.3	$ey\ gvl/ci^{D}$	—	28.8
$gvl\ ey^{R}\ gvl\ ey^{R}/ci\ ey^{R}$	211	24.6	$ci^{D}\ gvl/ey^{4}$	260	43.2
$ci\ ey^{R}\ gvl/gvl\ ey^{R}$	1852	36.1	$ci^{D}\ ey^{4}/gvl$	260	31.4
$gvl\ ey^{R}\ gvl/ci\ ey^{R}$	1110	25.6	$gvl\ ey^{4}/ci^{D}$	—	25.4
$ci\ ey^{R}\ gvl\ ey^{R}/gvl$	—	38.3	$ci^{D}\ gvl/ey^{2}$	763	37.4
$gvl\ gvl/ci\ ey^{R}$	863	21.4	$ci^{D}\ ey^{2}/gvl$	763	42.2
$ey^{D}\ ci^{D}/ci$	323	30.3	$gvl\ ey^{2}/ci^{D}$	—	20.4
$ey^{D}\ ci^{D}/+y\ sc\ w^{a}$	226	35.4	$ey\ gvl/gvl\ ey^{R}$	222	30.6
$ey^{D}\ ci^{D}/sv^{na}$	570	36.0	$gvl\ ey^{4}/gvl\ ey^{R}$	1174	25.2
$ey^{D}\ ci^{D}/+$San G.	457	41.4	$gvl\ ey^{2}/gvl\ ey^{R}$	1653	27.8
$ey^{D}\ ci^{D}/+$Bham	350	42.8			

Table 5 summarizes all the data available from triplo-IV experiments (of whatever type) in which as many as 200 flies were recorded. The final column gives the frequencies of the segregation-types concerned (that is, the values of table 3, twice those of table 4). These values have not been corrected, in cases where the viability correction is absent from the experiment itself; but the probability is that such corrections, were it possible to estimate them quantitatively, would be small in most cases.

Many of the values recorded in table 5 are strikingly different from the 33.3 percent that would result from random segregation. Analysis shows that these deviations are in general in agreement with the principle stated in discussing table 3. If A, B, and C represent the three chromosomes concerned, in the order of their position in the seriation, then the three types of segregation occur in the relative frequencies AB/C > AC/B > BC/A. More exact quantitative analysis will be discussed below.

Algebraic Analysis of the Data

Given three IV chromosomes (A, B, and C), the simplest assumption seems to be that when A and B pass to opposite poles at meiosis, C goes to the same pole as A in a fixed proportion of the eggs that is independent of the nature of C. It is clear, from even a cursory examination of the data, that the nature of C does influence the frequency with which A and B separate, so the absolute frequency of AC/B separation will be influenced by the nature of C; the assumption is that the nature of C does not influence the *relative* frequency of AC/B and A/BC.

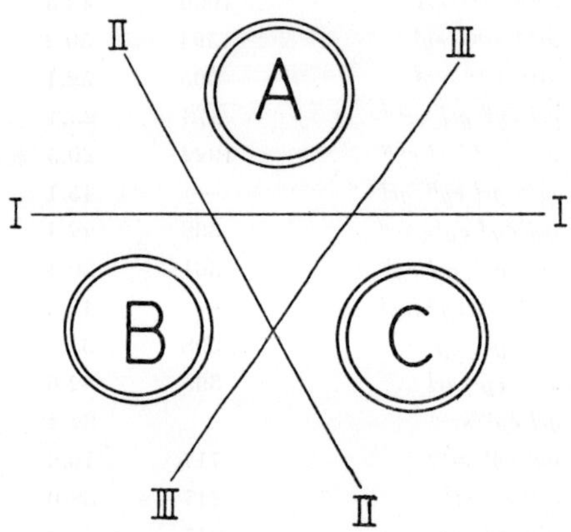

Figure 1. Three types of segregation of three IV chromosomes.

The testing of this assumption may be approached by an analysis of the three types of segregation into certain possible components. In figure 1 these three possible types are labelled I, II and III. The terminology used follows:

Let k = I + II (i.e., frequency of separation of A and B)
l = I + III (i.e., frequency of separation of A and C)
m = II + III (i.e., frequency of separation of B and C)

What will hereafter be referred to as the "preference ratios" may then be defined as follows:

r = proportion of k in which C goes with A $\left(\frac{II}{I + II}\right)$

s = proportion of l in which B goes with A $\left(\frac{III}{I + III}\right)$

t = proportion of m in which A goes with B $\left(\frac{III}{II + III}\right)$.

Then

$$I + II + III = 1$$
$$I = l(1 - s) = k(1 - r)$$
$$II = kr = m(1 - t)$$
$$III = ls = mt.$$

These equations may be shown to lead to the following results:

$$I = 1 - \left(\frac{r + s - 2rs}{1 - rs}\right) \quad (1)$$

$$II = \frac{r - rs}{1 - rs} \quad (2)$$

$$III = \frac{s - rs}{1 - rs} \quad (3)$$

$$t = \frac{s - rs}{r + s - 2rs} \quad (4)$$

The first three numbered equations are of interest in dealing with triplo-IV cases; (4) has been used in the analysis of the sc^{10-2} duplication series. The original assumption was that the values of r, s, and t were independent of the nature of the third element concerned—i.e., that r is a constant property of the pair A B, whether they be tested against C, D, or Z. Therefore equation (4) furnishes a method of calculating the preference between B and C when that between A and B and between A and C is known. The duplication series may be taken as giving such preference values directly, since one type of segregation is rare and does not enter into the calculations. If C (figure 1) be taken as the duplication, the rare type of segregation is III, and the definition of r is that it is equal to II/I + II. However, in most cases the occasional occurrence of III will give rise to two classes, one of which will be indistinguishable from one of those resulting from I or II. Nevertheless, since III is infrequent, the data of tables 1 and 2 may be taken as giving direct determinations of preferences, and equation (4) may be applied to them. (See discussion below.)

Table 6 shows the calculated values of t for each of the 26 tested chromosomes in combination with the five most thoroughly tested chromosomes. (See column headings.) From tables 1 and 2 were extracted all cases in which pairs of chromosomes entered in the cells of table 6 were each tested against a common third chromosome. From such combinations t was calculated (for example, sv/ey^D gave 52.2, $ci\ ey^R/ey^D$ gave 50.9; $sv/ci\ ey^R$ may be calculated from equation (4), substituting 52.2 for r, 50.9 for s). The

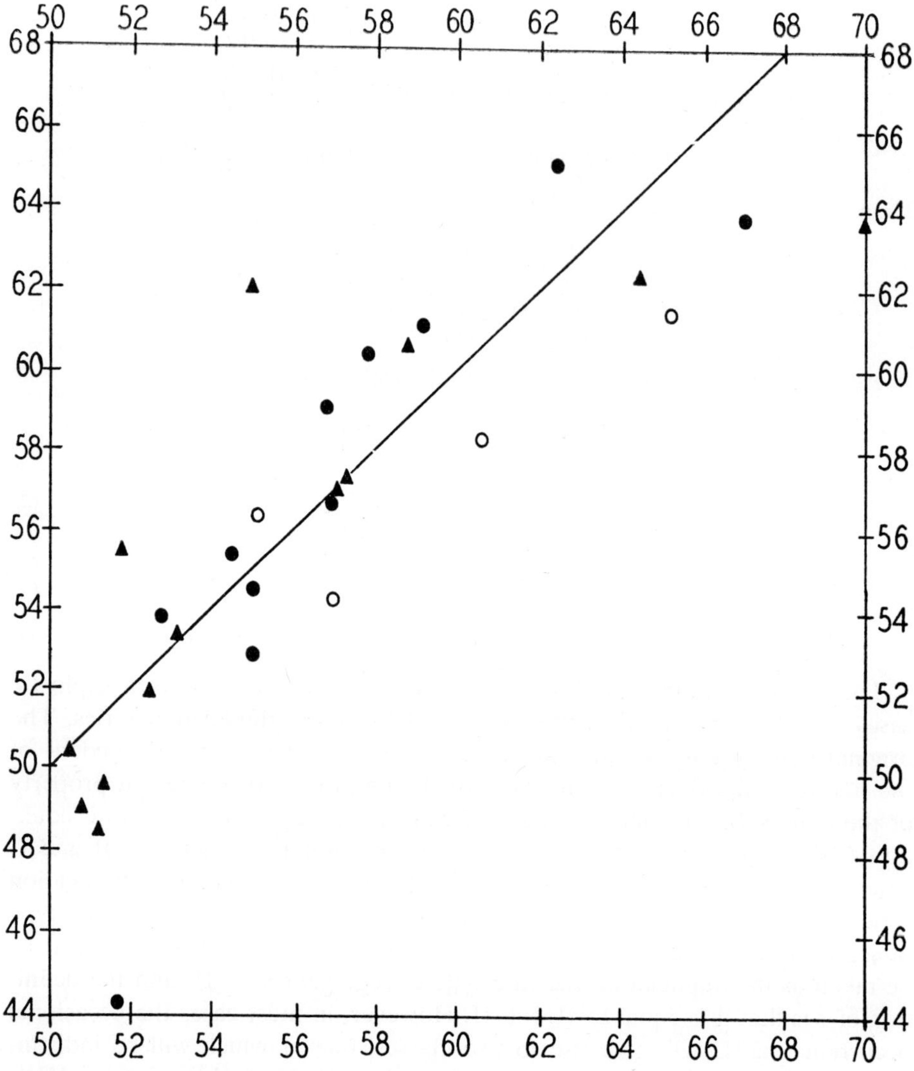

Figure 2. Comparison of observed (horizontal axis) and calculated (vertical axis) preference values for the scute-$^{10-2}$ duplication. Solid circles represent cases where the probable error of the observed value is 1.3 or less and the calculated value is an average of 9 or more determinations; solid triangles, cases where the probable error of the observed value is 1.3 or less, the calculated value the average of 5 to 7; open circles, cases where the probable error of the observed value is 1.4 or 1.5, the calculated value the average of 12 or more determinations.

various possible calculated values for each cell of table 6 have been averaged; averages of this type have been used directly in figure 2, but in table 6 the observed values from tables 1 and 2 have been averaged in, each being given the same weight as a single calculated value.

Table 6

Average of calculated preference values (directly observed values also included), based on tables 1 and 2.

OPPOSED CHROMOSOME	PREFERRED CHROMOSOME				
	ci ey^R	ey^D	sv^{na}	*gvl* ey^R	*gvl*
$+^{ci}$ ey^R	50.4	47.6	41.7	36.7	34.8
sv	48.1	49.6	44.9	36.0	36.4
ci ey^R	50.0	48.6	44.4	36.7	35.3
ey^D	51.4	50.0	45.5	38.1	36.0
ci	51.0	51.0	44.6	40.3	37.8
Xci	53.0	54.8	45.8	39.4	38.7
+*y sc* w^a	55.7	54.3	47.9	40.2	39.2
sv^{na}	55.6	54.5	50.0	41.2	39.6
+San G.	55.1	56.3	51.1	42.2	40.4
+Bham	57.2	58.5	52.9	44.3	42.7
bt	58.8	56.7	54.7	46.3	43.3
ci^D	59.2	58.4	54.4	45.5	43.7
+Tonto	58.2	60.9	52.1	45.9	45.0
bt^D	60.2	59.5	54.0	45.4	44.5
+Red. M.	59.1	60.4	54.2	46.1	45.5
+Arr. S.	59.9	59.4	55.4	46.7	45.1
+Fla.	60.5	61.3	53.6	46.7	45.3
ey	60.8	60.2	56.9	47.0	44.6
+Seattle	62.9	63.3	56.5	49.7	48.2
+Sch. X	62.4	63.2	57.2	50.2	47.8
gvl ey^R	63.3	61.9	58.8	50.0	47.2
ar	64.6	65.0	60.1	52.6	49.6
gvl	64.7	64.0	60.4	52.8	50.0
ey^4	67.1	66.1	62.2	53.8	50.8
M^4	67.7	67.3	64.0	57.7	51.0
ey^2	68.0	68.4	64.8	55.5	55.1

The values in this table are clearly self-consistent, each of the columns giving a regularly increasing series of values as one reads from top to bottom. The discrepancies that exist are all small in amount. In effect this table represents a "smoothing" of the raw data of tables 1 and 2. The regularity of the gradation of values is in part inevitable, since the final seriation is based on these data; the essential point is that it is possible to get so nearly a perfect gradation for all five columns simultaneously.

When either r or s is close to 50, equation (4) is approximated by the equation

$$t = 50 + r - s. \qquad (5)$$

That is, t differs from 50 by an amount equal to r − s. This relation is useful in checking calculations; and has also been made use of in applying arbitrary numerical values to the various chromosomes in table 7. This table in appearance, in method of construction, and in practical application, has a resemblance to the familiar "chromosome maps" that are based on crossover values. It is necessary to keep in mind, however, that no spatial relations are represented, and that the table concerns properties of whole chromosomes, not of loci.

Table 7

Preference table (Description in the text).

$+^{ci}ey^R$	20.6	+Bham	11.2	+Seattle	6.3
sv	19.8+	*bt*	10.4	+Sch. X	6.2+
ci ey^R	19.8	**ci**^{D}	10.1	**gvl ey**^{R}	6.2
ey^D	18.6	+Tonto	9.9	*ar*	4.2+
ci	17.4	bt^D	9.6	**gvl**	4.2
Xci	16.2	+Red. M.	9.3	ey^4	2.5
+*y sc* u^a	14.9	+Arr. S.	9.0	M^4	1.0
sv^{na}	14.2	+Fla.	8.8	**ey**^{2}	0.0
+San G.	13.5	**ey**	8.4		

The table is based primarily on the relations of the eight most-studied chromosomes; these are represented in bold-faced type. The values of *t* for each successive pair of these were calculated from the values of table 6—i.e., "second order" calculations were made and averaged. The differences between these values and 50 were then taken as representing the differences between the chromosomes concerned. The interval concerned was first estimated by comparing the totals of the rows in table 6, and assuming the seriation to be such as to give a regularly ascending series of such totals. Each of the remaining 18 chromosomes was then fitted into the scheme by making "second order" calculations (from table 6) for its relations to the two of the chosen 8 better-studied chromosomes between which it lies. When the sum of the values so obtained was not equal to the difference between the two chosen chromosomes (in no case did it differ markedly) the fitting was done by the method of proportional parts. These calculations resulted in a seriation of the chromosomes, with a set of relative differences. One of the terminal chromosomes (ey^2) was arbitrarily taken as the zero point, and numerical values were assigned to each of the other 25 chromosomes, corresponding to the sum of the differences appearing in each interval between them and the arbitrary zero point.

Table 7 may be used to give a prediction of the behavior of any three IV chromosomes when tested together. The differences between the values given in the table for any two chromosomes, when added to 50, give the preference ratio for those two. Three such values may be read off (for the possible combinations by two) for any three chromosomes; these, substituted for r, s, and t in equations (1), (2), and (3), give the required prediction. Furthermore, any IV chromosome not listed in table 7 may be entered in its proper

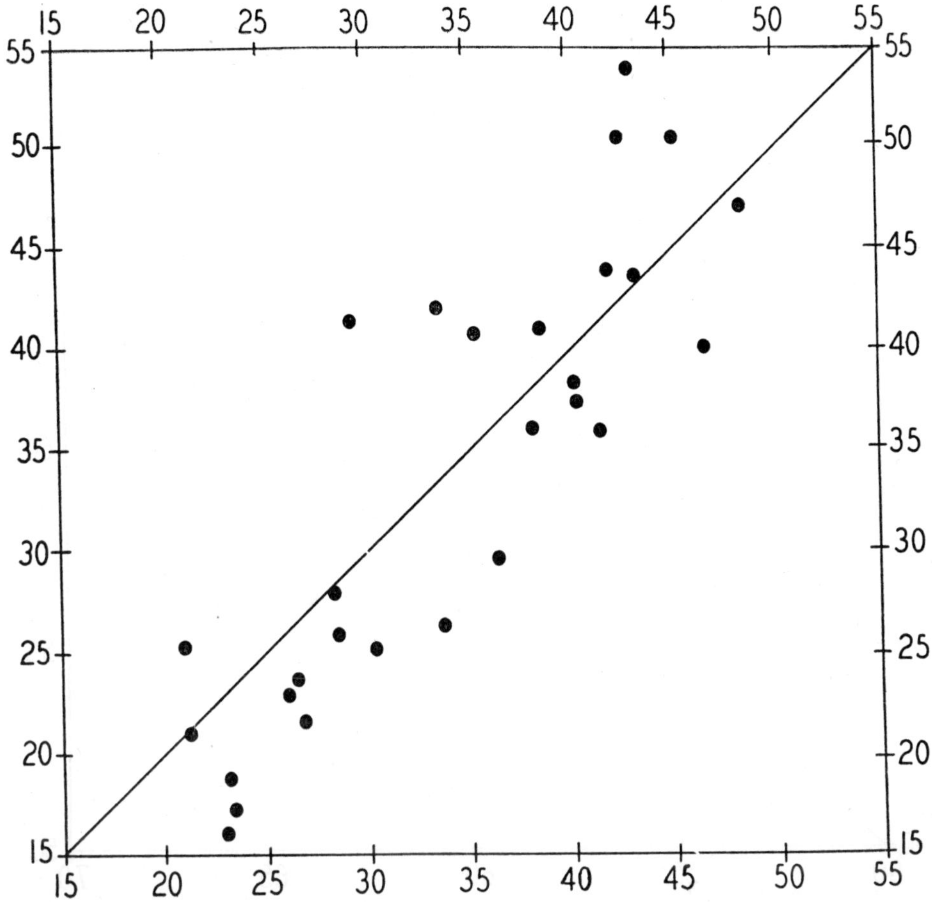

Figure 3. Comparison of observed (vertical axis) values for frequencies of segregation types in triplo-IV females with values calculated (horizontal axis) from the scute-$^{10-2}$ duplication experiments. Includes all cases where the observed value is based on 700 or more flies.

place by means of very few (theoretically by any single one) combinations with known chromosomes; it will then be possible to predict its behavior with any listed chromosomes.

Correlation of Duplication and Triplo-IV Data

The equations developed in the preceding section were assumed to apply both to the duplication experiments and to those with triplo-IV females. It is possible to test this directly, by calculating from one series the results to be expected from the other. Figures 3 and 4 show the results of such a procedure.

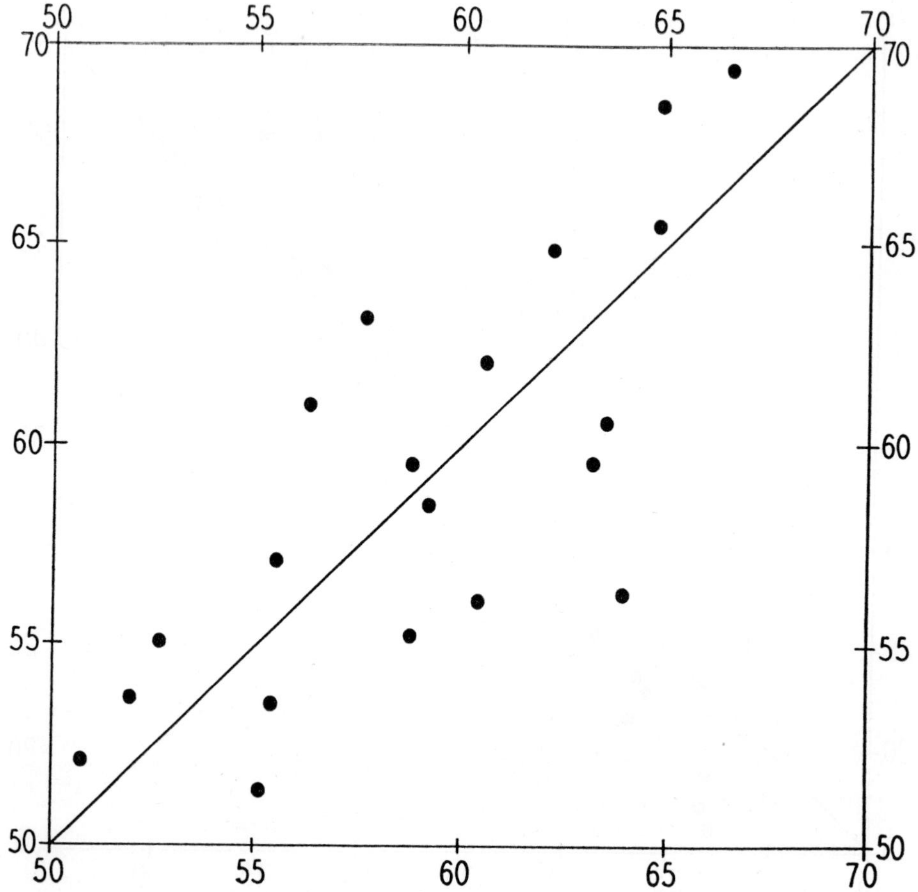

Figure 4. Preference ratios calculated from triplo-IV experiments (vertical axis) plotted against those calculated from scute-$^{10-2}$ duplication experiments (horizonal axis).

Equations (1), (2), and (3) make it possible to calculate the types of segregation expected from triplo-IV females, given the preference ratios of table 6. In some cases the necessary values are not included in table 6; such values were calculated by the same method as that used in deriving table 6. The observed frequencies are recorded in table 5. Figure 3 shows the relations here; all cases where the observed value is based on 700 or more flies are entered. The agreement is evidently satisfactory—calculations based wholly on scute-10 − 2 duplication data do in fact give good agreement with observations on triplo-IV.

There is, however, a suspicion of a systematic deviation in figure 3, such that observation gives a greater deviation from randomness than calculation would lead one to expect. The problem was therefore approached in another

Table 8

Preference ratios calculated from triplo-IV data of table 5.

CHROMOSOMES		NUMBER OF DETERMINATIONS	AVERAGE PREFERENCE RATIO
sv	*gvl* ey^R	1	56.1
sv	*gvl*	2	60.5
sv	ey^2	1	69.4
ci ey^R	sv^{na}	2	57.0
ci ey^R	ci^D	1	58.3
ci ey^R	*gvl* ey^R	2	59.5
ci ey^R	*gvl*	4	65.4
sv^{na}	*gvl* ey^R	3	55.1
sv^{na}	*gvl*	3	56.0
sv^{na}	ey^4	4	64.8
sv^{na}	ey^2	1	68.5
ci^D	*ey*	1	53.5
ci^D	*gvl*	4	61.0
ci^D	ey^4	1	63.0
ci^D	ey^2	1	62.1
ey	*gvl*	1	53.4
gvl ey^R	*gvl*	2	55.0
gvl ey^R	ey^4	2	59.4
gvl	ey^4	1	51.9
gvl	ey^2	2	51.2

way. From the definitions of r, s, and t it is possible to calculate them from triplo-IV data in cases where the frequencies of all three types of segregation are known. Table 5 includes a number of such cases. The preference values from these were calculated; where several determinations for the same two chromosomes were available these were averaged. The results are shown in table 8, and are plotted against the corresponding values of table 6 in figure 4. Here there is clearly no systematic deviation, so the indication of one in figure 3 is probably not significant. It may be noted that, if a table analogous to table 6 is constructed from the data of table 8, the same seriation of the nine chromosomes concerned will be found to fit, with surprisingly few discrepancies and these of small magnitude. If the seriation is deduced solely from the data of table 8, it is found to agree with that shown in table 7, with the exceptions that *sv* might be placed below *ci* ey^R, *ey* below *gvl* ey^R (only one comparison possible), and ey^4 below ey^2. In all these cases table 7 indicates only slight differences; the rearrangements suggested by the triplo-IV data would not seriously disturb them.

The calculations from the two series thus agree quantitatively as well as could be expected from the nature of the data.

Influence of Duplication on Segregation of Fourth Chromosomes

It has been assumed, up to this point, that in diplo-IV females carrying the scute 10 – 2 duplication the frequency with which the two IVs go to one pole and the duplication goes to the other is negligible. It was clear, when the experiments of table 1 were carried out, that this type of segregation does occur; since these experiments regularly gave a few haplo-IV flies (more than the occasional ones found in any experiment), and such flies were always not-yellow (that is, carried the duplication). These flies were recorded, but have not seemed worth dealing with, since the viability of haplo-IV is extremely low and is also quite variable. In one of the experiments of table 1, however, it is possible to identify the other (triplo-IV) product of such segregation, namely in the combination ey^D/ci^D/dupl. From this combination there were recorded 146 ey^D, 106 ci^D, 94y ey^D, 160 y ci^D, 11 y ci^D ey^D. The latter class results from the type of segregation under discussion; an equal number of (not-yellow) haplo-IV must have been produced but failed to survive. From these data one may deduce that the three types of segregation occurred with the following frequencies: ey^D dupl/ci^D = 57.9; ci^D dupl/ey^D = 37.8; ey^D ci^D/dupl = 4.2. It should be noted that the experiment suffers from the low viability of ey^D referred to before.

Several experiments were carried out in an attempt to determine the frequency of this type of segregation in another way. Females of the type studied in table 1 were mated to males of the constitution ey^D/M^4. M^4 is completely suppressed in triplo-IV flies; accordingly any wild type flies must have arisen by the production of a diplo-IV egg by the mother. Since both ey^D and M^4 have reduced viability, these experiments give maximum values, the haplo-IV eggs always giving less viable zygotes than the diplo-IV ones. The procedure adopted has been to double the number of + zygotes (since a corresponding number of haplo-IVs are lost), and add this number to that one of the two diplo-IV classes that was largest, in order to obtain a total number of zygotes. The data are given in table 9. The average value from this table is 4.55 percent non-disjunction, none having been found in the few offspring studied from females with no duplication. It may be noted that the highest frequency of non-disjunction was obtained from the combination involving the highest members of the IV chromosome seriation, as might have been expected. Taking this and the ey^D/ci^D/dupl experiment together, we may conclude that 4 percent represents a fair estimate of the usual frequency of this type of segregation.

If 4 percent is taken as the frequency of AB/dupl segregation, it is possible to estimate the error introduced by its occurrence in the experiments of tables 1 and 2. If A and B each carry a recessive gene (a, b), the calculated preference will differ according as the male used in the test is y a or y b. If r repre-

Table 9

Diplo-IV females × ey^D/M^4 males.

CONSTITUTION OF MOTHER	ey^D	M^4	+	CALCULATED NON-DISJUNCTION
sv/ey^4/dupl	90	67	1	2.2
+, *y sc* w^a/sv^{na}/dupl	129	99	6	8.5
sv^{na}/sv^{na}/dupl	204	231	7	5.7
sv^{na}/bt/dupl	102	100	3	5.6
+, San G./ey^2/dupl	90	67	1	2.2
gvl/gvl/dupl	188	160	3*	3.1
Average	—	—	—	4.55
sv/ey^4	191	149	0	0
sv^{na}/sv^{na}	85	80	0	0

* These flies, being *gvl/gvl/M*^{4}, were expected to show the grooveless character, as do *gvl/M*^{4} flies. No trace of grooveless was found; the constitution of one of the three flies was verified by tests of its offspring. This result is presumably related to the fact that the grooveless character is often slight in haplo-IV *gvl*.

sents the true preference ratio (for A over B), then in backcrosses to *y a* males the observed ratio will be r + 2/102; in backcrosses to *y b* males it will be r/102. These expressions give the following values:

Male	r = 50	r = 60	r = 65	r = 70
y a	51.0	60.8	65.7	70.6
y b	49.0	58.8	63.7	68.6

Table 10 shows the results from all experiments in which females of a given constitution were mated to both kinds of males. The average difference between the two types of experiments, instead of being 2.0 as expected, is only 0.2; evidently this source of error in tables 1 and 2 is too slight to be further considered.

Position of the Duplication in the Seriation

It is clear that the duplication may be placed in the same seriation as the other IV chromosomes. Since, as just shown, it almost always goes to the diplo-IV pole, it belongs at the top of the series, far above the other chromosomes dealt with. There is too much uncertainty about the values to enable one to place it numerically in relation to table 7.

IV Chromosome Translocations

Several reciprocal translocations between the IV and the II or III chromosomes have been tested for preference. They have not given con-

Table 10

Females A/B/dupl mated to males y a or y b.

a	*b*	BACKCROSS TO *a*: N	BACKCROSS TO *a*: CALC. PREFERENCE	BACKCROSS TO *b*: N	BACKCROSS TO *b*: CALC. PREFERENCE	*a* MINUS *b*
$ci\ ey^{R}$	*gvl*	625	67.2	583	68.5	−1.3
$ci\ ey^{R}$	sv^{na}	1979	56.0	361	55.1	+0.9
gvl	ey^{2}	821	48.2	203	50.8	−2.6
gvl	ey^{4}	378	50.2	587	48.6	+1.6
ey	*gvl*	200	58.0	802	54.6	+3.4
X*ci*	sv^{na}	505	52.3	216	48.1	+4.2
sv^{na}	*ey*	149	53.6	532	57.5	−3.9
sv^{na}	ey^{2}	201	64.1	306	65.4	−1.3
sv^{na}	*gvl*	174	58.6	678	57.5	+1.1
				Average		+0.23

sistent results, presumably because in all such cases there are types of gametes that are not recovered (or have greatly reduced viability). Until these complications can be eliminated it does not seem necessary to present what data were obtained.

No Influence of Age and Temperature

Two separate experiments were carried out to test the effects of age of female and of temperature on the preference ratios. In the first case the tested females were sv/ey^{D}, carrying the scute 10 − 2 duplication. Here controls (25°) and females treated for two days at 27°C immediately after emergence showed no clear differences from the usual value of 52 percent in daily transfer cultures run for 12 days. In the second series the females were ci^{D}/gvl, with the duplication. The temperatures used were 19°, 25°, and 28°; the first and last were applied for three days that included the emergence from the pupae. Again there were no evident deviations from the usual value (55 percent) in daily transfer cultures made for 12 days. These temperatures and ages are evidently without effect on the preference ratios.

Preference Tests in Males

The results described in the preceding pages have been based entirely on tests of females. The results obtained from tests of males are shown in tables 11 and 12. Table 11, dealing with scute-10 − 2 duplication males, suggests that there is no preference in the male, none of the values being

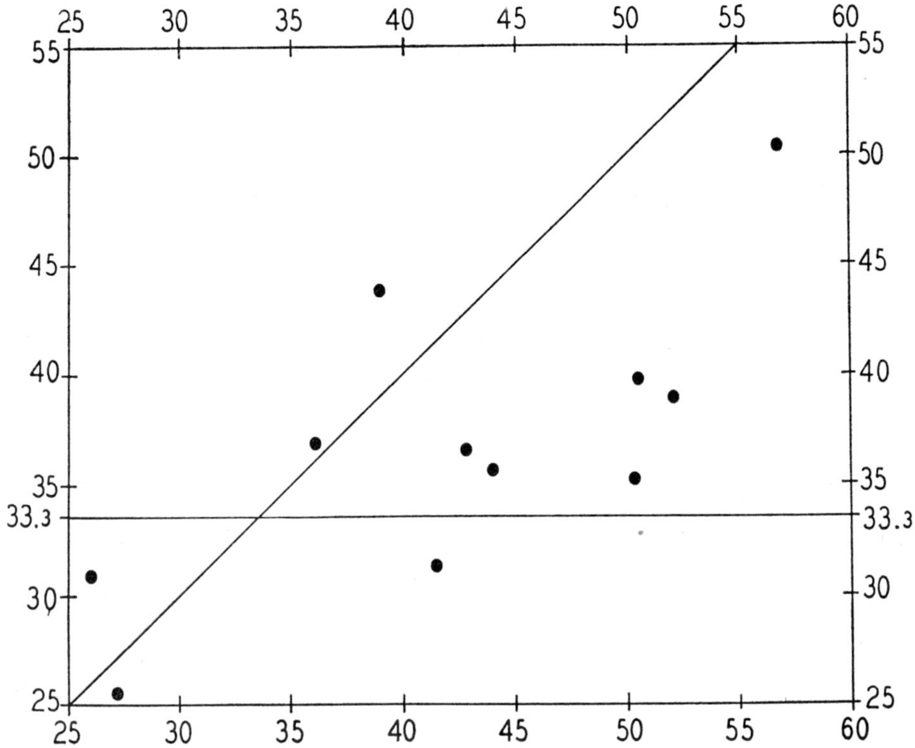

Figure 5. Frequencies of segregation types in triplo-IV males (vertical axis) plotted against the corresponding frequencies from triplo-IV females (horizontal axis).

clearly different from randomness (50.0). However, it is to be noted that the four combinations that give over 66.0 in females all give about 50.0 in males, while the four that give 57.0 or less in females all give less than 50.0 in males.

Table 12, showing the results from triplo-IV males, indicates the existence of a preference in males that is less than that in females. The correlation between comparable values for the two sexes is obvious in figure 5. The numbers of combinations are too few to warrant presentation of the results of application of the preference formulae to males; the deviations from randomness are so slight that very large numbers would be required to furnish an adequate test of these formulae.

Preference of the Duplication with Respect to X and Y

Tables 11 and 12 suggest that preference is more marked in triplo-IV males than in diplo-IV males carrying scute-10 – 2 duplication. One interpretation of this result might be that the unpaired X, having genes in

Table 11

Preference values for the scute-10 – 2 duplication in males and in corresponding females.

PREFERRED CHROMOSOME	OPPOSED CHROMOSOME	VALUE OBSERVED MALE	FEMALE
ci ey^R	*ar*	52.7 ± 1.6	70.0 ± 1.3
sv^{na}	bt^D	49.5 ± 1.4	56.6 ± 1.7
sv^{na}	ci^D	49.2 ± 2.1	57.0 ± 1.4
ci ey^R	*gvl*	52.6 ± 0.8	67.8 ± 0.9
ey^D	*gvl*	52.0 ± 1.1	67.0 ± 0.7
ci ey^R	sv^{na}	49.7 ± 1.7	56.8 ± 0.7
ey^D	sv^{na}	46.3 ± 2.1	55.0 ± 1.2
sv	ey^4	50.8 ± 1.9	66.4 ± 1.9

Table 12

Frequencies of segregation types in triplo-IV males and females compared.

	MALES		FEMALES	
	OFFSPRING	PERCENT	OFFSPRING	PERCENT
sv sv/ey^2	488	43.8	251	39.0
ey^D ci^D/*sv*	346	31.3	1020	41.5
bt^D ey^4/*ci* ey^R	343	25.6	484	27.3
sv^{na} sv^{na}/ey^4	300	35.4	1594	50.2
ey^D ci^D/sv^{na}	597	36.8	570	36.0
ey^D ci^D/+Bham.	382	36.7	350	42.8
ey^D ci^D/+Seattle	639	35.7	886	44.1
ey^D ci^D/ey^2	253	50.5	511	56.8
ey^D ci^D/*gvl*	384	39.7	782	50.6
sv^{na} sv^{na}/*ci* ey^R	592	31.1	1410	26.0
sv^{na} sv^{na}/ey^2	704	38.9	595	52.1

common with the duplication, affects its behavior by reducing its dependence on IV. If this were correct one might expect a relation of the duplication to the X-Y segregation in such males. The available data on this relation follow.

Male *y*, carrying scute-10 – 2 dupl

	$+^y$ ♀	*y* ♀	$+^y$ ♂	*y* ♂	Total
♀ *yy* (X's separate)	854	1010	772	1031	3667
♀ $\overline{yy}$ (X's attached)	252	350	314	368	1284

Analysis shows that in both series the duplication more often goes to the same pole as the Y (51.4 ± 0.6 percent and 51.7 ± 0.9 for the two series, 51.5 ± 0.48 for the two taken together). The deviation from randomness is very slight, and is doubtfully significant; since it is in the same direction in both series I am inclined to suppose that it is real.

Nevertheless the suggestion that such a relation is responsible for the difference between triplo-IV and duplication males cannot be taken as verified; for there is a more marked relation of the duplication to the sex-chromosomes in attached-X females, where its preference ratios for IV are unaffected. The data on this relation are as follows: $\overline{yy}$ dupl ♀ crossed with *y* ♂ gave + ♀ 1914, *y* ♀ 1793, + ♂ 1872, *y* ♂ 2162; preference of duplication for $\overline{XX}$ (as opposed to Y) = 4076/7741 = 52.6 ± 0.38 percent. That the result is not due to viability complications is shown by two facts. First, similar experiments using females with separate X's gave the same four classes as above in the frequencies 1099; 1178; 939; 1076. Here there can be no question of preference, since the mothers carried the duplication, whereas the fathers were responsible for the sex of the offspring. The percentage corresponding to the preference ratio above must then be 50.0 except for deviations due to viability. It is 50.6 ± 0.52. Second, other duplications, that have X spindle-attachments, similar in gene-content to the scute-10 – 2 duplication, give similar preference for $\overline{XX}$ as opposed to Y. My own observations in the duplication 118 give 307/555 = 56.4 ± 1.4; the data published by DOBZHANSKY (1934, p. 149) show similar relations for duplications 101, 106, 107, 118 and 136. The interpretation of these relations awaits further experiments.

Discussion

The algebraic analysis of the data from triplo-IV females given above has been tested and shown to be adequate. However, the IV chromosomes concerned are cytologically visible bodies, and a satisfactory description of their behavior must have a geometrical basis.

In the earlier account (STURTEVANT 1934) it was indicated that 66.7 was the limiting preference ratio, and that an interpretation might be approached from this observation as a starting point. The more extensive data here presented do not bear out the suggestion of such a limiting value; tables 1 and 6 suggest rather that the value is slightly exceeded and that there is no approach to a limit. If the duplication itself be considered as a IV chromosome, then the preference ratio for it as compared to any other IV chromosome is nearer 90 than 66.7.

One natural assumption is that two of the IV chromosomes pair and pass to opposite poles, leaving the third to go to either pole at random. The preference seriation would then be an indication of the strength of pairing affinity. This hypothesis alone is inadequate; for if, in *ABC*, pairing is always between *A* and *B*, there will result two types of segregation in equal numbers, and this value of 50 percent is the maximum that can be obtained. Since 50 percent is clearly exceeded in most of the duplication experiments and in several of those with triplo-IV, this hypothesis alone is inadequate.

Perhaps the most attractive of the remaining possible hypotheses is that all three chromosomes conjugate, as in figure 1, and preference is due to non-random orientation of the resulting hexad on the first meiotic spindle. One may suppose that the orientation is always such that one chromosome is directed toward one pole of the spindle, two toward the other. On this basis, a chromosome near the bottom of the preference seriation is one that is more likely to be directed toward a pole; it may be looked upon as being in some sense the equivalent of more than a single one of any type of chromosome lying above it in the seriation. It is probable that a workable model of this type could be imagined in terms of electrical charges.

The scute-10 – 2 duplication has nothing in common with the other IV chromosomes except its inert material and its spindle-attachment. The data on crossover IV chromosomes (of which more are needed) suggest also that these regions of normal IVs may be especially important in determining preference properties. Since the segregation relations of X and Y are also related to inert material and spindle-attachments, it seems likely that they are essential in the relations described in this paper. The spindle-attachment alone cannot be solely responsible, since the crossovers show that more than one locus must be involved. Since BRIDGES (1935) has shown that there is inert material at both ends of the IV chromosome, this demonstration does not weaken the argument for its importance.

Summary

1. Segregation in triplo-IV females of *Drosophila melanogaster* usually gives two chromosomes to one pole, one to the other.

2. The three resulting types of segregation do not usually occur with equal frequencies.

3. The IV chromosomes studied may be arranged in a definite seriation, such that any chromosome will, in a triplo-IV female, pass to the haplo-IV pole more often than will any chromosome that lies above it in the seriation (when the two are tested in like experiments or in the same experiment).

4. If two chromosomes, *A* and *B*, are tested against any third chromosome, *C*, then in those cases in which *A* and *B* separate, *C* will go with *A* in a proportion of the cases, r, that is independent of the nature of *C*—that is, r is a constant property of the pair *AB*.

5. If the preference ratio for *A* as opposed to *B* is called r, that for *A* as opposed to *C* is called s, and that for *B* as opposed to *C* is called t, then

$$t = \frac{s - rs}{r + s - 2rs}.$$

6. From this and related equations it is possible to show that the data

form a consistent whole, and to predict the results that will be given by any combination of three tested chromosomes.

7. The segregation in triplo-IV males is also non-random. It deviates from randomness in the same direction as that in corresponding females, but less markedly.

8. More than one locus is concerned with the preference properties of a IV chromosome.

LITERATURE CITED

BRIDGES, C. B., 1935: Cytological data on chromosome IV of *Drosophila melanogaster*. Trans. Dynamics Development **10:** 463–473.

DOBZHANSKY, T., 1934: Studies on chromosome conjugation. III. Behavior of duplicating fragments. Z. i. A. V. **68:** 134–162.

STURTEVANT, A. H., 1934: Preferential segregation of the fourth chromosomes in *Drosophila melanogaster*. Proc. Nat. Acad. Sci. **20:** 515–518.

Author's Note, 1961

Numerous crossover fourth chromosomes later became available (see paper no. 32), but unfortunately by this time the duplication that was used in the present experiments had been lost. Some tests of the new crossovers were carried out, using the triplo-IV technique. This technique is, as shown in this paper, relatively insensitive; the new data have never been published, for this reason. They do, however, serve to confirm the conclusion that the preference properties are influenced by more than one region in the fourth chromosome.

22

THE RELATIONS OF INVERSIONS IN THE X CHROMOSOME OF *DROSOPHILA MELANOGASTER* TO CROSSING OVER AND DISJUNCTION[1] †

Introduction

Studies of chromosome aberrations such as polyploidy and translocations have contributed much to the understanding of the meiotic behavior of chromosomes. One of the commonest types of structural difference in chromosomes within a species is that in which a segment of a chromosome has been inverted. These cases have not contributed as much as might have been expected toward an understanding of chromosome mechanics. It has been apparent for some time that they needed systematic study, and that the series of X-chromosome inversions accumulated in X-ray experiments furnished the necessary material. The present paper represents the results of a study made with these points in mind.

Recent papers on inversions illustrate the difficulties encountered in such studies. GERSHENSON (1935) and STONE and THOMAS (1935) found that the chromatids resulting from single crossing over are not recovered, and by making egg counts showed that there is no detectable mortality that can be invoked to account for them. They concluded that single exchange is so rare as to be negligible. SIDOROV, SOKOLOV, and TROFIMOV (1935) showed by the use of attached-X females that single exchanges do occur with a high frequency; but they appear to have been unaware that egg counts fail to show any corresponding mortality. The dates show that at least the second and third papers mentioned and our own preliminary note (BEADLE and STURTEVANT, 1935) were all sent to press independently. Our account, of which the present paper is the full presentation, included both the proof that single exchanges occur (using the same method as the Russian investigators), and the proof that there is no corresponding egg mortality. We were faced with a seeming

---

[1] The cost of the accompanying tables and illustrations is paid by the Galton and Mendel Memorial Fund.

† In collaboration with G. W. Beadle. Reprinted from Genetics, Vol. 21, 1936, pp. 544–604.

paradox, the only escape from which was the assumption that the single crossover chromatids are produced but are not included in the egg nucleus. We developed a scheme that gives this result (pp. 233–238, and figs. 6 and 7), and that is in good quantitative agreement with the data in other respects.

Inversions Studied

There follows a descriptive catalogue of the inversions we have studied. Figure 1 shows the lengths of these inversions. The data on the extent of each inversion are largely from the experiments with females heterozygous for two inversions, which are described below. The egg mortalities associated with the inversions are given in a later section.

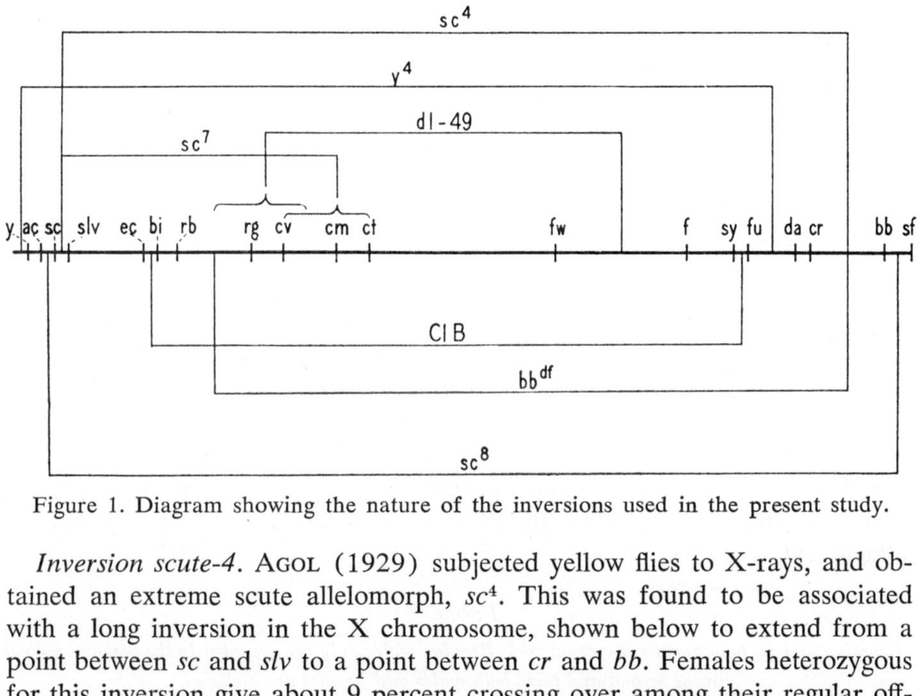

Figure 1. Diagram showing the nature of the inversions used in the present study.

Inversion scute-4. Agol (1929) subjected yellow flies to X-rays, and obtained an extreme scute allelomorph, sc^4. This was found to be associated with a long inversion in the X chromosome, shown below to extend from a point between *sc* and *slv* to a point between *cr* and *bb*. Females heterozygous for this inversion give about 9 percent crossing over among their regular offspring, and also produce about 6 percent of patroclinous sons, as shown in table 1. All the recovered crossovers are doubles.

Our best experiments are those in which *y*, *cv*, *v*, and *f* are followed, since here it is probable that few or no undetected doubles occur. The table shows a few crossovers entered as singles; these are evidently all really doubles in which the second crossover has occurred in the unmarked region between *f* and the inversion point. In another experiment, in which *cr* and *bb* were present, but the rest of the chromosome was not so well controlled, two such apparent singles were tested and found in fact to represent *cr-bb* crossovers as well.

Table 1 shows a total of 108 crossover males to 63 patroclinous from XX females; the ratio between these classes, 1.71:1, will be discussed below.

Table 1

Tests of females heterozygous for inversion sc-4. In this and following tables the crossover classes are labelled according to the standard sequence of loci, unless otherwise indicated. In cases where two contrary classes are entered under one heading, the right-hand or lower one of the two carries the mutant gene at the leftmost locus concerned.

MOTHERS	♀		REGULAR ♂♂															EXC. ♂♂
	REG.	EXC.	0		1		2		3		1, 2		1, 3		2, 3		TOTAL	
sc^3 *cv v f/y* sc^4	344	0	115	150	1	2	2	1	1	0	1	0	3	5	4	6	291	18
y^2 *cv v f/y* sc^4	331	0	154	86	3	0	0	0	1	2	1	0	4	7	5	3	266	14
v/y sc^4 *cv f*	801	1	294	291	2	3	7	6	0	2	3	2	10	11	6	4	641	31
y^2 *cv v f/y* sc^4/Y	397	11	144	105	3	1	4	3	1	4	3	2	9	9	4	7	299	52

The total crossover percentage among regular sons is 9.3 from the XX females, 16.7 from the XXY females. The latter value needs correction, however. Half of the exceptional gametes die, so the totals should be taken as regulars plus twice the exceptional females; this gives 50:299 + 22, or 15.6 percent. It seems clear that the presence of a Y significantly increases the frequency of recovered double crossovers.

Crossing over in sc-4/sc-4 is approximately normal. Three experiments are recorded (table 2). The third experiment also gave a total of 14 reversions of Bar and 6 occurrences of double-bar (in 8,523 flies), all of them *f-fu* crossovers. In many of these cultures the *B* flies were not classified for *f-fu*. This serves to confirm the report of MULLER and WEINSTEIN (1933), based on sc-8 experiments, that unequal crossing over occurs only between non-sister strands even when the *B* locus is far removed from the spindle attachment.

Table 2

Crossing over in homozygous inversion sc-4. Regular males only are recorded in the first experiment; in the second and third both males and females are included.

MOTHERS	0		1		2		3		1, 2		1, 3		2, 3		TOTAL
cv cr/w^e *f*	66	73	14	8	43	42	9	16	2	3	1	1	1	2	281
v f/cv	93	94	31	20	24	37			3	0					302
B/f B fu	2707		103												2810

Inversion scute-7. DUBININ (1930) has described scute-7. It was obtained by X-raying apricot (w^a) flies, and is a recessive scute allelomorph resembling but definitely different from scute-1.

Females heterozygous for scute-7 have shown no crossing over for any loci to the left of *ct*, and a reduction of crossing over for the interval from *ct* to *lz*. To the right of *lz* substantially normal values have been obtained (table 3). Tests of crossovers have shown that the decrease of crossing over is due to something lying to the left of *sn*, in the region most affected.

*Table 3**

LOCI	sc-7/+		sc-7/sc-7	
	N	PERCENT	N	PERCENT
sc-w	6213	0	—	—
w-cv	3103	0	—	—
fa-sn	—	(0.4)	603	2.8
cv-ct	1030	0	—	—
cv-sn	1395	0.4	—	—
cv-v	938	7.5	—	—
sn-lz	1395	1.6	292	5.1
lz-v	—	(5.5)	292	3.4
v-f	678	24.9	350	22.6

* These values are all from XX females. Less extensive data are available for XXY females in the case of sc-7/+; they show no significant differences from the above values.

Owing to the complete absence of crossing over to the left of *ct* it was at first impossible to test this region in homozygous sc-7. However, in a sc-7 chromosome a mutation occurred to an allelomorph of facet, found in an X-ray experiment by Mrs. C. E. Ruch. The results obtained with this chromosome, tested against genes introduced by crossing over (table 3), show that the reduction of crossing over in heterozygous sc-7 is due to an inversion that includes *fa* but not *sn*.

The extent of this inversion has not been determined accurately by genetic methods, since it does not give viable crossovers with any other inversion we have used, unless special methods are used. Salivary gland chromosomes have, however, been seen to give typical heterozygous inversion figures that show the existence of a short terminal uninverted piece. Dr. J. Schultz has studied such preparations more carefully, and we are indebted to him for the information that the left inversion point lies close to the right of scute, the right one not far from the locus of crossveinless.

Inversion scute-8. Sidorov (1930) subjected apricot (w^a) flies to X-rays and obtained a bristle mutant described as a new scute allelomorph. The relations of this to the scute and achaete series are peculiar; but it is most convenient to retain the name scute-8 for it. There is an associated inversion which has been studied by several investigators (Patterson and Stone 1935; Stone and Thomas 1935). These authors have described its properties; the data given here are in essential agreement with their account.

As will be shown below, the inversion extends from a point between *ac* and *sc* to a point to the right of *bb*. Its right end has been shown by GERSHENSON to be located in the inert region of the X.

The crossing over shown by sc-8/+ is illustrated in table 4. Our other experiments are less satisfactory in that fewer or less well-spaced loci are concerned, but they agree in indicating that these are about the usual values. As in other cases, the classes listed as single crossovers are obviously really doubles, with the second crossover between forked and the right inversion point.

No adequate data are available for crossing over in XXY sc-8/+ females.

Table 4 shows that crossing over is of about the normal frequency in sc-8/sc-8, though (as briefly stated by OFFERMAN and MULLER, 1932) there are local differences from the standard values. The w^a-*cv* interval gives 6.1 percent, as opposed to the usual value of about 12 percent, whereas *f-cr* gives 9.5 as opposed to about 6 percent. Thus in both cases the same section gives more crossing over when it lies far from the spindle attachment.

Inversion scute-8 deficiency. The scute-8 inversion reaches from a point between *ac* and *sc* to a point to the right of *bb*. We have found, as NOUJDIN (1935) has recently reported, that any series of sc-8, in which appropriate tests are made, produces occasional *y-ac* deficiencies, that is, the left inversion point involves an unstable union of parts. The resulting chromosome is deficient for *y*, *ac*, and probably *Hw*, and for no other known loci. All the remaining known loci are in reverse sequence. We have used this "Df sc-8" in some of our inversion experiments.

As in the cases of sc-4/+ and sc-8/+, the apparent single crossovers are evidently all really doubles, in which the second crossover was to the right of *f*. The relative numbers of double crossover males (30) and patroclinous males (31) need correction for comparison with other series, since only half the crossovers survive while the other half carry the deficiency. As thus corrected the ratio becomes 60:31 = 1.9:1.

Tables 4 and 5 indicate that there are about twice as many crossovers recovered from Df (sc-8)/+ as from sc-8/+. Analysis of a few other crosses, not here reported in detail because they include fewer genes or for similar reasons, suggests that the difference, if present, may be less than indicated. More experiments are needed.

Females of the constitution Df (sc-8)/sc-8 have all their genes in the same sequence, but are heterozygous for a terminal *y-ac* deficiency. Table 6 gives the crossing over observed.

In the case of *cv v* the data suggest a difference; other experiments (involving fewer loci) with sc-8/Df (sc-8) give values of 18.7 (in 215 flies) and 16.6 (in 380 flies). Both series are therefore to be taken as giving not far from 18 percent.

Inversion ClB. The "ClB" chromosome was first briefly described by

Table 4

Crossing over in inversion sc-8 females.

CLASSES	In/+		In/In			
	sc-8 w^a *cv v f* / y^2	*ct* / sc-8 w^a *cr*	w^a *ct cr* / *cv f*	+ / w^a *cv v f*	w^a *cv* / *v f*	w^a *v* / *cv f*
0	764	341	490	407	173	78
	913	277	509	257	123	79
1	0	1	33	26	13	9
	0	1	47	31	11	5
2	4	1	29	97	52	18
	3	4	42	75	27	21
3	9		345	110	54	25
	14		326	94	43	20
4	0		74			
	4		71			
1, 2	0	6	0	1	0	0
	0	14	1	1	1	0
1, 3	0		13	3	4	4
	0		13	5	2	4
1, 4	0		6			
	1		9			
2, 3	2		9	18	10	3
	3		10	5	2	1
2, 4	8		6			
	12		2			
3, 4	20		12			
	13		15			
1, 3, 4	0		0			
	0		1			
2, 3, 4	0		0			
	0		1			
Exc. males	57	21				

MULLER (1928), and has been very extensively used since in special experiments. The most detailed account of its properties is that given by GERSHENSON (1935). PAINTER (1934) has figured the inversion that is visible in the salivary gland cells of ClB heterozygotes. As described by MULLER, this chromosome has an inversion (shown below to extend from a point between *ec* and *bi* to a point between *sy* and *fu*), and carries a lethal and

*Table 5**

$$\frac{sc\ cv\ v\ f}{\text{Df (sc-8)}\ w^{a}} \times y^{2}\ cv\ v\ f$$

INTERVAL	MALES	FEMALES	
0	211	245	245
1	2 }	4	5
2	2 }		
3	7	16	9
4	1	1	6
1, 3	0 }	1	1
2, 3	0 }		
1, 4	1 }	2	5
2, 4	7 }		
3, 4	10	8	7
Total	241	555	
Ex.	31		

* Df sc-8 is lethal, so that each crossover class of males is represented only once. The females could not be classified for w^{a}, so intervals 1 and 2 were not separable in them. In the female row the not-yellow class is entered to the left in each case. Egg counts made by Miss M. Groscurth show that the males carrying Df sc-8 die in the egg stage.

the mutant genes *sc, t, v, sy,* and *B*. The leftmost of these, *sc,* lies outside the inversion and separates from it by crossing over occasionally. The other mutant genes lying in the inversion are also occasionally lost by double crossing over. We should interpret the "reinverted" ClB chromosome described by GERSHENSON (1930) as having arisen from a triple crossover—a double within the inversion and a simultaneous single to the left of it. GERSHENSON (1935) records a total of 1 percent crossing over for the whole chromosome (*sc* to *cr*), a value that is, if anything, slightly higher than general experience would lead one to expect. About three-fourths of this crossing over is made

Table 6

Crossing over in $sc^{8}B$/Df (sc-8) *cv v f* ♀♀.*

0	1	2	3	4	1.2	1.3	2.3	TOTAL
89 62	26 16	8 6	2 5	0 0	29 2	25 27	7 9	332

* This table gives the results of a cross to *y cv v f*; it includes both sexes, and contrary classes are therefore not equivalent because of the lethal nature of the deficiency. In reality the sequence is Df *B f v cv*; the classes above, renumbered on this basis, become, in the order given: 0; 1; 1.4; 1.3; 1.2; 4; 3; 3.4).

From table 6 values for comparison with sc-8/sc-8 have been calculated:

	w^{a}-*cv*	*cv-ct*	*cv-v*	*v-f*	*f-B*	*f-cr*	*B*-Df
sc-8/sc-8	6.1	4.8	17.4	21.3	—	9.5	—
sc-8/Df (sc-8)	—	—	24.1	22.6	0	—	19.0

up of singles to the left of the inversion. Our own data are less extensive than GERSHENSON'S, and need not be presented since they add nothing of importance.

GERSHENSON has also studied crossing over in XXY females heterozygous for ClB. He found, as have we, that the exceptional females produced are practically always non-crossovers. After correcting his observed crossover values by adding the non-disjunctional eggs to the observed non-crossover regulars, he arrived at the conclusion that approximately the same amount of crossing over occurs as in XX females. His values are 1.29 percent for crossing over to the left of the inversion, .07 percent for doubles within the inversion, the differences from the XX values being opposite in sign and insignificant in amount; the percentages are, however, too small to be useful for such comparisons.

GERSHENSON has added one other essential bit of evidence, concerning a relation we have not studied. He determined (the method used is not stated) the frequency of XX and XXY females among the regular daughters of XXY ClB females. His table IV shows that among the regulars carrying the ClB chromosome there were 47 XXY:49 XX; among those carrying the non-inverted chromosome there were 68 XXY:63 XX. Evidently, in the regular eggs the Y is distributed at random.

A similar relation was recorded by BRIDGES (1916) for XXY females carrying no inversion, half the regular daughters being shown to be XXY, the other half XX. KIKKAWA (1932) reports that in *D. virilis,* significantly less than half are XXY. This result is based on cytological observations and

*Table 7**

$♀ \frac{\text{dl-49 } cm^2 \, bb}{v \, cr} \times ♂ v \, B \, bb$

	XX ♀♀		XXY ♀♀	
Regular (*B*) ♀				
0	220	201	134	255
1	30	40	152	57
Exc. (+) ♀	0		578	
Reg. (not-*B*) ♂				
cm or *cr* (*v* or not-*v*)	419		567	
v	19		40	
Exc. (*v B*) ♂	1		480	

* In the male classes *bb* cannot be identified, and *cm* and *cr* are not separable. The *v-cr* crossing over can only be determined by doubling the *v* class of males, the contrary crossover, *cm cr*, not being identifiable. In the female classes the crossing over detected is that between *v* and *bb*; in the case of the XXY females *bb* is suppressed by the Y present in some of the regular daughters so the observed classes are misleading. Using *bb* classes only the numbers are: non-crossovers 134, crossovers 57.

seems to us to be doubtful. The Y of *virilis* is not visibly different from the X's or from eight of the autosomes; under these circumstances wholesale counts made on the minute oögonial chromosomes seem questionable.

Inversion delta-49. Inversion dl-49 was first described in an abstract by MULLER and STONE (1931). The left break was shown to lie between *rb* and *cm*, the right between *fw* and *f*. According to PAINTER (1934) unpublished data of these workers show that the right break lies between *fw* and *g*. PAINTER (1934) showed from a study of salivary chromosomes of females heterozygous for inversion dl-49 that the left break occurred between *rb* and *cv*. Since the method that we have used for determining the ends of inversions could not (with the material at our disposal) be used with dl-49, we can add no further information.

Females heterozygous for Inversion dl-49 give few or no crossovers within the inversion among their offspring. We have not collected any extensive series of data showing this to be true since it seemed quite unnecessary. Inversion dl-49 is extensively used in "balanced" stocks and, so far as we know, no case of double crossing over within the inverted segment has occurred in heterozygous females; the opportunities for detecting such, had they occurred, have been abundant.

Females homozygous for inversion dl-49 show approximately normal crossing over both within the inversion and outside it, according to STONE and THOMAS (1935).

Females heterozygous for inversion dl-49 do give recoverable crossovers outside the limits of the inversion, those to the left occurring with considerably lower frequency than those to the right. Among 533 progeny of the cross inversion dl-49 $cm^2\ f/y\ ec\ cv\ ct^6\ v\ g^2\ f \times y\ ec\ cv\ ct^6\ v\ g^2\ f$, 3 or 0.56 percent were crossovers between *y* and *ec*.

The data of table 7 show, from XX females, $v\ cr = 8.7$ percent, $v\ bb = 14.3$ percent. The latter value may be taken as giving the amount of crossing over between the inversion and the spindle attachment; it is in reasonable agreement with the value of 13.4 obtained by SCHULTZ (quoted by L. V. MORGAN 1933).

Table 7 also shows the crossing over from XXY females heterozygous for inversion dl-49. After correcting for the inviable non-disjunctional gametes the values are $v\ cr = 4.5$, $v\ bb = 10.1$. Another series of XXY females, of the constitution dl-49 $cm^2\ bb/y^2$, gave (corrected) $y^2\ cm^2 = 0.5$, $y^2\ bb = 13.4$, based on 275 regular males and 70 *bb* regular females, respectively. The conclusion seems warranted that the Y somewhat decreases crossing over between the spindle attachment and the inversion.

Inversion yellow-4. According to DUBININ and FRIESEN (1932), the y-4 inversion was found by SEREBROVSKY. It presumably arose as the result of X-ray treatment and is inseparably associated with a mutation of the yellow gene to an allelomorph very closely resembling the original y^1. As shown below, the leftmost inversion break is located very near to or at the yellow locus (to the left of it according to MULLER and PROKOFJEVA 1935); the rightmost break is located between the genes *fu* and *da*.

Females heterozygous for inversion y-4 give, among their regular offspring, about 2.7 percent of double crossovers within the inversion; about 2.4 percent of the sons of such females are patroclinous. The data from which these percentages are obtained are presented in table 8. Certain of the crossovers appear as singles within the inversion; they are presumably doubles with the second crossover in the short uncontrolled region between *f* and the end of the inversion. The ratio of recovered double crossovers among male offspring of XX mothers (57) to patroclinous males (51) is 1.12:1.

The data from XXY females (table 8) can be compared with those from XX females, since the XXY mothers were sisters of approximately half of the XX mothers and since the two lots of XX data gave similar results (2.8 and 2.3 percent of males patroclinous, 2.7 and 2.0 percent of regular males crossovers). The percentage of crossovers among regular sons of XXY mothers is 6.1, or 5.8 when corrected for exceptional offspring. As in the case of inversion sc-4, the frequency of recovered double crossovers is increased by the presence of a Y chromosome.

Table 8

Crossing over in In y-4/sc cv v f and in sister In y-4/sc cv v f/Y females by B or sc B males.

	XX FEMALES		XXY FEMALES	
B ♀♀	2106		548	
+ ♀♀	1		39	
B ♂♂	51		56	
Regular males				
0	1019	1044	287	250
1–2	2	3	3	3
1–3	16	11	9	9
1–(4)	0	1	1	0
2–3	11	8	7	6
2–(4)	3	1	0	0
3–(4)	0	1	0	0
Total regular males		2120		575

Inversion bobbed deficiency. From X-ray treated males SIVERTZEV-DOBZHANSKY and DOBZHANSKY (1933) got an X chromosome carrying a deficiency for the proximal third of the somatic metaphase X chromosome, and a small-wing (*sl*) allelomorph. Females heterozygous for this and a normal X chromosome gave very low crossover values, a result ascribed by the above workers to the presence of the deficiency in heterozygous condition. We have obtained clear evidence that the bobbed deficiency chromosome carries an inverted segment extending from between *rb* and *rg* to between *cr* and the spindle attachment.

Sivertzev-Dobzhansky and Dobzhansky published data on crossing over in females heterozygous for Df (bb). Our data are substantially the same (table 9). The *y-ec* interval lies outside the inversion and shows (in XX females), 0.8 percent of recovered single crossovers. This is a marked reduction as compared with the standard crossover value for this interval. The region from *ec* to *ct* is partly outside and partly within the inversion. To get the frequency of recovered double crossovers within the inversion, apparent singles in this region (four in number) are assumed to be actually doubles since the greater part of the region must lie inside the inversion. Recovered doubles within the inversion, assuming apparent singles to be doubles with the second single in the unmarked *f*-spindle attachment region, constitute 3.9 percent of the regular males. The relation between double crossovers and patroclinous males will be considered in connection with the mechanism of disjunction in inversions.

Table 9

Data from the cross sc sl^2 Df(bb)/ec ct^6 g^2 f × w B. The XX and XXY females were sisters.

	XX females		XXY females	
B ♀♀	1244		1622	
+ ♀♀	0		255	
w B ♂♂	21		292	
Regular males				
0	443	601	695	774
1	5	4	4	10
2–(6)	1	3	5	2
3–(6)	12	10	13	17
4–(6)	1	2	4	1
5–(6)	1	0	2	0
2–3	0	0	1	0
2–4	0	0	0	1
2–5	0	1	2	1
3–4	1	6	3	0
3–5	3	1	5	3
4–5	1	0	0	0
1–3–(6)	0	0	0	1
Total regular males		1096		1544

Females heterozygous for Df (bb) and carrying a Y chromosome gave about the same frequency of double crossovers within the inversion as did their XX sisters, 3.7 (corrected for non-disjunction) as compared with 3.9 percent. There is no indication here of an increase in crossing over in the presence of a Y chromosome such as that shown by the sc-4 and y-4 inversions.

Crossover data from XXY females homozygous for Df (bb) are given in table 10. (XX homozygotes do not survive, as shown by SIVERTZEV-DOBZHANSKY and DOBZHANSKY.) The *y-f* interval gives a crossover value of 37.1 percent which is higher than is given by this segment in its normal position in the chromosome (8 plus 10 units). The remaining intervals show less than normal crossing over, with the decrease becoming more marked toward the spindle attachment. Presumably we are here dealing with the so-called spindle attachment effect, that is, segments moved away from the spindle attachment show increased crossing over; distal segments moved near the spindle fiber show a decrease in crossing over (OFFERMAN and MULLER 1932; BEADLE 1932).

Table 10

Data from the cross $y^2 f v Df(bb)/sc g^2 ct^6 Df(bb)/Y \times w B$.

		B ♀♀	2269		
		+ ♀♀	69		
		w B ♂♂	84		
Regular males					
0	478	434	1–4	15	17
1	304	312	2–3	0	2
2	72	107	2–4	3	5
3	46	52	3–4	1	1
4	33	37	1–2–4	1	1
1–2	23	19	1–2–3	1	0
1–3	25	27			
			Total regular males		2016
		Region	Percentage of crossovers		
		1	37.1		
		2	11.6		
		3	7.7		
		4	5.7		

Combinations of Different Inversions

In females carrying overlapping inversions, single crossovers within the region common to the two inverted segments should give chromatids with single spindle attachments, in contrast to the chromatids with two or with no spindle attachments resulting from single crossing over within the inverted segment in a female heterozygous for a single inversion. If two overlapping inversions are not too different in length such single crossovers should be viable in the heterozygote. Actually we know this to be the case in several combinations of X-chromosome inversions. GERSHENSON (1932) has reported a bobbed-deficiency chromosome resulting from single crossing over between In sc-4 and In sc-8.

Crossovers between different inversions will of course give different results depending on the relative positions of the inversion points. Thus, representing the normal sequence of segments of a chromosome such as *A B C D E F*, inversions differing only at one end, $\frac{A\ B\ D\ C\ E\ F}{A\ B\ E\ D\ C\ F}$, will give crossovers *A B D C F*, a single deficiency and *A B E D C E F*, a single net duplication (fig. 2). If both ends differ in position, there are two more possibilities. Thus $\frac{A\ B\ D\ C\ E\ F}{A\ E\ D\ C\ B\ F}$ gives *A B D C B F*, a duplication for *B* and a deficiency for *E*, and the complementary duplication-deficiency *A E D E F* (fig. 3). The third possibility $\frac{A\ B\ E\ D\ C\ F}{A\ D\ C\ B\ E\ F}$ gives the double duplication *A B B D C B E F* and the double deficiency *A D C F* (fig. 4). We shall consider examples of all these possibilities.

The recoverable single crossovers with either one or two deficient segments are useful in determining genetically the position of the inversion points. Thus if such a crossover shows a deficiency for gene *B* but not for genes *A* and *C*, we can say that one inversion end is located between genes *A* and *B*, the other between *B* and *C*. The precision of this method is limited only by the extent and accuracy of the genetic map and by the fact that only recessive mutant effects are available for deficiency tests.

From many combinations of two inversions we have collected data on non-disjunction (table 14) which are discussed below.

Table 11 gives a summary of the available information concerning the properties of the chromosomes derived from single crossing over in the region common to two inversions. Where we have recorded a given crossover chromosome as being inviable, this is to be understood as meaning "under ordinary conditions." It is quite likely that some of the types in question could be brought to maturity by special culture techniques, which we have not used in any case.

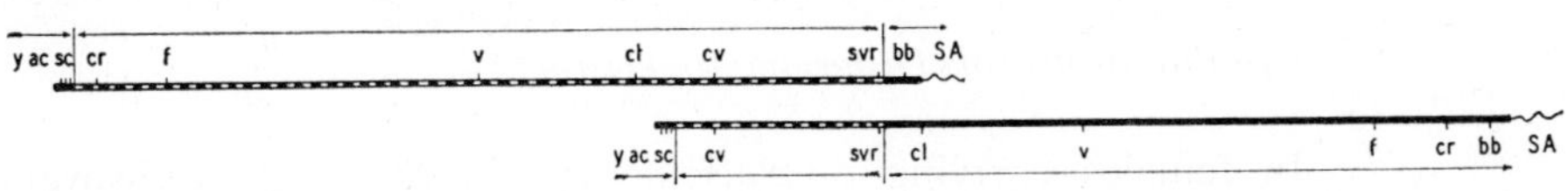

Figure 2. In sc-4 (above)/In sc-7 (below). Diagram showing conjugation of the common inverted regions. The arrows point in the direction in which the loci are arranged in "normal" chromosomes, reading from the distal end to the spindle attachment.

Inversion sc-4/ Inversion sc-7. In this case both left inversion points lie very close to *sc*, and to the right of it; they must be nearly or exactly at the same level. The right point is much further from this level in sc-4. Singles within the common inverted region would be either duplications or deficiencies for the long section from near *cv* to *cr* (fig. 2), and it is accordingly in agree-

Table 11

Chromosomes resulting from single crossing over within common inverted regions.

SOURCE OF LEFT END	SOURCE OF RIGHT END	DUPLICATION FOR	DEFICIENCY FOR	♀ HETEROZYGOUS FOR NORMAL CHROMOSOME	♂
sc-4	sc-7	*ct-cr*	none	Inviable	Inviable
sc-4	sc-8	*sc*	*bb*	Normal; fertile	Normal; fertile; inviable without Y
sc-4	ClB	*fu-cr*	*slv-ec*	Inviable	Inviable
sc-4	dl-49	*g-cr*	*slv-rb*	Inviable	Inviable
sc-4	y-4	*y-sc, cr*	none	Normal	Normal
sc-7	sc-4	none	*ct-cr*	Inviable	Inviable
sc-7	sc-8	*sc*	*ct-bb*	Inviable	Inviable
sc-8	sc-4	*bb*	*sc*	Fertile; legs often abnormal	Nearly completely lethal; sterile; extreme sc
sc-8	sc-7	*ct-bb*	*sc*	Inviable	Inviable
sc-8	ClB	*fu-bb*	*sc-ec*	Inviable	Inviable
sc-8	dl-49	*g-bb*	*sc-rb*	Inviable	Inviable
sc-8	y-4	*y-ac, cr-bb*	none	*Hw* effect of sc-8; fertile	*Hw* effect of sc-8
ClB	sc-4	*slv-ec*	*fu-cr*	Inviable	Inviable
ClB	sc-8	*sc-ec*	*fu-bb*	Inviable	Inviable
ClB	dl-49	*g-sy*	*bi-rb*	Inviable	Inviable
ClB	y-4	*y-ec*	*fu*	Fertile; abnormal eyes, wings, hairs	Inviable
ClB	Df(bb)	none	*bi-rb, fu-bb*	Poorly viable; sterile; minute bristles	Inviable
dl-49	sc-4	*slv-rb*	*g-cr*	Inviable	Inviable
dl-49	sc-8	*sc-rb*	*g-bb*	Inviable	Inviable
dl-49	ClB	*bi-rb*	*g-sy*	Inviable	Inviable
dl-49	y-4	*y-rb*	*g-fu*	Inviable	Inviable
y-4	sc-4	none	*y-sc, cr*	Poorly viable; fertile; minute bristles	Inviable
y-4	sc-8	none	*y-ac cr-bb*	Poorly viable; fertile; minute bristles	Inviable
y-4	ClB	*fu*	*y-ec*	Inviable	Inviable
y-4	dl-49	*g-fu*	*y-rb*	Inviable	Inviable
y-4	Df(bb)	none	*y-rb, cr*	Inviable	Inviable
Df(bb)	ClB	*bi-rb, fu-cr*	none	Fertile; wings slightly narrowed	Sterile; wings notched at tip
Df(bb)	y-4	*y-rb, cr*	none	Fertile; wings narrow, bristles abnormal	Inviable

ment with expectation that they were not recovered in crosses to normal males. The only other type of crossover that might be expected to appear is the double within this *cv-cr* section, and a few of these were obtained.

From *y* sc^4 *cv f*/sc^7 w^a ♀ (× various ♂♂) were obtained 459 regular

♀♀ (no crossovers were observed in those 2 of the 4 cultures where they could have been identified, but detailed counts were not recorded), 228 *sc* w^a ♂♂, 176 *y sc cv f* ♂♂, 8 *sc* w^a *f* ♂♂, 3 *y sc cv* ♂♂.

By mating sc-4/sc-7 heterozygotes to translocation 1,2–7 (break between *rb* and *cv* in X and attached to the right of *sp* in II) or translocation 1,3–3 (break between *rb* and *cv* in X, attached to the right of *ca* in III) we were able to save the crossovers in the common inverted segment which are deficient for the long segment from the right end of the sc-7 inversion (near *cv*) to the spindle attachment (fig. 2). From the cross *y* sc^4 *v f cr*/sc^7 w^a fa^2 *sn v* by T 1, 2–7, 455 normal ♀♀, 2 + ♂♂, 181 *y* sc^4 *v f cr* ♂♂, 184 *sc* w^a *fa sn v* ♂♂ were obtained. In addition there were 16 males carrying the deficiency crossover plus the proximal X segment from the translocation. Of these 13 were *sc*, 1 was *sc* w^a *fa* and 2 were *sc fa*. All were strong scutes, otherwise normal. Of these 11 were tested; none was fertile (they are of course expected to be XO). The 16 males constitute 4.05 percent of the regular males, which would indicate, since the contrary class is lost, a frequency of 8.1 percent crossing over in the common inverted segment. It is assumed that the translocation males produce four types of gametes in equal numbers.

Females of the same constitution as above mated to T1,3–3 males gave essentially similar results. There were 407 regular non-crossover males and 6 crossovers of the type considered above. Here the frequency of single crossovers, corrected for the class not recovered, is 2.9 percent. Three of these males were tested and, as expected, were sterile.

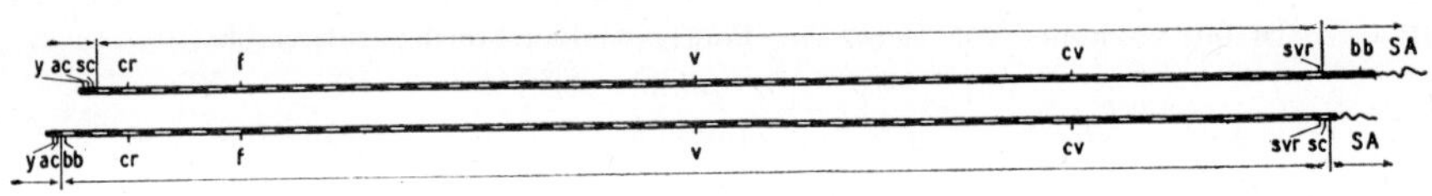

Figure 3. In sc-4 (above)/In sc-8 (below). Conjugation of the common inverted regions.

In both of the above cases the percentages of recovered crossovers indicate minimum values of crossing over in the common inverted segment. Since the recovered individuals carry a net duplication (difference between sc-7 right break and translocation break) they are probably lower in viability than non-crossover males. Likewise if the translocations give more regular than non-disjunctional gametes the observed frequency of crossing over will be lower than the real value.

Inversion sc-4/Inversion sc-8. The sc-8 inversion is slightly longer than the sc-4 one at both ends, the resulting single crossovers carrying either a deficiency for *sc* and a duplication for *bb* or a duplication for *sc* and a deficiency for *bb*. The conclusions as to their limits depend in part on the results obtained from the heterozygote here under discussion, so the argument may now be presented.

GERSHENSON (1932) has already described the crossover that receives the left end of sc-4 and the right end of sc-8 (fig. 3), showing that it acts as though it carried a lethal allelomorph of *bb*, both the lethal and the bobbed effects being suppressed by a Y. This shows that the *bb* locus is absent, that is, that it is present in the inverted portion of sc-8 and in the uninverted portion of sc-4, both of which are absent in this crossover. GERSHENSON has also studied the deficient chromosome cytologically in the oögonia of heterozygous females; he finds it to be reduced in length by about one quarter. This can only mean that a large section (not far from half) of the inert region is included in In sc-8 and not in In sc-4. The latter presumably includes little or no inert region. Finally, GERSHENSON showed that this chromosome did not carry a deficiency for *cr*. It follows that In sc-4 has its right break between *cr* and *bb*, while In sc-8 has it to the right of *bb*, conclusions which our own data confirm.

The other crossover, that must carry two *bb* loci, is usually lethal in the male. We have found that it does not carry a deficiency for *ac*, *slv*, or *br*. These three loci must thence be alike in the two inversions, that is, either in both or outside of both, since the crossover studied by GERSHENSON also carries no deficiency for them. When tested against scute allelomorphs this chromosome behaves as an extreme scute, but so does sc-4 itself, so this is not a critical result. However, we have been able to obtain a few males carrying this chromosome. Occasionally they emerge but only live a few hours. Examination of their bristles shows that they have many fewer than sc-4 males; the only named ones observed were the inner verticals, posterior supra-alars, and the dorsocentrals, that is, the "achaete" as opposed to the "scute" bristles, and even these were frequently absent. Dr. J. SCHULTZ informs us that a study of the salivary gland chromosomes also indicates that the sc-8 break is to the left of the *sc* band, the sc-4 one to the right of it. It must be concluded that these males represent the occasional survival of specimens in which the scute locus is wholly absent.

Females heterozygous for this chromosome often have some of their legs abnormal. The abnormality, which is most frequent in the posterior pair, may consist in bifurcation, shortening and twisting, or basal fusion of the two members of one pair.

The above results show that In sc-4 runs from a point between *sc* and *slv* to a point between *cr* and *bb*; In sc-8 from a point between *ac* and *sc* to a point between *bb* and the spindle attachment.

Table 12 shows the crossing over from sc-4/sc-8. Since one type of single crossover is practically lethal, the simplest way of calculating crossing over seems to be to use in each case only the larger of the two contrary classes. If this is done the values become: total 201; *y-f*, 19.9; *f-v*, 17.9; *v-cv*, 21.4; *cv*-w^a, 9.0. These values are, as expected, not different from those for sc-8/sc-8 or sc-8/Df sc-8. Classification for sc^8 was not attempted. The *y*

Table 12

$\frac{f\ v\ cv\ w^a\ \text{(sc-8)}}{y\ sc^4}$ × various ♂♂

♀					
Except. ♂♂	456		1, 2	1	3
Regular ♂♂	0		1, 3	7	3
0	52	76	1, 4	1	1
1	0	28	2, 4	2	3
2	1	37	3, 4	3	1
3	0	30	1, 2, 4	0	1
4	0	10	Total ♂♂	260	

males were typical sc^4; the *y* w^a were nearly wild-type for bristles, carrying both sc^4 and sc^8.

Both of the crossovers recovered from sc-4/sc-8 have been tested for crossing over and disjunction, and have given the expected results. The one with the left end of In sc-8 (the sc deficiency) when tested against a normal chromosome gave the following results: Df (sc) *f*/*y* w^a ♀ × various males: 453 regular ♀♀, no exceptions; 337 *y* w^a ♂♂, 7 *sc* (extreme) *f* ♂♂, 17 *y* w^a *f* ♂♂, 1 extreme *sc* ♂, 1 *y* ♂, 47 exceptional ♂♂. (In some cases sc^1, sc^{10}, or *slv* were used instead of *y*.) Calculating the frequency of exceptional males by doubling the *y* class (since the deficiency is nearly lethal), there were 47/767 or 6.1 percent exceptions. In other words this chromosome gives results both as to crossing over and as to disjunction comparable to those shown by the inversions from which it was derived. This result is confirmed by a small series in which the crossover was tested against In sc-8. Here there was about 32 percent crossing over between *sc* and *f*, and only 0.7 percent exceptional males were produced.

The other crossover (*sc* duplication, *bb* deficiency) behaves similarly. Females of the constitution *y* w^a *cv Df* (*bb*)/f gave 347 regular sons (of which 10 were crossovers, 3 clearly doubles and the others presumably so) and 11 exceptional sons. The totals for all experiments of this type show 12/391 = 3.1 percent exceptions. This same chromosome was also tested against In sc-4. There resulted 826 regular ♀♀, no exceptional ♀, 657 regular ♂♂, 4 exceptional ♂♂. Among the regular males crossing over could be checked, and the following values were obtained: *sc*-w^a, 0.5; w^a-*cv*, 7.0; *cv*-*v*, 18.3; *v*-*f*, 21.3. Here again the crossover, as expected, behaves much like the inversions from which it was derived.

Inversion sc-4/Inversion ClB. The ClB inversion lies wholly within that of sc-4. ClB/+ gives very few crossovers, so that doubles would be expected to be rare here, and none was found. Of the singles within the common inverted region, one type should be a deficiency for *slv-ec* and a duplication for *fu*; the other should be a duplication for *slv-ec*, and a deficiency for *fu*. Neither was obtained; evidently both are inviable in males and also in heterozygous females.

From ClB (*sc v B*)/*y* sc^4 *cv f* ♀ × y^2 *cv v f* ♂ were obtained 254 *v B* ♀♀, 238 *y cv f* ♀♀, 2 *sc B* (exceptional) ♀♀, 182 *y sc cv f* ♂♂, 1 y^2 *cv v f* (exceptional) ♂. An XXY ♀ of the same constitution gave 27 *v B* ♀♀, 33 *y cv f* ♀♀, 61 *sc B* ♀♀, 24 *y sc cv f* ♂♂ and 52 y^2 *cv v f* ♂♂.

Inversion sc-4/Inversion dl-49. Scute-4 inversion includes the segment from just to the right of *sc* to between *cr* and *bb*, the greater portion of the chromosome (fig. 1). Delta-49 extends from between *rb* and *cv* to between *fw* and *f* and is consequently entirely included within the sc-4 inversion. From the cross *y* sc-4/dl-49 *y-Hw* cm^2 m^2 g^3 × *t v m* g^2, 1,752 males were recorded, all non-crossovers. This result might have been expected from our knowledge of the behavior of females heterozygous for each of these inversions separately. Doubles within the sc-4 inversion and outside In dl-49 either do not occur or are very rare. Since heterozygous dl-49 gives single crossovers within the inversion (p. 229), it is probable that in the sc-4—dl-49 combination some singles occur in the inverted segment common to the two inversions. Such crossovers result in either a *slv-rb* duplication and *f-cr* deficiency or the complementary deficiency-duplication. Both of these products would be expected to be inviable.

XX females heterozygous for In sc-4 and In dl-49 give few or no exceptional daughters and a frequency of exceptional sons not significantly higher than normal.

Inversion sc-4/Inversion y-4. Inversion sc-4 runs from a point between *sc* and *slv* to a point between *cr* and *bb*; In y-4 from a point near *y* (to the left of *ac*) to a point between *fu* and *cr*. One of the single crossovers is deficient for *y* (?) *ac-sc* and *cr*; the other is a duplication for both these sections. The former is lethal in males but survives occasionally in heterozygous females as a minute-bristled type. The latter (duplication) crossover is viable in offspring of both sexes.

Owing to the presence of *y* in both inverted chromosomes and of an extreme *sc* allelomorph in sc-4, the tests for deficiencies for these two loci are inconclusive. Fully satisfactory tests have shown, however, that the minute crossover is deficient for the loci *ac* and *cr*, not for *rst*, *pn*, *sy*, nor for *od*. The other (duplication) crossover showed no deficiency effects for any of these loci. That is, the inversions differ with respect to *ac* and *cr*; In sc-4 was shown above to include *cr* but not *ac*. Hence y-4 includes *ac* but not *cr*. Both include *rst* and *pn*, as follows from this analysis and from the direct test. In sc-4 includes *sy* and *od*; therefore In y-4 has its right break between these two and *cr*; other data show that it is also to the right of *fu* which is 0.3 units to the right of *od* and *sy*.

The crossing over in In sc-4/In y-4 is similar to that in In sc-4/In sc-8. Here again one single crossover class of males dies, and the data have been treated by using the larger member of each pair of contrary classes. The results are then: total 433; *sc-f*, 20.3; *f-v*, 7.9; *v-cv*, 11.3; *cv*-w^a, 10.2; w^a-*sc*, 1.2.

The w^a *sc* value appears here and not in the corresponding series for In sc-4/In sc-8 because In y-4 carries a *sc* gene that is clearly dominant to sc^4, whereas the allelomorph present in In sc-8 gives with sc^4 a variable type not always clearly separable from sc^4.

Data from XXY In sc-4/In y-4 show the same type of crossing over and roughly the same amount; the experiments however include too few flies to be valuable for detailed comparisons.

Inversion sc-7/Inversion sc-8. This combination of inversions is essentially the same as sc-4/sc-7 already discussed. Females heterozygous for the two inversions give progeny carrying chromosomes derived by crossing over within the common inverted segment. As in the case of the sc-4/sc-7 both of these crossovers are inviable. Doubles within the sc-8 inversion but outside the short sc-7 inversion are recovered.

From the cross sc-8 w^a/sc-8 *v*/Y by various males the following offspring were recovered:

Regular	♀♀	589	
Exceptional	♀♀	70	
Exceptional	♂♂	103	
Regular	♂♂		
0		247	246
2–(3)		11	10

In addition there was one exceptional male expected to be y^2 *cv v B* which did not show *y* but which did show a hairy wing effect. This male was sterile. Evidently he carried a duplication for *y*.

From crosses of sc-7/sc-8 to T 1, 2–7 or T 1, 3–3 males (as in the case of sc-4/sc-7) males were recovered which carried the crossover deficient for the long segment from the right break in sc-7 to *bb*. These were enabled to survive by the proximal X segment from the translocation males. Such males differ from those obtained from sc-4/sc-7 (fig. 2) in not carrying a *bb* duplication and in carrying a duplication for the *sc* locus. From the cross sc-8 w^a *cv v f*/sc-7 w^a fa^2 *sn v* by translocation males the following flies were recorded:

		♂♂ T 1,2–7	♂♂ T 1,3–3
Regular females		844	867
Matroclinous	♀♀	1	0
v minute	♀♀	0	1
Patroclinous	♂♂	4	5
Regular	♂♂	796	728
sc w^a	♂♂	17	19
sc w^a *fa*	♂♂	0	1

The *v* minute female recorded above carried the long crossover chromosome, tandem attached-X chromosomes deficient for the common inverted segment of the parent inversions. This deficiency was partly covered by the distal X segment from the translocation.

The *sc* w^a and *sc* w^a *fa* males carry the short crossover chromosome. The actual frequency was 2.1 and 2.7 percent of regular males. These males carry both the sc^7 and sc^8 genes and were intermediate between sc^7 and sc^8 males for the scute character. Their wings were spread. All of them that were tested (seven) were sterile as expected and were presumably XO.

Assuming that the translocation males produce four types of gametes in equal numbers, and that the above mentioned males have normal viability the true percentages of crossovers in the common inverted segment would be obtained by doubling the crossover male classes (the contrary crossover is not recovered) which would give values of 4.1 and 5.2 percent. Since these males are almost certainly of lower viability than the regular males, these percentages represent minimum values. The real value is probably considerably higher.

Inversion sc-7/Inversion ClB. These two inversions overlap in the rather short region from *bi* to near *cv*. Singles would be expected to be rare within this region and would be inviable. None was recovered. From the behavior of each when heterozygous for a normal chromosome it is inferred that few crossovers of any kind would be recovered. In fact, sc-7 w^a/ClB *sc v B* females gave 327 regular sons with one crossover which was sc^7 w^a *v*; that is, a double within that part of In ClB not common to In sc-7. Similar XXY females gave 270 regular males with no crossovers.

Inversion sc-7/Inversion dl-49. The right break in In sc-7 is near *cv*, the left break in dl-49 between *rb* and *cv*. In the absence of more accurate information we cannot say definitely whether or not these two inversions overlap although it seems more probable that there is a short overlapping segment. The data from sc-7 w^a/dl-49 cm^2 XX females by *w B* males are limited; 157 males show no crossovers but the region to the right of the dl-49 inversion where one might expect a low frequency of crossovers is not under control. From XXY mothers, similarly marked, 206 non-crossover males were recorded.

Inversion sc-7/Inversion Df(bb). In the combination of In sc-7 and In Df(bb) there is probably a short overlapping segment between *rb* and *cv*. From the cross sc-7 w^a/*y* sl^2 Df(bb) × *sc ec ct* g^2 *sl*, 1651 regular males were recorded of which 19 or 1.15 percent were apparent single crossovers between w^a and *sl*. Presumably all of these were actually doubles with the second crossover in the uncontrolled region between *sl* and *bb*. Several sc-7 w^a sl^2 males and *sc sl* females were tested and found not to carry Df(bb) showing that they were double crossovers. Presumably an appreciable number of undetected double crossovers occurred in the rather long unmarked region between the right end of In sc-7 and the locus of *sl*.

In the above cross 22 patroclinous males were recorded. Exceptional females could not be distinguished from one crossover class but since there were only 18 females in this class and 12 in the contrary crossover class, the number of exceptional females could not have been large.

Inversion sc-8/Inversion ClB. Inversion ClB is wholly within the limits of In sc-8, and few or no recovered crossovers of any kind are to be expected—the case being very similar to that of In sc-4/In ClB. In fact none was obtained among 21 regular sons of XX females or 160 sons from XXY females.

Inversion sc-8/Inversion dl-49. This combination is essentially similar to the combination of In sc-4 and In dl-49 already considered; the discussion given there applies in the present combination. From the cross sc-8 w^a/dl-49 cm^2 by *w B*, 1742 non-crossover regular males were recorded and 6 patroclinous males. From 5 additional cultures not recorded in detail, 2 apparent w^a *cm* males were obtained. These proved, on testing, to be the result of mutation of an eye color gene in the dl-49 chromosome rather than of crossing over. The locus of the mutation was not determined. From XXY females of the above constitution mated to *w B* males 576 non-crossover regular males and 444 patroclinous males were recorded.

Stone and Thomas (1935) also studied this combination. They obtained one double crossover (outside of the dl-49 inversion, inside of the sc-8 one) in experiments carried out at 30°C.

Inversion sc-8/Inversion y-4. Inversion y-4 extends to the left further than In sc-8 by the locus of *ac* (and *y*?); In sc-8 extends further to the right by the *cr-bb* section. No crossovers are to be expected outside the inversions; of the singles within the common inverted region, one is a duplication for *ac* (and *y*?) and for *cr-bb*; the other is a deficiency for both these sections. The latter is lethal in males, the former survives; in heterozygous females the former (duplication) is fully viable, the latter gives a minute-bristled individual that has reduced viability.

Tests against recessives show that the "minute" crossover is deficient for *ac* and *cr*, not for *slv*; the other one (duplication) is deficient for none of these loci. Therefore the inversions differ in that one includes *cr*, the other does not; likewise they differ for *ac*; both or neither include *slv*. These results are in agreement with the conclusions from sc-4/sc-8 and sc-4/y-4, which show that *slv* is included in both, *cr* in sc-8 but not in y-4, *ac* in y-4 but not in sc-8.

The crossing over tests for this combination show results similar to those from sc-4/y-4 but they are too scanty to permit detailed comparisons.

The duplication crossover, tested against a normal chromosome gave no exceptions among 314 daughters, 18 among 332 sons (5.4 percent). These values, as expected, are comparable to those from the In sc-4–In sc-8 crossovers.

Inversion ClB/Inversion dl-49. The delta-49 inversion is entirely included within the ClB inversion. From the cross ClB *sc v sl B*/dl-49 cm^2 bb^x × *sc cv v f cr*, 629 regular non-crossover males (cm^2) and 2 patroclinous males were obtained. Among 1266 females, 3 (0.24 percent) were crossovers between *sc* and the left break of ClB; one was of the constitution *sc B*, presumably a primary exception equational for the scute gene. XXY females of this combination were not studied.

Single crossovers in the inverted segment common to ClB and dl-49 presumably occur but are evidently inviable.

Inversion ClB/Inversion y-4. The right inversion points here are very similar, differing only in that In y-4 includes *fu*; In ClB does not. At the left In y-4 is considerably longer. Crossing over might occur between the inversions and the spindle attachment, but tests have not been made. Of the singles within the common inverted region, one gives a long deficiency for (*y*?) *ac* to *ec*, and a short duplication for *fu*, the other has the corresponding duplication and deficiency. Both crossovers are lethal in males. The one with the long (*ac-ec*) deficiency is also lethal in heterozygous females; the other is viable. This latter crossover usually carries the *B* gene of ClB, and in that case the resulting *B*/+ females have very narrow bar eyes similar to those of *B*/*B*. They also have irregularly arranged acrostical hairs, and their wings are slightly reduced in size and are less convex than is normal on the posterior margin. Tests of this crossover chromosome show that it carries a deficiency for *fu*, not for *y*, *ac*, *br*, *w*, *ec*, *f*, *vb*, *sy*, *od*, *cr*. The negative results could all have been predicted from conclusions already established in this paper; the positive case constitutes our proof that *fu* is in the inverted section of y-4, not in that of ClB.

From y-4 w^a/ClB *sc v B* only one of the 280 regular sons was a crossover. This one, y^4 w^a *v*, was a double within the common inverted region. The females from these same mothers (excluding the mating to *fu* ♂ because of the low viability of *fu*/Df) gave 218 broad bar non-crossover ♀♀, 237 not-bar

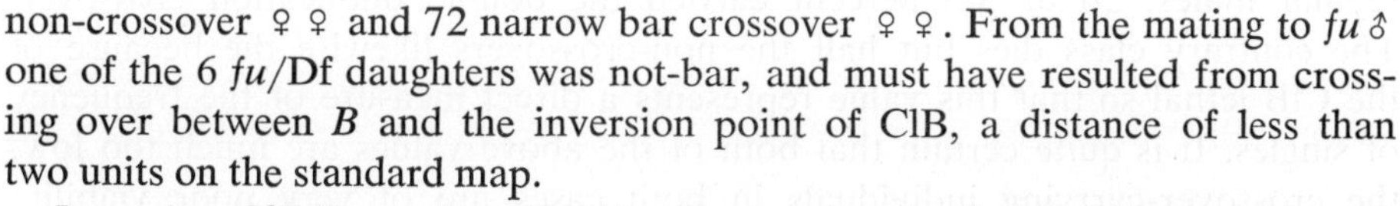

Figure 4. In Df(bb) (above)/In ClB (below). Conjugation of the common inverted regions.

non-crossover ♀♀ and 72 narrow bar crossover ♀♀. From the mating to *fu* ♂ one of the 6 *fu*/Df daughters was not-bar, and must have resulted from crossing over between *B* and the inversion point of ClB, a distance of less than two units on the standard map.

Inversion ClB/Inversion Df(bb). The left break of ClB is between *bi* and *ec*, the right between *sy* and *fu*. The left break of Df (bb) is between *rb* and *rg*, the right between *cr* and *bb*. That these statements are correct will be shown below from studies of single crossovers between ClB and Df (bb). Both breaks of Df (bb) are to the right of those of ClB. Consequently one single crossover should give duplications for the *bi-rb* and *fu-cr* segments (fig. 4). The contrary crossover should give a deficiency for these two segments plus the deficiency for bobbed from the Df (bb) chromosome (fig. 4). Both of these crossovers are viable and can be recovered in heterozygous females. The duplication chromosome is viable in the male. Such males are small with wings having a less convex outer margin than normal and usually with one or more notches at the tips; they are sterile. Dissections by Professor DOBZHANSKY show that the testes are collapsed like those of very old males. Females heterozygous for the double deficiency survive as extreme minutes with wings of a characteristic shape. They have normal ovaries as shown by dissections made by Professor DOBZHANSKY but according to many tests are sterile. Males carrying this deficiency chromosome are inviable.

Tests of the deficiency-carrying crossover which gets the right end from Df (bb) and the left end from ClB give positive evidence that the *rb* and *cr* loci are absent. From crosses of ClB/Df (bb) to *bi* and *fu* males no deficiency heterozygotes were obtained. We conclude that both of these loci are included in the deficiency crossover. The results from ClB/y-4 establish this for *fu*. The number of flies examined was in each case adequate to have given many deficiency crossover females were they not inviable. Similar tests have shown that the *ec*, *rg*, *f*, *bb*, and *sy* loci are not included in these deficiencies. These results confirm the conclusion already drawn that the right break of ClB is between *sy* and *fu*. The left break must lie between *ec* and *bi*. These conclusions confirm and extend those of PAINTER (1934) derived from studies of salivary chromosomes. In a similar way it is clear that the left break of Df (bb) lies between *rb* and *rg* and that the right one lies to the right of carnation.

Frequency and Distribution of Single Crossovers

Among 2010 regular females from ClB/Df(bb) mothers, 48 or 2.4 percent carried the deficiency crossover. The contrary crossover could not be classified accurately in the females; if they were of equal frequency, the recovered single crossovers would be 4.8 percent of the total. Among 749 regular males, 26 or 3.5 percent carried the double duplication crossover. The contrary class dies but half the non-crossovers likewise die because of the ClB lethal so that this value represents a direct measure of the frequency of singles. It is quite certain that both of the above values are much too low; the crossover-carrying individuals in both cases are of very poor viability and no precautions were taken to prevent overcrowding in the cultures.

It is of interest to determine how the single crossovers in the common inverted segment are distributed. In all experiments *v* was carried by the ClB chromosome. We can then separate crossovers in males into those which occurred between *rg* and *v* and those which occurred between *v* and *sy*. When this is done, the following results are obtained:

CROSSOVER INTERVAL	STANDARD MAP LENGTHS	NUMBER OF CROSSOVERS
rg to *v*	22+	111
v to *sy*	27.2	141

The ratios of standard map lengths of these intervals is 1:1.24− ; that of singles within these regions 1:1.26. We can conclude that the distribution of single crossovers within the common inverted segment is approximately normal.

Crossing Over in Heterozygotes for the Duplication Crossover

Crossing over in females heterozygous for the Df (bb)-ClB crossover has been studied in two experiments. Only one of these is reported here. The other, although involving larger numbers of individuals, is not as well controlled for crossing over in different regions. The data from females carrying the duplication chromosome and a bobbed deficiency chromosome are given in table 13.

Table 13

Data from the cross y^2 f v dupl/*sc g^2 ct^6 Df (bb)* ♀♀ by *w B* ♂♂. *The females of this experiment were sisters of those used in the experiment summarized in Table 10, homozygous Df (bb) of the same constitution.*

		B ♀♀	1471		
		+ ♀♀	0		
		w B ♂♂	5		
*Regular ♂♂					
0	91	411	1–5	0	6
1	99	14	2–3	0	2
2	60	14	2–4	0	3
3	55	7	3–4	0	1
4	31	6	3–5	0	1
5	7	0	1–3–4	1	0
1–2	0	8	1–2–4	1	0
1–3	3	14			
1–4	0	8	Total	844	

* In each case the smaller of the two contrary classes represents the males carrying the duplication.

In comparison with the Df (bb) chromosome, the double duplication crossover chromosome carries one net duplication, namely a segment including *rb-bi-f-cr* from the ClB chromosome. This segment together with the bb segment is simply added to the Df (bb) chromosome which of course carries the *bi-rb-cv-f* segment (no difference in arrangement) at the left end. The data of table 13 show the following crossover values as measured in the classes not carrying the duplication:

y-f	19.4
f-g	10.6
g-v	10.5
v-ct	6.4
ct-dup	2.1

For these same regions in homozygous Df (bb) sister females with a Y chromosome the values for the first four regions above (Table 13) are 37.1, 11.6, 7.7, 5.7 respectively. The fifth region cannot be measured in Df (bb)/Df (bb); its standard map length is 10 units.

It is clear from the above that the duplication crossover chromosome crosses over freely with Df(bb), one of the inversions from which it was derived. As compared with Df(bb)/Df(bb)/Y, crossing over is reduced in the *y-f* interval but is the same in the other intervals which can be compared. The reduction in the *y-f* interval is presumably the result of the duplication which is of course homologous with a segment included in the *y-f* interval.

Inversion dl-49/Inversion y-4. From the cross y-4 w^a/dl-49 *y Hw* m^2 g^3 ♀♀ by *t v m g* ♂♂, 1506 regular sons showed no crossovers. There were two patroclinous males. There were 1,651 regular females, all non-crossovers, and five exceptional females, four from 1 of the nine cultures.

From a cross of XXY females of the above constitution with *w* or *w B* males, 428 regular non-crossover, 487 patroclinous males, 446 regular females and 506 exceptional females were recorded.

It is clear that recoverable crossovers are practically absent in females of this combination, unless they occur between the right break of y-4 inversion and the spindle attachment. Data from attached-X females heterozygous for y-4 (p. 207) indicate that singles in this interval are very rare.

Single crossovers in the segment common to the two inversions presumably occur and are lethal both in heterozygous females and in males.

Inversion y-4/Inversion Df (bb). The location of the breaks in both y-4 and Df(bb) have already been discussed; both breaks in Df (bb) are to the right of those of y-4. Consequently single crossovers in the common inverted segment of these two inversions will give either a double duplication or a double deficiency. The latter is inviable both in males and in heterozygous females; the former is viable in females heterozygous for a normal chromosome. Such a female usually has stubby outer verticals, disarranged scutellars, and outer wing margins less convex than normal. If such a duplication female is heterozygous for *B*, the eyes are usually as narrow as those of a female homozygous for *B*. These crossover females are fertile but produce very few offspring. Their viability is good considering the number of loci carried in the two duplications.

From the cross y-4 w^a *cv v* s^2/*sc* sl^2 Df(bb) by *sc B* males there were recorded one exceptional female (+), 330 regular ♀♀ (*B* and *sc B*), 4 *sc B* ♂♂, 123 *sc sl* ♂♂, 85 *y* w^a *cv v s* ♂♂ and 38 duplication ♀♀. The frequency of crossover females in percent of regular females is 10.7. From sister females of those used in the cross above, but XXY in constitution, mated to *sc B* males, the following offspring were obtained: 264 + ♀♀, 390 *B* ♀♀, 236 *sc B* ♀♀, 272 *sc B* ♂♂, 159 *sc sl* ♂♂, 198 *y* w^a *sv v s* ♂♂, and 144 duplication ♂♂. Here duplication females constitute 18.7 percent of the regular females. Since only one crossover is recovered here the true percentage of singles from these data will be 18.7 for XX and 19.8 (corrected) for XXY. The frequencies are not significantly different.

Since females heterozygous for the duplication crossover produce very few offspring, few studies of them were made. It is known that crossing over between the duplication chromosome and a normal chromosome is very low and that a few patroclinous males are produced. These results are expected since the crossover chromosome is in effect an inversion plus an intercalated duplication.

Non-Disjunction

Table 14 is a summary of the available data on the production of matroclinous females and patroclinous males.

In many of these experiments the exceptional females could not be distinguished and only the males are recorded, in others the male exceptions were known to have very low viability and only the females are recorded. It follows that the numbers of individuals in the two sexes from a given combination are often not comparable. ClB and Df(sc-8) are lethal in the males; accordingly in all series involving these the observed number of regular males has been doubled in calculating the recorded total, a point to be remembered in judging the significance of the values given. In many of the combinations of two inversions, single crossovers between the two inversions occur; some of these are lethal and others have reduced viability. No corrections have been made for this; therefore in several of these cases it is certain that the totals are too low and the percentages of exceptions too high.

In the case of ClB/+ we have added our own data to those recorded by Gershenson (1935) though in the XXY experiments we obtained somewhat higher values than he records. We have excluded the males from his XX experiment in which the father was bb^l, since the exceptions (having no Y) would be inviable. We have also excluded one unexplained XX culture of our own that gave 9 exceptional females to 223 regulars and 11 exceptional males to 133 regulars. We have observed in some other combinations a suspiciously high frequency of cultures that gave more than one exception when

Table 14

Summary of non-disjunction data.

	XX MOTHERS						XXY MOTHERS					
	FEMALES			MALES			FEMALES			MALES		
	TOTAL	EXC.	% EXC.	TOTAL	EXC.	% EXC.	TOTAL	EXC.	% EXC.	TOTAL	EXC.	% EXC.
sc-4/+	6287	1	0.02	5861	337	5.75	933	40	4.3	817	109	13.4
sc-4/sc-4	953	0	0.00	600	0	0.00						
sc-7/+	5386	5	0.09	4919	14	0.28	1518	195	12.9	1697	256	15.1
sc-7/sc-7	1610	0	0.00	1370	1	0.07						
sc-8/+	4703	1	0.02	5138	164	3.20	1310	130	9.9	1847	255	13.8
sc-8/sc-8	574	0	0.00	481	0	0.00	310	4	1.3	252	5	2.0
Df(sc^8)/+	641	2	0.31	1053	47	4.46	85	9	10.6	80	12	15.0
sc-8/Df(sc^8)	631	0	0.00	504	0	0.00						
ClB/+	5693	14	0.25	3438	16	0.47	7478	2729	36.6	7172	2712	37.8
dl-49/+	3238	0	0.00	3168	6	0.19	4355	1985	45.6	4145	1747	42.2
dl-49/dl-49							126	5	4.0	99	12	2.0
y-4/+	2007	1	0.05	2171	51	2.34	587	39	6.8	631	56	8.9
y-4/y-4	206	1	0.49	169	0	0.00						
Df(bb)/+	1244	0	0.00	3437	67	1.95	1877	255	13.6	1836	292	15.9
Df(bb)/Df(bb)							2338	69	2.9	2100	84	4.0
sc-4/sc-7	459	0	0.00	422	7	1.66	341	34	10.0	280	43	15.3
sc-4/sc-8	456	0	0.00	264	0	0.00						
sc-4/dl-49	2084	0	0.00	1755	3	0.17						
sc-4/y-4	1187	3	0.26	954	0	0.00	413	33	8.0	224	30	13.4
sc-7/sc-7	439	0	0.00	350	12	3.43	659	70	10.6	617	103	16.7
sc-7/Df(sc^8)	523	0	0.00	531	7	1.32						
sc-7/ClB	754	0	0.00	658	4	0.61	894	322	35.8	824	284	34.5
sc-7/dl-49	169	2	1.18	157	0	0.00	355	138	38.8	346	140	40.5
sc-7/y-4	479	0	0.00	297	1	0.34						
sc-7/Df(bb)				1673	22	1.31						
sc-8/dl-49	1959	5	0.26	1748	6	0.34	1146	541	47.3	1020	444	43.5
sc-8/y-4	316	0	0.00	240	1	0.42	906	70	7.7	968	134	13.8
ClB/dl-49	1270	4	0.32	1260	2	0.16						
ClB/y-4				561	1	0.18	101	4	4.0	57	5	8.8
ClB/Df(bb)	1557	2	0.13	1200	6	0.50	1622	520	32.0	1351	561	41.5
dl-49/y-4	1656	5	0.32	1508	2	0.13	952	506	53.2	915	487	53.2
y-4/Df(bb)	370	1	0.27	194	4	2.06	1034	264	25.5	629	272	43.2

others of the same constitution gave none. In no case were the resulting frequencies high enough to be interpreted as due to the presence of an unsuspected Y. The frequencies are about those that result from the presence of a short duplication carrying the X spindle attachment, but we have not studied the descendants of such females with this possibility in mind.

The results of STONE and THOMAS (1935) for sc-8/+ and dl-49/+ have not been included in the table.

Presence of Single Crossovers

Representing a normal X chromosome schematically as *B C D E a* and an homologous chromosome with the *CD* segment inverted as *B D C E a* (*a* in both cases representing the spindle attachment), then single crossovers in the *CD* segment will give the products (1) *B C D B* (duplication for *B*, deficiency for *E*, and having no spindle attachment) and (2) *a E C D E a* (deficiency for *B*, duplication for *E*, and having two spindle attachments). Product (1) would be expected to be lost because of its lack of a spindle attachment. Product (2), because of its two spindle attachments, should form a tie between the two poles of the first meiotic division. It is known from the cytological studies of McClintock (1931, 1933) on Zea, Mather (1934) on Vicia, Stone (1933) on Tulipa, and Smith (1935) on Trillium that, for these plants, single crossovers do occur between segments relatively inverted and that the results are as described above. In order to understand the mechanism of disjunction in inversion heterozygotes in Drosophila it is essential to know whether such crossovers occur in this organism and, if so, with what frequency. The most direct method of answering these questions, namely, cytological examination as used in the cases cited above, is very difficult in the case of oögenesis in *Drosophila melanogaster*. We have resorted to less direct genetic methods.

From the data already presented on single crossing over in combinations of two overlapping inversions it seems highly probable that single crossovers occur within the inverted segment in females heterozygous for a single inversion. Inversion scute-8 represents an inversion of the entire X chromosome with the exception of the *y* and *ac* loci and the spindle attachment. It can be considered as representing essentially a transfer of the spindle attachment from the right to the left end of the chromosome.

As regards crossing over it should behave essentially like a normal chromosome. In the heterozygote of sc-8 and sc-4 crossing over is practically normal as has already been shown. We can therefore argue that in the heterozygote sc-4/+, single crossovers should be of approximately normal frequency. Similarly in the combination sc-7/sc-8 we have shown that single crossing over occurs in the common inverted segment. From this we can conclude that single crossing over occurs in the inverted segment of sc-7/+. Here the data from the combination sc-7/sc-8 and also from sc-4/sc-7 suggest that the frequency is reduced below that for the sc-7 inverted segment normally arranged. The same general kind of an argument can be made for several of the other combinations considered in the previous section.

In a female heterozygous for a single X-chromosome inversion, crossing over can be more or less directly measured by using attached-X females, (see also Sidorov, Sokolov, and Trofimov 1935). By selecting the appropriate crossover from a triploid of the constitution y-4/$\widehat{XX}$ we obtained an attached-X female heterozygous for the y-4 inversion. In such a female, exchange in the inverted segment will give either (1) a closed chromosome carrying a duplication for *cr-bb* and a deficiency for the small segment to the left of *y*, or (2) a chromatid with two spindle attachments plus a chromatid with none (fig. 5). The Y chromosome that is usually present disjoins from the attached-X chromosomes at the first division. The results described above and shown in figure 5 take place during the second division. The types and

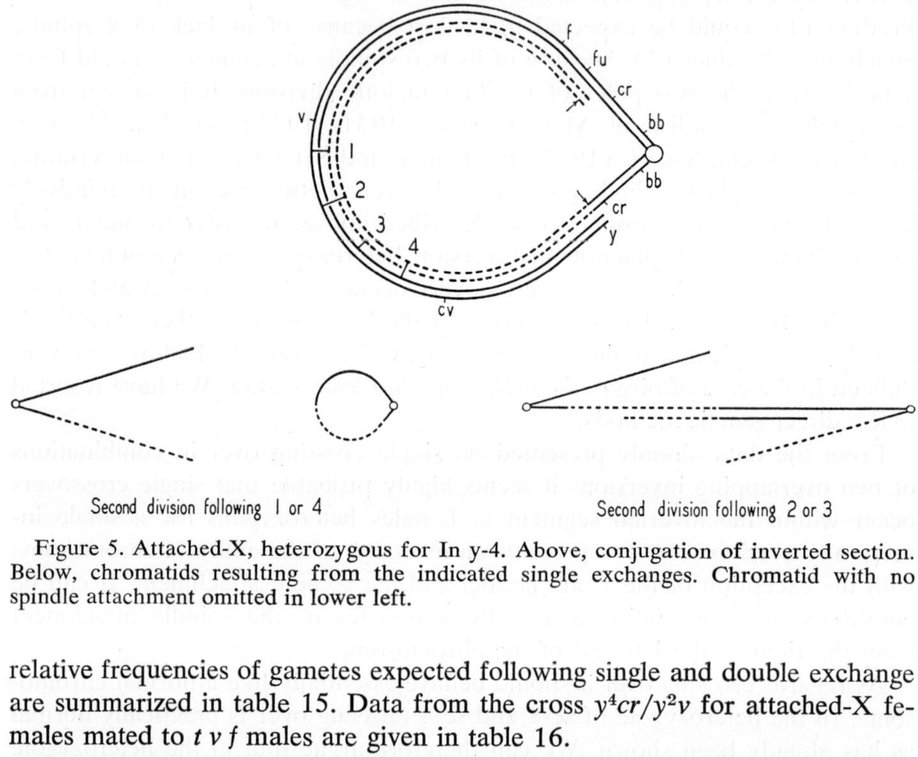

Figure 5. Attached-X, heterozygous for In y-4. Above, conjugation of inverted section. Below, chromatids resulting from the indicated single exchanges. Chromatid with no spindle attachment omitted in lower left.

relative frequencies of gametes expected following single and double exchange are summarized in table 15. Data from the cross y^4cr/y^2v for attached-X females mated to *t v f* males are given in table 16.

The daughters showing *y* and *cr* carry attached-X chromosomes. Those which show neither *y* nor *cr* carry the closed chromosome derived by single crossing over in the inversion. From the fact that the sex ratio approaches 2 males to 1 female we infer that those cases in which a double spindle attachment arises by crossing over within the inversion result in lethal eggs. In such cases the X chromosomes are not simply eliminated for we should then expect a ratio of 2+ males to 1 female. Males carrying the closed chromosome are almost completely inviable. The *y cr* male recorded in table 16 apparently carried such a chromosome. This male had narrow wings, other characteristics of duplication-carrying males, was *y* (not y^2) and was sterile. We have not made a cytological study of this closed-X; this has been done by SIDOROV, SOKOLOV, and TROFIMOV (1935) for the closed-X they obtained

Table 15

EXCHANGE	$\widehat{XX}$ GAMETES: NON-CROSSOVER	$\widehat{XX}$ GAMETES: RECIP-ROCAL	$\widehat{XX}$ GAMETES: EQUA-TIONAL-a	$\widehat{XX}$ GAMETES: EQUA-TIONAL-b	X^c GAMETES	SECOND DIVISIONS WITH CHROMATID TIE	Y GAMETES
None	1						1
Single	1				1	2	4
Double* 2–s	1	1	1	1			4
3–s a					2	2	4
3–s b			1	1	2		4
4–s						4	4
Total Doubles	1	1	2	2	4	6	16

* Double exchanges are designated by the number of strands that undergo crossing over. Thus, in a 2-strand (2-s) double exchange, the same two strands cross over at the two levels

in the same way (using an unspecified inversion). They have published a drawing of one metaphase plate showing the closed-X.

From the data given in table 16 and the information summarized in table 15, we can make an approximate calculation of the frequencies of single and double exchanges within the inverted segment. Doubles are measured by equationals for genes within the inverted segment (*v* and *f*). Since equationals for one chromosome only are detected, there is one chance in eight of detecting a crossover from a double exchange tetrad. Thus the twelve equationals indicate that there were actually 96 double exchange tetrads. Since a part of the females are eliminated, a corrected total must be used and the most direct way of getting this is to double the number of males. This gives 2196 as a corrected total. The frequency of double exchanges is then 4.4 percent. Using the information in table 15 we can subtract from the observed numbers of individuals those which carry products of double exchange. The remainder should give a measure of single exchanges. Single crossovers are directly recovered as closed chromosomes. The frequency of single exchanges is also measured by the deficiency of females as compared

Table 16

Progeny of the cross $y^4cr/y^2v\,f\,cr$ attached-X females by t v f males. Four egg-laying periods; 3, 2, 2 and 3 days.

$\widehat{XX}$ FEMALES		$+/X^c$ FEMALES		MALES	
y^2cr	304	*f*	137	*t v f*	1098
$y^2v\ cr$	11	*v f*	197	*y cr*	1
$y^2v\,f\ cr$	1	*v*	3		
Total	316		337		

with males. The average of these two measures of single exchange is 90.8 percent (assuming 50 percent as the value given by recovered closed chromosomes). The summary of the above operations is as follows:

	$\widehat{XX}$ FEMALES	X/X^c FEMALES	MALES	EXCESS OF MALES OVER FEMALES
Observed	316	337	1098	445
Double exchange (4.4%)	36	24		36
Remainder	280	313		409
Single exchange (90.8%)	298.7	248.4		454.9

The two measures of single exchange agree with one another only approximately. However it is quite clear that the frequency of single exchange is high and approaches the frequency characteristic of the segment normally arranged. The distribution of the single crossovers in the inversion heterozygote is indicated by the data, and for the regions *y-v* and *v-f,* is approximately the same as that for normal chromosomes.

Crossing over in the segment between the right break of the inversion and the spindle attachment is very low as indicated by the low frequency of forked equationals. The one y^2 *v f cr* female recorded in table 16 is assumed to have resulted from double exchange within the inverted segment rather than from single exchange to the right of it.

In the case of $\widehat{XX}$ females heterozygous for short inversions, the closed chromosomes resulting from crossing over within the inverted segment are inviable in heterozygous condition and so cannot be recovered unless the duplications and deficiencies are compensated for in some way. However it is clear from the above discussion of attached-X In y-4 heterozygotes that the distortion of the sex ratio in itself can be used as a measure of crossing over.

Early experiments with $\widehat{XX}$ females heterozygous for In sc-7 consistently gave an excess of males over females. The results of three such experiments are summarized in table 17. These experiments indicated relatively high frequencies of exchange in the inverted segment and approximately normal crossing over between the inversion and the spindle-attachment, the latter result being in substantial agreement with those from free-X In sc-7 heterozygotes (p. 200). Since no particular precautions were taken to insure that male and female offspring were of comparable viability in these experiments, stocks more satisfactory with respect to the mutant genes used were made up and the experiments repeated. To decrease viability differences, relatively short egg-laying periods were used. The results are summarized in tables 18 and 19. These data are in agreement with those from the first experiments in showing that crossing over to the right of the inversion is about normal (the sc-7 chromosome of the females whose progeny are summarized in table 19 apparently carried a semi-lethal mutant of unknown origin and the raw data must be corrected accordingly). However these experiments indicate a lower exchange frequency within the inverted segment than did the earlier ones. The inversion involves a segment 13 to 20 map units long. The data of tables 18 and 19 give crossover values of 9 and 10 percent (one half exchange frequencies).

Table 17

Progeny of $\widehat{XX}$ females heterozygous for inversion sc-7 mated to various males.

CONSTITUTION OF DAUGHTERS*	CONSTITUTION OF MOTHERS		
	sc-7 w^a v f / y^2 w^a ec f	sc-7 w^a f / y^2 w^a ec f	sc-7 w^a v B / y^2 w^a ec f
+	267	206	155 (B/+)
sc-7	6	45	10 (B/+)
y^2 ec	53	24	18 (B/+)
sc-7 v	53		14 (B/+)
y^2 ec f			4
sc-7 v, B/B			5
B/B			2
f			1
y^2 ec, B/B			1
Total females	379	275	210
Corrected total females†	385	296	216
Total males	493	483	309
Exchange (%)	29.2	51.6	40.1

* Constitutions are given only with respect to genes heterozygous in the mother.

† Corrected total females obtained by adding to non-crossover phenotype, twice the number of equationals for the sc-7. The correction is for the indicated lower viability of equationals for y^2.

Experiments were made with In dl-49 using the same technique as for the later experiments with In sc-7. The results of two experiments with controls are summarized in table 20. The extent of the distortion of the sex ratio in the two series is not the same. The control experiments indicate that the difference is due to the difference in relative viability of the two kinds of males used. Making appropriate corrections of the number of males, the two series indicate exchange values for the inverted segment of 11.5 and 12.5 percent. Exchange in the segment between the right inversion break and the spindle attachment is measured by equationals for genes within the inversion limits. The values indicated are 3.8 and 10.4 percent for the two series. The cause of the rather large difference is not known. Exchange in the segment to the left of the inversion is measured by equationals for *y* and by equationals for the gene *v* which are not equational for *y*. About all that can be said about the exchange frequency for this terminal segment is that it is low (less than 2 percent).

Table 18

Progeny of the cross sc-7 w^acr/y^2ct v f cr attached-X females by wild type males. Egg-laying periods; 3, 2 and 2 days.

FEMALES								TOTAL*	MALES	PERCENT EXCHANGE
cr	sc-7 w^a*cr*	*y ct cr*	*y ct v cr*	*y ct v f cr*	*f cr*	*v f cr*	*v cr*			
629	124	27	79	45	2	3	1	937	1088	18.5

* Total corrected for indicated lower viability of sc-7 w^a.

Table 19

Progeny of the cross sc-7 w^a/y^2v f cr attached-X females by wild type males. Egg-laying periods; 3, 2 and 2 days.

FEMALES										TOTAL*	MALES	PERCENT EXCHANGE
+	sc-7 w^a	*y*	*y v*	*y v f*	*y v f cr*	*v f*	*f cr*	*f*	*v*			
393	2	14	66	8	1	1	1	1	1	575	679	20.4

* Total corrected as in table 18.

From the evidence considered above it is clear that single crossovers do occur between segments of chromosomes inverted relative to one another. The frequency of such crossovers evidently depends on the length of the inverted segment and its position in the chromosome. These relations will be discussed in more detail in another connection (page 237). In any case we can say that long inversions such as In sc-8, In sc-4, and In y-4 show, with a normal chromosome, a high frequency of single crossing over between

Table 20

Progeny of the cross dl-49 cm^2/y^2 v f cr attached-X females × Bar and + males and from control crosses of y^2 v f cr/y^2 v f cr attached-X females by Bar and + males. In each case, three egg-laying periods; 3, 2 and 2 days.

PROGENY	INVERSION ×*B*	INVERSION ×+	CONTROL ×*B*	CONTROL ×+
+ ♀♀	984	1672		
cm ♀♀	11	46		
y ♀♀	0	1		
y v ♀♀	2	10		
y v f ♀♀	4	28		
y v f cr ♀♀	5	18		
v f cr ♀♀	0	1		
Total ♀♀	1006	1776	590	841
♂♂	1004 (1101)*	2076 (1960)*	538	891

* Numbers in parentheses are corrected totals obtained by multiplying the observed numbers of males by the ratio of females to males in the appropriate control.

the inversion segments. These frequencies are of the same order of magnitude as those characteristic of these same segments arranged in the normal way.

Egg and Larval-Pupal Mortality

We have shown in the preceding section that single crossovers occur with a relatively high frequency in inversion heterozygotes. The question that we shall consider now is whether or not such crossovers result in inviable zygotes. This question can of course be directly answered by determining the amount of mortality in the progeny of heterozygotes for inversions known to give a high frequency of single crossing over within the inverted segment.

Technique. The method that we have used in determining the amount of mortality is essentially the same as that commonly used by other workers (e.g., L. V. MORGAN 1933). Certain modifications were found useful. Paper spoons have usually been used as containers for the medium on which the eggs are collected. They have two disadvantages: (1) the surface of the medium is usually not flat and (2) the depth of the medium varies which often results in drying out around the edges. To overcome these disadvantages small metal (nickel has been found satisfactory) containers about 38 mm long, 17 mm wide and 3 mm deep were made. A handle of the same material about 10 mm wide was soldered to the bottom so that it projected about 3 cm.

The standard cornmeal-molasses-agar medium with the addition of animal charcoal (to increase the contrast between eggs and medium) was liquefied, pipetted into the containers and allowed to cool. The flat surface was then painted with a rather heavy yeast suspension. The addition of fermented banana, alcohol or wine was found to be of no advantage. A single female was allowed to deposit eggs on the medium for a period of 24 hours. The container was then removed from the vial and replaced with one containing fresh medium. After removing the container the food was removed from it with a strip of cardboard of appropriate size. The eggs, including those already hatched, were then counted and recorded. The food was placed in a vial at 25°C in a moist incubator for 28 hours after which time the unhatched eggs were counted. The food block was then placed in a standard half pint culture bottle and the flies allowed to develop to maturity. Unless care is taken to have the outside of the food container dry there is danger in error from eggs deposited on the sides of the container. The larvae from these eggs may hatch and crawl onto the food block. There is also some error in losing or killing a few larvae in handling the food blocks. The magnitude of these errors can be kept reasonably low with careful manipulation.

Our experience indicates that the percentage of egg mortality is dependent upon the genetic constitution of the mother as well as upon the genetic constitution of the eggs themselves. Thus females from inbred stocks or females homozygous for several recessive genes generally give relatively high mortality regardless of the type of males to which they are mated. Because of this fact, strictly comparable controls cannot be had. To minimize this "residual" egg mortality, crosses between more or less unrelated stocks were made wherever possible and the F_1 females from not overcrowded cultures were used in the egg-laying experiments. Normal controls more or less comparable in genetic constitution were run simultaneously with the experiments on inversion heterozygotes.

Results. The results of our experiments on In sc-4, In sc-8, In y-4, and In dl-49 heterozygotes are summarized in table 21. The answer to the question that we set out to study is quite clear: single crossovers do not give rise to inviable zygotes. In the cases of In sc-4, In sc-8 and In y-4 exchange is approximately normal in frequency. If the distribution of the four strands of a tetrad were random at meiosis, we should expect about half the products to be single crossovers or their equivalent. Since such crossovers are not recovered in the viable zygotes they would have to be eliminated as inviable zygotes. However, it is evident that inviable zygotes do not approach 50 percent in frequency. Since these inversion heterozygotes produce an appreciable number of patroclinous males (about 3 to 5 percent of the viable zygotes) we know that there should be a corresponding frequency of inviable zygotes (no-X eggs fertilized by Y-carrying sperms). In these cases crossing over could be followed sufficiently well to know that the inversion heterozygotes were giving the usual results. In addition the frequency of patroclinous males was determined and found to be approximately "normal" for the inversion heterozygotes under consideration. When the data are considered in connection with the frequencies of patroclinous males produced (page 236) and with the controls, and when allowance is made for the difference in genetic constitution between inversion heterozygotes and controls, we can conclude that the only zygotes whose death is the direct result of the presence of the inversion in the parent females are those corresponding to patroclinous males and differentiated from them by the sperm.

Stone and Thomas (1935) have also published egg counts for In sc-8/+ and In dl-49/+. The mortality indicated is higher than in our data, as it is also for their controls. They have also not distinguished between egg mortality and larval-pupal mortality. It seems clear that the lowest adequately established mortality is the one that gives the best picture of the case; and it is to be noted that Stone and Thomas conclude, as do we, that single crossovers are not responsible for any detectable portion of the observed mortality.

Data from XXY In sc-8 heterozygotes are included in table 21. Here both egg and larval-pupal mortality is relatively high. This is of course expected from the fact that here secondary non-disjunction occurs with an appreciable

Table 21

Egg and larval-pupal mortality data for inversion heterozygotes.

MATING	TOTAL EGGS	INVIABLE	PERCENT INVIABLE	HATCHED EGGS	ADULT FLIES	PERCENT EMERGENCE OF HATCHED EGGS
y sc-4 *v f cr*/+ × *B*	2839	214	7.5	2625	2187	83.4
sc-8 w^a *cv v f*/y^2×*B*	4190	284	6.8	3906	3745	96.0
y^{2S}/B (control for above) × *B*	3947	52	1.3	3895	3741	96.2
sc-8 *B*/w^a *v*/*Y* × y^{2S}	1313	184	14.0	1129	990	87.6
+/w^a *v* (control) × y^{2S}	1018	26	2.5	992	991	99.9
dl-49/+ × *B*	1645	51	3.1	1594	1525	95.7
y-4/*v* g^3 × *w*	884	68	7.7	816	753	92.4
y-4/y-4 × *w*	570	113	19.8	457	375	82.2

frequency (page 224). The frequency of inviable eggs is of the same order of magnitude as that of exceptional males (YY or YX zygotes, depending upon the sperm) and larval-pupal mortality is of the same order of magnitude as exceptional females (XXX or XXY zygotes, depending upon the sperm).

In connection with studies of sister-strand crossing over, SCHWEITZER and KALISS (1935) have made extensive determinations of egg mortality in inversion heterozygotes. Their results are in agreement with the conclusion we have drawn that single exchanges between inverted segments do not result in inviable zygotes.

The Mechanism of Disjunction in Inversion Heterozygotes

It has been shown in preceding sections that single exchanges occur within the inverted segment of inversion heterozygotes, and further, that the crossover products of single exchange are not recovered and do not result in inviable zygotes. It is evident that we must assume that such single crossover chromatids are selectively eliminated during the meiotic process. It has also been shown that X-chromosome inversion heterozygotes give rise to patroclinous males among their progeny. We have implied that the frequency of such exceptional males is a function of double exchange.

The problem that we shall consider is how these two results, (1) elimination of single crossover chromatids and (2) the production of no-X eggs, are brought about. We know from cytological studies on plants (Zea, MCCLINTOCK 1933; Tulipa, STONE 1933; Vicia, MATHER 1934; Trillium, SMITH 1935) that double spindle attachment chromatids resulting from crossing over between segments inverted with respect to one another pro-

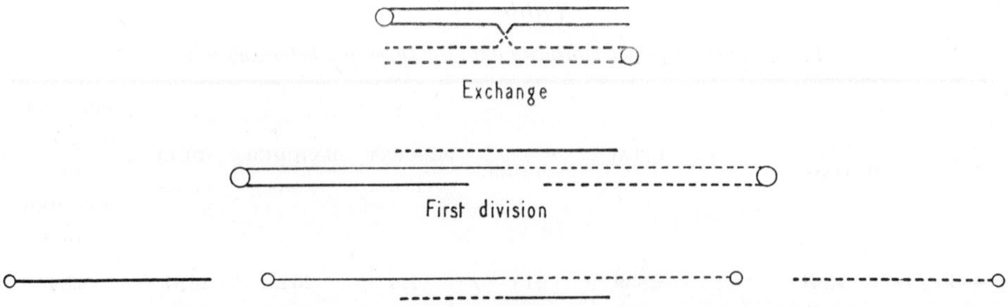

Figure 6. Single exchange within a heterozygous inversion. The upper figure represents the two X's of a female in which one chromosome is practically wholly inverted. At the first meiotic division there results a chromatid tie; this leads to an orientation of the second division such that the two terminal nuclei receive only non-crossover chromatids; one of these is the egg nucleus. The result is the total loss of all single crossover chromatids to the polar body nuclei.

duce chromatin ties between the two poles of the first meiotic division (or under certain conditions to be considered below, between poles of second meiotic spindles). Knowing that the four nuclei resulting from meiosis in the Drosophila egg lie approximately on a single straight line (HUETTNER 1924), we are prompted to propose the following scheme for the X chromosome of Drosophila.

1. A single chromatid tie at the first meiotic division results in orientation of the spindle attachments in such a manner that only chromatids with a single spindle attachment get into the terminal nuclei, one of which will become the egg nucleus (HUETTNER 1924).

2. A double chromatid tie results in the formation of end nuclei with no X chromosome, and a no-X egg will result.

The behavior of various types of crossover tetrads expected according to this scheme is shown diagrammatically in figures 6 and 7.

As to the precise nature of the orientation of single exchange tetrads or their equivalent we have insufficient information; we know only the end result. It seems reasonable to suppose that the orienting influence of a double attachment chromatid is mechanical. However, we do not know whether or not such chromatids in Drosophila break during the division as they are known to do in the plants mentioned above. If they break, the orientation probably results from the retardation prior to breakage. Single crossover chromatids without spindle attachments are probably not included in either daughter nucleus but lost in the cytoplasm during division as is known to be the case in plants (McCLINTOCK 1933).

From the diagrammatic representation of the suggested scheme (figs. 6 and 7), it is evident that certain quantitative relations should hold. These should serve as tests of the assumptions we have made. The types and frequencies of gametes expected to result from non-, single, double, and triple

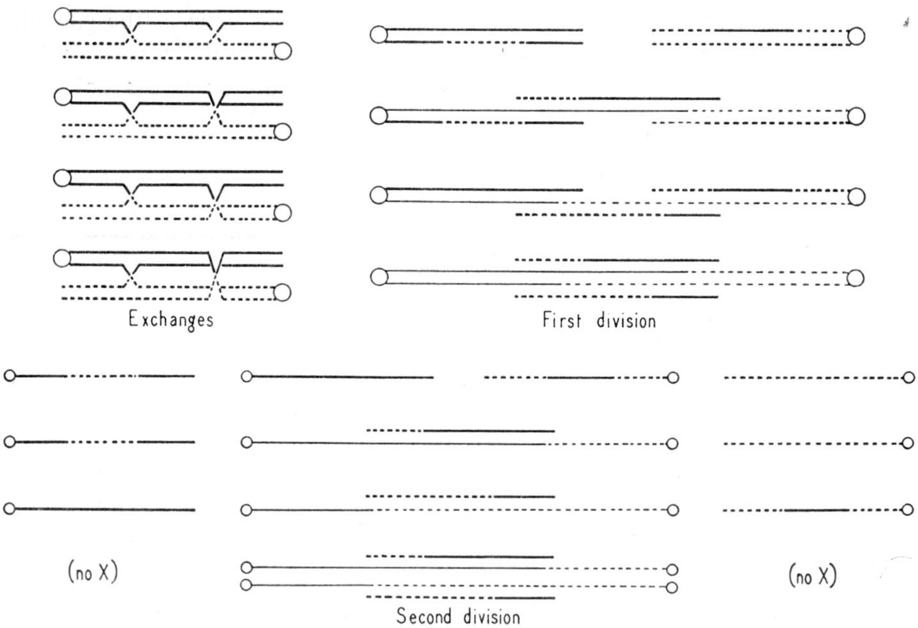

Figure 7. The four possible types of double exchange within a heterozygous inversion. The two-strand exchange (upper row) leads to equal numbers of non-crossover and double crossover chromatids in the terminal nuclei, each of which will be the egg nucleus in half the cases. Three-strand doubles (second and third rows) result in chromatid ties at the first division, and also lead to equal numbers of non-crossover and double crossover chromatids in the egg nuclei. Four-strand double exchange (fourth row) leads to a double tie at the first division, and to no-X egg nuclei.

exchange tetrads are summarized in table 22. It is seen that from double exchange tetrads, double crossover and no-X gametes are expected to occur in the ratio of 3:2. This 3:2 ratio was also found experimentally by STONE and THOMAS (1935) for In sc-8 and for another long inversion that we have not studied. For the longer inversions some triple exchanges presumably occur and these give double crossover and no-X gametes in the ratio of 21:4. Since we have no way of measuring the relative frequencies of double and triple exchanges in inversion heterozygotes (because triple crossover chromatids must be eliminated) we cannot predict precisely what the ratio of double crossover to no-X eggs should be. However, if we make the assumption that the frequency of triple relative to double exchanges in inversion heterozygotes does not exceed that of normal X chromosomes (about 1:10), then the ratio of double crossover to no-X gametes should lie between 3:2 and 3.4:2.

The numerical data showing the relation of double crossovers to patroclinous males are given in table 23. Half the no-X eggs are lost (fertilization with Y sperm) and in the cases recorded in the table, half the double crossovers were not detected since only male offspring were used to measure crossing over. Hence the zygotic ratios expected are the same as the gametic ratios mentioned above and given in table 22. The observed relative frequencies approach rather closely the ratio 3:2. In no case is the deviation

Table 22

Relative frequencies of types of gametes produced following single, double, and triple exchange within the inverted segment of an X-chromosome tetrad.

EXCHANGE	DESIGNATION*	NON-CROSSOVER	DOUBLE CROSSOVER	NO-X EGG
None		1	0	0
Single		1	0	0
Double	2 (1)	1	1	
	3 (2)	2	2	
	4 (1)			2
	Total	3	3	2
Triple	2, 3, 3 (4)	4	4	
	2, 4, 4 (2)		4	
	2, 2, 2 (1)	2		
	3, 4, 3 (2)			4
	3, 2, 3 (2)	1	3	
	3, 3, 4 (4)		8	
	4, 2, 4 (1)		2	
	Total	7	21	4

* Doubles are designated by the number of strands involved in the two exchanges; triples in the same way by considering them as three doubles, taking successive exchanges a, b and c in combinations of two in the order a-b, a-c, and b-c, and in this case, disregarding the direction (a-b-c is equivalent to c-b-a). Relative frequencies of different types among doubles or among triples are indicated by numbers in parentheses.

The frequencies of gametes (totals of last three columns) must in each case be proportional to the frequencies of occurrence of the types of exchange (numbers in parentheses). This is true even in the case of two-strand doubles, where the potentially good chromatids are twice as numerous as in the case of three-strand doubles, since each tetrad gives rise only to a single gamete. The same principle applies also to triples.

from 3:2 statistically significant. The totals approach very closely a ratio of 3:2; the actual observed small deviation is in the direction expected to result from triple exchanges. The inversion heterozygotes known to give approximately normal crossing over (sc-4, sc-8 and Df sc^8) actually are the ones that give the higher ratios. The results expected on the proposed scheme of disjunction in inversion heterozygotes are thus in quantitative agreement with the experimental data.

The scheme proposed should enable one to predict quantitatively the results from closed-X heterozygotes. These have been studied by L. V. MORGAN (1933). Her results differ from those expected according to the

Table 23

Relative frequencies with which double crossovers and patroclinous males are recovered from inversion heterozygotes. In all cases double crossovers as recovered in males only are recorded.

CONSTITUTION	DOUBLE CROSSOVER MALES	PATROCLINOUS MALES	CALCULATED (3:2)		ACTUAL RATIO
sc-4/+	108	63	102.6	68.4	3.4:2
sc-8/+	93	57	90.0	60.0	3.3:2
Df (sc-8)*/+	60	31	54.6	36.4	3.9:2
y-4/+	57	51	64.8	43.2	2.2:2
Df (bb)/+	93	66	95.4	63.6	2.8:2
sc-4/sc-7	11	7	10.8	7.2	3.1:2
	422	275	418.2	278.8	3.1:2

* Number of double crossovers corrected, owing to lethal nature of Df (sc-8).

scheme formulated from our knowledge of inversion heterozygotes in two important respects:

1. Among the progeny of X/X^c, X is recovered more frequently than X^c.
2. Egg mortality is too high relative to the frequency of recovered double crossovers.

The inequality of X and X^c among the progeny was ascribed by L. V. MORGAN to differential viability. The egg mortality data led her to conclude that single exchanges result in inviable zygotes. Fortunately a second closed-X chromosome was found by Mr. R. D. BOCHE of this laboratory. This closed-X has an advantage over the original X^c used in the experiments of L. V. MORGAN in that it has less effect on viability. We have made several experiments with X^{c-2} heterozygotes set up especially to test the scheme proposed in this paper. The results of these experiments are to be presented in another paper but we can say here that both discrepancies mentioned above appear now to be viability effects. The results obtained from X/X^{c-2} and X^{c-2}/X^{c-2} females are in as good agreement with those predicted from the assumptions we have made in this paper as could reasonably be expected.

We have pointed out that single exchange between two segments inverted relative to one another does not result in inviable zygotes. In the case of attached-X inversion heterozygotes we have pointed out that following certain types of single exchange a chromatid tie is formed during the second meiotic division. From the numerical data we concluded that the condition in which the X-chromosome spindle attachment is tied to a spindle attachment in the nucleus lying next in line does result in an inviable egg. There is another case in which single exchange within the inverted segment should result in a chromatid tie at the second division. This is where a single exchange within the inversion is accompanied by a second exchange outside the inversion of such a nature that the two exchanges make a three-strand double exchange. SMITH (1935) has observed this result cytologically in Trillium. In the inversions dealt with in our experiments, lethal eggs from this source would be very infrequent except for the case of In sc-7. This inversion is the only one that we have used in which exchange in the heterozygote is frequent both inside and outside the inverted segment. Here, however, the mortality has not been studied.

Effects of Inversions on Frequency of Crossing Over

a. Homozygous inversions. The data are in agreement with earlier conclusions (STURTEVANT 1931) that homozygous inversions show about the same total amount of crossing over as do homozygous normals. They are also in agreement with the conclusions of BEADLE (1932) and of OFFERMAN and MULLER (1932) that the distribution of this crossing over is altered by relation to the spindle attachment, a given section giving less crossing over if it is near the attachment, more if it is near the free end of the chromosome.

b. Heterozygous inversions. The effects on crossing over are, as might be expected, different for sections within the inversion and those outside it, and are also dependent on the length and position of the inversion concerned.

Crossing over within the inversion is evidently decreased in heterozygotes for In sc-7 and In dl-49 (pp. 228–230), and is probably decreased in In ClB since so few doubles are recovered. The other inversions studied here, all of them longer than these, seem to have much less effect on crossing over within the inversion, though the data are scarcely adequate to permit the conclusion that there is no effect.

Crossing over in sections outside the inversion is regularly reduced. In sc-4 and In sc-8 do not leave any sections uninverted in which crossing over occurs in normal flies and can be measured in inversion heterozygotes. The same is true for the sections to the left of In sc-7 and In y-4 and for that to the right of In Df (bb). The remaining seven uninverted sections all show a reduction in crossing over, localized close to the inversion itself to the right of In sc-7, relatively slight to the right of In dl-49, and very marked in the other cases, namely, on both sides of In ClB, to the left of In dl-49 and of In Df (bb), and to the right of In y-4.

These data are in approximate agreement with expectations from the competitive pairing hypothesis of DOBZHANSKY (1931). A short inversion may be supposed to have its pairing more interfered with by the uninverted sections than does a long inversion which has shorter uninverted sections. Conversely, a long inversion may be supposed to interfere more seriously than a short one with the pairing of the uninverted sections.

One other relation is suggested here, as it is by the data on autosomal inversions (STURTEVANT 1931); namely that an inversion is more effective in suppressing crossing over in segments distal to itself than in proximal segments. This relation is difficult to analyze, and of the present series of inversions dl-49 seems to be the only one favorable for its study. What is needed is a comparative study of a larger series of more diverse types of inversions than we have used. This need is supplied in part by STONE and THOMAS (1935), who reach conclusions similar to the one just suggested.

The Effects of the Y Chromosome on Crossing Over

The most extensive series of data on the effects of a Y chromosome on crossing over is that of BRIDGES and OLBRYCHT (1926). DR. BRIDGES informs us that the XXY females there recorded gave, in addition to the published results, 246 exceptional offspring (2.5 percent). Correcting the data (by adding twice the number of exceptions to the non-crossover class in the case of XXY females) gives the following comparisons:

	sc – *ec*	*ec* – *cv*	*cv* – *ct*	*ct* – *v*	*v* – *g*	*g* – *f*	total	N
XX	6.7	8.8	8.2	14.4	11.3	12.3	54.5	11325
XXY	6.5	9.1	8.3	14.2	10.1	9.7	50.8	9461
XX/XXY	1.03	.97	.99	1.01	1.12	1.27	1.07	

Evidently the Y has no effect on crossing over in the region from *sc* to *v*, reduces *v*-*g*, and reduces *g*-*f* still more. This is in agreement with the results on duplications described by DOBZHANSKY (1934), since the Y is homologous only with the right end of the X, and reduces crossing over only in the portions of the X near this homologous section.

The results recorded in this paper for comparable XX and XXY females can also be interpreted in terms of the hypothesis of "competitive pairing" (DOBZHANSKY 1931). In Df(bb)/+ gives no effect of the Y, as might be expected, since Df(bb) presumably carries little or no material homologous to the Y. In the cases of sc-4/+ and y-4/+ there is an increase of double crossing over within the inversion, which may be looked upon as due to interference of the Y with pairing of the attachment ends of the X's, this in turn leading to less interference of these attachment ends with full pairing of the inverted segment. In the case of dl-49/+ the crossing over studied is that between the inversion and the spindle attachment; the data are in agreement with the analysis just given in that they indicate a decrease in crossing over due to the Y. The one remaining case in which we have comparable data is that of y-4/Df(bb), where the frequency of singles within the common inverted region seems to be unaffected by the presence of a Y, as would have been expected.

Secondary Non-Disjunction

XXY females of all kinds give segregation of two chromosomes to one pole, one to the other (X-XY or XX-Y). The inversions affect the relative frequencies of these two types, and therefore a fuller understanding of the meiotic behavior of inversions should throw light on the mechanism of secondary non-disjunction.

If p be taken as the frequency of XX-Y segregation, then assuming random fertilization by X sperm and Y sperm and death of the XXX and YY classes, the frequency of recovered exceptions, q, will be $\frac{p}{2-p}$.

The earlier analyses of secondary non-disjunction (BRIDGES 1916, ANDERSON 1929, GERSHENSON 1935) have been based on the assumption that the maximum frequency of XX-Y segregation occurred when one X separated from Y and the other X went to either pole at random. This gives $p = .5$, $q = .333\%$. GERSHENSON himself obtained, from In ClB/+, $q = .353 \pm .0040$ in the female classes, a deviation about five times the probable error. Adding our data the value becomes $.366 \pm .0038$, a deviation of .033 or nearly 9 times the probable error. In the case of In dl-49/+ the female data of table 14 give $q = .456 \pm .0051$, a deviation of .123, 24 times the probable error.

There can be no doubt, then, that XX-Y segregation can occur with a frequency greater than 0.5. As a matter of fact the dl-49/+ value for q (.456) gives $p = .626$.

It becomes necessary, therefore, to search for a new interpretation of secondary exceptions. As pointed out by BRIDGES (1916), nearly all the exceptional females from +/+/Y mothers are non-crossovers, carrying the same two X's as their respective mothers. The same relation holds for In/+/Y exceptions, as shown by GERSHENSON (1935) and by our own data. In both types of experiments occasional exceptions are found with crossover chromosomes; but these are little, if any, more frequent than are such crossover exceptions from XX mothers; they may safely be disregarded in analyzing the effect of the Y on non-disjunction. Secondary exceptions, then, carry two unlike non-crossover chromatids. The failure of separation must take place at the first meiotic division, rather than the second, since the latter would give two *like* chromosomes. This is true provided Drosophila agrees with plants in having the first division reductional for spindle attachments. Indirect evidence as well as direct cytological evidence (KAUFMANN, 1935) indicates that this assumption is correct. It may be concluded also that secondary exceptions result from X-*tetrads* in which no crossing over occurred, for otherwise one would have to assume that the orientation of sister chromatids on the second meiotic spindle was not random. This is contrary to what is known in other cases; and even this assumption would not suffice to account for 3-strand or 4-strand doubles.

If secondary exceptions arise from non-crossover X-tetrads, the next step is to determine the frequency of such tetrads in various kinds of females and to compare this with the frequency of non-disjunction in such females. This can be done best in the case of In dl-49. As shown above, XXY dl-49/+ females gave about 1 percent exchange to the left of the inversion, 12 percent in the inversion, and 20 percent to the right of the inversion. There are probably not over 3 percent multiple crossovers here, so 30 percent is not far from the true value for the crossover X-tetrads. (Three-strand multiples where one crossover is in the inversion and one is outside it should give rise to inviable eggs; the data presented above show that these are negligible in frequency.) Therefore, among the 70 percent non-crossover X-tetrads, 62.6 or 90 percent give rise to non-disjunctional gametes. If we use the frequency of exceptional females actually observed in the same experiment in which crossing over was studied, we find that 66/79 or 83 percent

Table 24

Comparison of species.

SPECIES	SECONDARY EXCEPTIONS		TOTAL MAP LENGTH OF X	
	%	AUTHORITY	UNITS	AUTHORITY
melanogaster	4.3	Bridges 1916	66	Bridges, unpubl.
simulans	2.9	Sturtevant 1929	70	Sturtevant 1929
willistoni	1.7	Lancefield and Metz 1921	84	Lancefield and Metz 1922
virilis	0.5	Kikkawa 1932	182	Kikkawa 1932
pseudoobscura	0	Schultz and Redfield, unpubl.	170	Lancefield 1922

of the non-crossover X-tetrads gave XX-Y segregation. This does not take into account exchanges within the inversion. There can be little doubt that 90 percent is too low rather than too high a value. If one assumes that this same proportion holds in all cases the resulting deduced frequencies of complete non-crossover X-tetrads (for example 9 percent for +/+/Y) seem not unreasonable. In any case, the proportion .667 suggested by BEADLE and STURTEVANT (1935) by analogy with the fourth chromosome is clearly incorrect for dl-49/+.

Table 14 shows that the frequency of secondary exceptions rises as the total frequency of crossing over decreases in the various inversion combinations. Changes in the reverse direction have not been recorded in *D. melanogaster,* but other species give more crossing over; that is, they have longer crossing over maps and presumably fewer non-crossover tetrads. The available data are shown in table 24. The map lengths given are probably too short, especially in *willistoni* and *pseudoöbscura,* owing to fewness of available loci for study. Other species have been omitted because this element of uncertainty is even greater. It is clear that the table is in agreement with expectation.

The frequency of secondary exceptions thus shows strong negative correlation with the frequency of tetrad crossing over. Since the latter value is not greatly affected by the presence of a Y, whereas the former is, it may be concluded that the frequency of secondaries is dependent on the occurrence of non-crossover tetrads, rather than the reverse.

Normal Disjunction of X's

In the case of dl-49/+ the data show about 14 percent crossing over (28 percent exchange) between the spindle attachment and the inversion. The attached-X data show about 12 percent exchange within the inversion. These results are from XXY females; in the cases of other inversions the presence of a Y has been shown to give a slight increase in crossing over within the inversion. There is a small percentage (about 1 percent) of exchange between the inversion and the free end. The indicated total frequency of exchange is thus 41 percent. There is a fairly large probable error attached to this value; but, since there are probably some double exchanges involved, it seems safe to conclude that at least half the tetrads undergo no exchange.

The data of table 14 show, from this combination, no matroclinous females in a total of 3238 daughters. It follows that exchange is not necessary for normal disjunction. This conclusion can be avoided only by supposing that undetectable exchanges occur between the known genes and the attachment end of the chromosome. This supposition has no evidence in its support and is made unlikely by the absence of matroclinous females from In Df(bb)/+ and their presence only in numbers similar to those given by +/+ in the cases of In sc-8/+ and In Df(sc-8)/+. These three cases all involve inversions that upset the homology well within the inert region, and might well be expected to interfere with crossing over in the region concerned.

The results here reported may, then, be taken as supporting the conclusion, which is probable on other grounds, that crossing over is not a necessary requirement for regular disjunction of the X chromosomes of *Drosophila melanogaster*.

Population Mechanics of Inversions

The scheme for inversion crossovers here developed should apply in all cases of inversions that do not include spindle attachments, since singles within such inversions should always give ties between first meiotic nuclei. The resulting selective eliminations of crossover chromatids may be expected in any case where three of the four products of meiosis are eliminated, the effective one being terminal (in terms of the orientation of the second division spindles). These conditions hold in the oögenesis of most animals and in the megasporogenesis of most seed-plants. In such forms as the Ascomycete Neurospora, where all the products of meiosis are potentially functional but are still arranged in a line, inversions of the type under discussion should lead to non-functioning of "inner" nuclei in much higher proportions than terminal ones. In plants the result will be the production of numerous inviable pollen grains, but no increase in egg inviability. There will therefore be no decrease in fertility, a circumstance that must prevent the rapid elimination of inversions through a reduced rate of reproduction. In animals the aberrant sperm will presumably be viable and functional, but will lead to the production of inviable zygotes and therefore to reduced fertility. In Drosophila this result is not brought about because of the absence of crossing over in the male.

A mechanism that increases the number of gametes carrying a single complete haploid set of chromosomes exists also in the case of heterozygotes for reciprocal translocations, where there is a higher frequency of "regular" than of "irregular" gametes in most cases. Here, however, there is no marked sexual difference, and the frequency of irregulars is high enough to produce an appreciable decrease in fertility in most (probably in all) cases. These relations are probably responsible for the observation that, within a given species of Drosophila, wild populations carry inversions far more frequently than translocations.

Inversions that include the spindle attachment cannot produce a chromatid tie, and will therefore decrease fertility if single exchanges occur within them. This is probably the explanation of the fact that no such inversions have been found in wild populations of Drosophila, though they do occur as a result of X-ray treatment.

Summary

1. Seven inversions are discussed. Their nature is illustrated in figure 1.

2. The results obtained from females heterozygous for two inversions are described. The properties of the chromosomes produced by single crossing over within the common inverted sections are summarized in table 11.

3. The frequencies of matroclinous females and patroclinous males from the combinations studied are shown in table 14.

4. Females carrying attached X's, in one of which there is an inversion, give rise to closed X's by single crossing over within the inversion.

5. Egg counts show that the mortality from inversion heterozygotes can all be accounted for as due to the fertilization of no-X eggs by Y

sperm. This is very much less than the indicated frequency of single crossover chromatids.

6. Since single crossovers are produced but are not recovered, they must be eliminated from the egg at meiosis, leaving a non-crossover chromatid in the reduced egg.

7. A scheme for such oriented divisions is shown in figures 6 and 7. This is based on cytological observations on plants and on the observed geometrical relations of the meiotic divisions in the Drosophila egg.

8. According to this scheme, crossover chromatids with two spindle attachments form ties between two nuclei at the first meiotic division, resulting in the tied chromatids failing to pass to either terminal pole; or at the second division, resulting in death of the egg when the egg nucleus is concerned.

9. This scheme results in several numerical predictions, which are borne out by the data:

(a) matroclinous females from XX mothers are not increased in frequency by inversions.

(b) patroclinous males are to recovered double crossover males as 2:3.

(c) egg mortality is practically equal to the frequency of patroclinous males.

10. Inversions, and also the presence of a Y chromosome, decrease crossing over in accordance with the hypothesis of competitive pairing.

11. Females (XX) heterozygous for inversions may give many no-exchange tetrads; these segregate normally, with the production of no significant number of XX gametes.

12. In XXY females that are dl-49/+, 90 percent or more of the eggs in which no exchange occurs give XX-Y segregation. Similar frequencies are probable in all cases studied.

LITERATURE CITED

AGOL, I. J., 1929: Step-allelomorphism in *Drosophila melanogaster*. J. Exp. Biol. (Russian) **5:** 84–101.

BEADLE, G. W., 1932: A possible effect of the spindle fibre on crossing over in Drosophila. Proc. Nat. Acad. Sci. **18:** 160–165.

———, and STURTEVANT, A. H., 1935: X-chromosome inversions and meiosis in *Drosophila melanogaster*. Proc. Nat. Acad. Sci. **21:** 384–390.

BRIDGES, C. B., 1916: Non-disjunction as proof of the chromosome theory of heredity. Genetics **1:** 1–52, 107–163.

———, and OLBRYCHT, T. M., 1926: The multiple stock "Xple" and its use. Genetics **11:** 41–56.

DOBZHANSKY, T., 1931: The decrease of crossing over observed in translocations, and its probable explanation. Amer. Nat. **65:** 214–232.

1934: Studies on chromosome conjugation III. Behavior of duplicating fraghents. Z. i. A. V. **68:** 134–162.

DUBININ, N. P., 1930: Step-allelomorphism in *Drosophila melanogaster*. Allelomorphs scute^{7} achaete^{2}-scute^{10}, achaete^{1}-scute^{11}, achaete^{3}-scute^{13}. J. Exp. Biol. (Russian) **6:** 4.

———, and FRIESEN, H., 1932: Die Unmöglichkeit einer Erklärung des Falls der Treppenallele scute vom Standpunkte der Goldschmidtschen physiologischen Theorie der Vererbung. Biol. Zbl. **52:** 147–162.

GERSHENSON, S. M., 1930: Phenomenon of reinversion in the sex-chromosome of *Drosophila melanogaster*. Reports IV. Congr. Zool., Kiev (Russian). P. 7.

1932: Studies on the genetically inert region of the X chromosome of Drosophila. I. J. Genet. **28:** 297–313.

1935: The mechanism of non-disjunction in the ClB stock of *Drosophila melanogaster*. J. Genet. **30:** 115–125.

HUETTNER, A. F., 1924: Maturation and fertilization in *Drosophila melanogaster*. J. Morph. Physiol. **39:** 249–265.

KAUFMANN, B. P., 1935: Cytological evidence of pre-reduction in *Drosophila melanogaster*. Rec. Genetics Soc. Amer. **4:** 71–72.

KIKKAWA, H., 1932: Contributions to the knowledge of non-disjunction of the sex chromosomes in *Drosophila virilis*. I. General Problems. Cytologia **3:** 340–349.

LANCEFIELD, D. E., 1922: Linkage relations of the sex-linked characters in *Drosophila obscura*. Genetics **7:** 335–384.

LANCEFIELD, R. C., and METZ, C. W., 1921: Non-disjunction and the chromosome relationships of *Drosophila willistoni*. Proc. Nat. Acad. Sci. **7:** 225–229.

1922: The sex-linked group of mutant characters in *Drosophila willistoni*. Amer. Nat. **56:** 211–241.

MATHER, K., 1934: The behavior of meiotic chromosomes after X-radiation. Hered. **19:** 303–322.

MCCLINTOCK, B., 1931: Cytological observations of deficiencies involving known genes, translocations and an inversion in *Zea mays*. Mo. Agr. Exper. Sta. Res. Bull. **163,** 30 pp.

1933: The association of non-homologous parts of chromosomes in the midprophase of meiosis in *Zea mays*. Z. Zellf. mikr. Anat. **19:** 191–237.

MORGAN, L. V., 1933: A closed X-chromosome in *Drosophila melanogaster*. Genetics. **18:** 250–283.

MULLER, H. J., 1928: The measurement of gene mutation rate in Drosophila, its high variability, and its dependence upon temperature. Genetics **13:** 279–357.

———, and STONE, W. S., 1931: Analysis of several induced gene rearrangements involving the X chromosome of Drosophila. Anat. Rec. **47:** 393 (Abstract).

———, and PROKOFYEVA, A. A., 1935: The individual gene in relation to the chromomere and the chromosome. Proc. Nat. Acad. Sci. **21:** 16–26.

———, and WEINSTEIN, A., 1933: Evidence against the occurrence of crossing over between sister chromatids. Amer. Nat. **67:** 64–65.

NOUJDIN, N. I., 1935: An investigation of an unstable chromosome in *Drosophila melanogaster* and of the mosaicism caused by it. Zool. J. (Russian). **14:** 317–352.

OFFERMAN, C. A., and MULLER, H. J., 1932: Regional differences in crossing over as a function of the chromosome structure. Proc. Sixth Int. Cong. Genetics **2:** 143–145.

PAINTER, T. S., 1934: The morphology of the X chromosome in salivary glands of

Drosophila melanogaster and a new type of chromosome map for this element. Genetics **19:** 448–469.

PATTERSON, J. T., and STONE, W. S., 1935: Some observations on the scute-8 chromosome of *Drosophila melanogaster*. Genetics **20:** 172–178.

SCHWEITZER, M. D. and KALISS, N., 1935: Does sister-strand crossing over occur in *Drosophila melanogaster?* Genetics **20:** 581–585.

SIDOROV, B. P., 1930: A case of allelomorph scute rising and evoking simultaneously the hairy wing character. Proc. Fourth Cong. Zool. (Russian) 149–150.

———, SOKOLOV, N. N., and TROFIMOV, I. E., 1935: Forces of attraction of homologous loci and chromosome conjugation. Nature **136:** 108–109.

SIVERTZEV-DOBZHANSKY, N. P. and DOBZHANSKY, T., 1933: Deficiency and duplications for the gene bobbed in *Drosophila melanogaster*. Genetics **18:** 173–192.

SMITH, S. 1935: Chromosome fragmentation produced by crossing over in *Trillium erectum*. J. Genet. **30:** 227–232.

STONE, L. H. A., 1933: The effect of X-radiation on the meiotic and mitotic divisions of certain plants. Ann. Bot. **47:** 815–825.

STONE, W., and THOMAS, I., 1935: Crossover and disjunctional properties of X chromosome inversions in *Drosophila melanogaster*. Genetica **17:** 170–184.

STURTEVANT, A. H., 1929: The genetics of *Drosophila simulans*. Pub. Carnegie Instn. **399:** 1–62.

1931: Known and probable inverted sections of the autosomes of *Drosophila melanogaster*. Pub. Carnegie Instn. **421:** 1–27.

1934: Preferential segregation of the fourth chromosomes in *Drosophila melanogaster*. Proc. Nat. Acad. Sci. **20:** 515–518.

Author's Note, 1961

There is an error in Table 15 that was pointed out by Dr. E. Novitski. The second row of "Doubles, 3-sa," should show no X^c gametes and 4 second division ties. The last row, "Total Doubles," should be corrected accordingly, giving the numbers:

1, 1, 2, 2, 2, 8, 16

The discussion in the text is not affected by this correction.

23

AUTOSOMAL LETHALS IN WILD POPULATIONS OF *DROSOPHILA PSEUDOÖBSCURA* †

Introduction

It has been recognized for some years, following the work of Muller, that lethals are especially convenient material for the study of mutation rates. This is because they occur with a frequency that is great enough to be measured, and because their occurrence can be detected by a technique that is independent of the personal equation of the observer. These same two advantages apply to the use of lethals in the study of the constitution of wild populations. Two other advantages are also to be noted in this field. Owing to the extensive studies on the mutation rates of lethals, there is available a large body of data on the frequency of the occurrence of new lethals under a variety of conditions. On the other hand, the rate of elimination of lethals from a population furnishes the minimum possible difficulty of quantitative estimation. It is, therefore, not surprising that there is a rapidly increasing literature in the field—to which the present paper belongs.

Material and Methods

A large series of wild strains of *Drosophila pseudoöbscura* have been studied in this laboratory (see, for example, Dobzhansky, 1935; Dobzhansky and Boche, 1933; Sturtevant and Dobzhansky, 1936). Tests have been carried out on a series of these, to determine the frequency with which autosomal lethals occur. Most of the work has been with the third chromosome, though a few tests on the second will be described below.

Wild males, or a single son from each of a series of wild females (i.e., females already fertilized when trapped), are mated individually to females carrying the recessive gene for orange eyes. From each such mating a single son is mated to females of the multiple stock orange Scute (dominant, bristle reduction) purple (recessive, eye color). All the Sc, not-or, offspring of this

† Reprinted from the Biological Bulletin, Vol. 73, 1937, pp. 542–551.

mating carry one particular third chromosome from the wild specimen; when they are mated together $\left(\frac{+}{\text{or Sc pr}} \times \frac{+}{\text{or Sc pr}}\right)$ the not-Sc offspring are homozygous for one of the two third chromosomes of the original wild male. If this chromosome carries a lethal, the not-Sc class is absent.

The difficulty with the method is that, as just outlined, it neglects the possibility of crossing over. This possibility exists only in the females used in the final test generation, since everywhere else the tested chromosome is kept in males, in which crossing over is absent (or at most negligible in frequency). Even in the females in question, it happens that in the majority of cases crossing over is not a serious source of error. As reported by Sturtevant and Dobzhansky (1936), there is a wide variety of sequences in this chromosome. The or Sc pr stock has the Standard sequence; and females heterozygous for Standard and any other sequence give few crossovers. The presence of the three mutant genes in the experiment makes it possible to distinguish between Standard, on the one hand, and all the remaining sequences, on the other hand, in the tested third chromosome. In case the tested chromosome does not have the Standard sequence (and the majority of them do not have it) the presence of a lethal is at once apparent from the absence or extreme rarity of wild-type flies in the test generation. Most of the sequences do give occasional crossovers in one region or another when tested against Standard; for this reason special tests are required to distinguish between semi-lethals and lethals that cross over with Sc. These have not, in general, been carried out in the present experiments; but it is clear that very few semi-lethals are present, since in most cases the wild-type class is either wholly absent or approximately half as numerous as the heterozygous Scute class. This results from the fact that few of the lethals happened to lie in the sections that give crossovers with Scute.

If the test chromosome has the Standard sequence, lethals can still be detected, though it is necessary to make more extensive counts to be certain no lethal is present. This is due to the fact that Sc lies near the middle of the chromosome (about 30 units from the left end, 40 from the right), so that any lethal shows appreciable linkage with it. Here, however, the distinction between lethals remote from Sc, and low-viability genes close to it, requires special tests. The test, applied in most cases, has been to use a chromosome carrying the dominant Emarginate (eye shape) and the Arrowhead sequence. From the mating $\frac{\text{Em}}{+} \times \frac{\text{Em}}{+}$, results comparable to the Sc test outlined above are obtained.

In the case of the second chromosome, the test procedure is similar. Wild males or single sons of wild females are mated to glass (recessive, eye structure) females, and a single male from each such mating is mated to Bare (dominant, bristle form) glass females. The resulting Bare offspring

$\left(\frac{\text{Ba gl}}{+}\right)$ are mated together, and counts are made of their offspring. This test is less efficient than that used for the third chromosome, since there is more crossing over. No inversions are present in the material tested, and Bare is about 60 units from the left end of the chromosome, 40 from the right. Lethals even at this distance do cause a disturbance of the ratios; and all suspected lethals were verified by making a series of matings of Ba × Ba from sibs. In every case these tests agreed in showing a lethal to be present. However, there is no test for distinguishing lethals and semi-lethals; and the technique is more laborious than that used for the third chromosome. For these reasons relatively few tests were made on the second chromosome.

The mutant strains available did not permit reasonably efficient tests for lethals in chromosomes IV or V, and none was attempted. In the case of the X chromosome no detailed tests were made, but observations on the sex ratio from wild females and their daughters indicated that no lethals were present in any of the wild strains studied.

It should be pointed out that what is being studied here is lethal-bearing chromosomes, rather than lethals. That is to say, if a chromosome carries more than one lethal the technique used does not make the fact evident. With frequencies of lethals as high as those found it seems likely that such duplications of lethals in a single chromosome do in fact occur. The net result will be that the frequencies recorded are too low. Further complications arise in connection with the tests for identity of lethals, but since few cases of identity were found this does not seem a serious source of error.

The results obtained are shown in summary form in Table I.

Table I

Frequency of lethals

	Chromosomes tested	Lethals	
		Number	Percentage
Chromosome II			
1936	21	4	19 0
Old stocks	16	3	18.8
Chromosome III			
1935	67	13	19.4
1936	120	23	19.2
Old stocks	29	7	24.2

The material studied in 1935 came largely from the Rocky Mountains and from the Mexican plateau; that in 1936 largely from southern California. The "old stocks" were, many of them, from the Pacific Northwest. These latter had been kept in the laboratory for at least a year before the lethal tests were begun. The striking thing about the table is that, in spite of these differences, there was little variation in the frequencies found for a given chromosome or even as between the second and third chromosomes. Table II, giving the data for the most-studied individual localities (included

Table II

	Chromosomes tested	Lethals	
		Number	Percentage
Chromosome III			
San Gabriel Canyon, Azusa, Calif....	27	8	29.6
Julian, Calif........................	21	3	14.3
San Gorgonio Mt., Calif.............	17	3	17.6
Santa Cruz Isd., Calif...............	16	3	18.8

in Table I) shows a similar relative constancy of the frequency of lethals. All these results are from Race A; 5 Race B third chromosomes (all from the Olympic Peninsula of Washington) were tested, and one lethal was found, giving a frequency of 20 per cent.

As reported by Sturtevant and Dobzhansky (1936), the third chromosomes of this species exist in a wide variety of sequences. This phenomenon does not appear to be related to the occurrence of lethals. Lethals have been found in the Standard, Arrowhead, Pikes Peak, Santa Cruz, Chiricahua, Cuernavaca, and Klamath sequences. Their frequency seems to be essentially the same in regions such as southern California or southern Mexico, where several different sequences exist in fairly large numbers within each population, and in the area near the common corner of Utah, Colorado, Arizona, and New Mexico, where only the Arrowhead sequence is found. Fifteen third chromosomes tested from this pure Arrowhead area gave 3 lethals—exactly 20 per cent. The data on Chromosome II also indicate that diversity of sequences is not important, since all the strains from the regions concerned in these tests had the Standard Race A sequence in that chromosome.

Tests for Allelomorphism

If two lethals, different in origin, are kept in stocks of the form $\frac{\text{lethal}}{\text{or Sc pr}}$, it is easy to test their identity by crossing the two strains. If two distinct lethals are concerned, wild-type offspring will appear in numbers approximately equal to half the size of the heterozygous Scute class; if the two lethals are identical, no wild-type offspring will be produced except as a result of crossing over in the female used, and this will be infrequent if the lethal-bearing chromosome does not have the Standard sequence. Numerous tests of this kind have been carried out.

Fifteen of the 1935 and "old stock" lethals were tested in all possible combinations; the result showed 14 lethals to be present. One from Metaline Falls, Washington (1934) was identical with one from Florence, Texas (1935). Thirteen of the 1936 lethals were tested against each other, and were found to include 12 different ones. One of three tested from Julian was found to be identical with one of eight from San Gabriel Canyon. The Metaline 1934 lethal was tested against the 12 different 1936 lethals, and was found to be allelomorphic to one from Barton Flats, on the slopes of Mt. San Gorgonio, Calif. The San Gabriel-Julian duplicate was different from all the seven surviving members of the 1935 series with which it was tested. These tests, with a few miscellaneous others, total 225 crosses between separately found lethals, with 3 cases of allelomorphism. That is, an average of about $\frac{3}{225} = 1.3$ per cent of the lethals found may be expected to be allelomorphic to any given one selected at random. Or, since the total frequency of lethals in this chromosome in wild stocks is just under 20 per cent, the average frequency of a given lethal may be estimated at about one-fourth of 1 per cent.

It may be observed that no cases of recurrence have been found within a locality. That is to say, the same lethal has not been recovered from any two specimens collected in the same place. As shown above, one lethal was found both at Julian and at San Gabriel Canyon, in southern California, in 1936. While the two localities are only about 100 miles apart, the populations of their respective mountain ranges (San Jacinto and Sierra Madre) appear to be distinct in that the former contains the Santa Cruz sequence while in the latter Tree Line occurs (along with Standard, Arrowhead, and Chiricahua, common to both regions—see Dobzhansky and Sturtevant MS in press—*Genetics*).[1] The other recurrent lethal was found in stocks from Metaline Falls, Washington (1934), Florence, Texas (1935), and Mt. San Gorgonio, Calif. (1936). It seems clear in the latter case, and probable in the former, that we are dealing with recurrent mutations rather than with the persistence of a lethal gene in heterozygous form.

Rate of Origin of New Lethals

A preliminary lethal-accumulation experiment has been carried out. The technique used is not essentially new, and the results are not yet extensive. It seems sufficient at present, therefore, to record that 2 new

[1] *Genetics* **23:** 28–64.

lethals occurred in a total of 120 chromosome-generations, giving a frequency of 1.7 per 100—with a large probable error.

Properties of Non-Lethal Chromosomes

The frequency of lethals found in the tested wild stocks is unexpectedly high. A possible explanation of this result seemed to be that the species is in a condition approximating that of balanced lethals. That is to say, that most or all its third chromosomes are so constituted that flies homozygous for any one of them cannot compete with heterozygotes. Such an assumption amounts to supposing that the phenomenon of heterosis is well developed within a single pair of chromosomes. Under these conditions it might result that the rate of elimination of a lethal would be greatly decreased, since lethals are eliminated only when homozygous, and the assumption is that homozygotes play little part in the perpetuation of the species, so that their properties are unimportant.

The simplest way to test this hypothesis is to study the properties of individuals homozygous for a series of non-lethal chromosomes, from the same wild populations as those that contained lethals, in comparison with individuals heterozygous for two such chromosomes. There is another question that can conveniently be investigated at the same time—are the lethals completely recessive? This latter question is of importance in any calculation of the probable rate of elimination of lethals from wild populations.

Only one characteristic of the flies concerned has been investigated—namely their viability. It is clear that fertility, length of life, resistance to unfavorable conditions, and other properties, would all be of importance in determining the frequencies found in natural populations; but the labor involved in studying a series of combinations for such a variety of characteristics has seemed prohibitive. It also seems probable, *a priori,* that there will be, in general, a rough correlation between all these properties, such that data on one of them will give a satisfactory picture of the general situation.

The technique used was the same as that used in the detection and testing for identity of lethals—in fact many of the cultures served both purposes. For the study of lethals partial counts were adequate, and were used in many cases. However, if complete counts are made on a culture there results a measure of the viability of the wild-type chromosome concerned. From the mating together of $\frac{+}{\text{or Sc pr}}$, in the absence of crossing over, three types are expected in the ratio 1 : 2 : 1. The homozygous or Sc pr is usually present in rather small numbers, and has been neglected in making comparisons, since the numbers seem to be very much influenced by minor environmental differences and are often so low as to be inconveniently sensitive to the error of random sampling. The index of viability actually used is the number of wild-type ÷ the number of heterozygous Scute × 100—i.e., the percentage that the wild type is of the Scute. Evidently, with equal viability of all classes the index will be 50.

Several objections may be raised to this method. The most serious is that each of the classes with which we are concerned is compared, not with a common standard type, but with a $\frac{+}{\text{or Sc pr}}$ class that varies in constitution from one test to another. In the case of a test of a homozygote the comparison concerns a heterozygote of that same chromosome; in the test of a heterozygote the $\frac{+}{\text{or Sc pr}}$ class is made up of heterozygotes for each of the two + chromosomes, in approximately equal numbers. In a few cases tests were repeated, and the result shows a definite, though far from complete, correlation between successive tests of a given combination. The use of this technique can, therefore, not be considered as giving more than an indication of the situation, and accordingly it does not seem desirable to present the results obtained in detail.

In general, the homozygotes for non-lethal third chromosomes from wild stocks are viable and fertile in both sexes, with the result that homozygous stocks are easily established. The test outlined above indicates that some homozygotes are definitely below par, others apparently at no disadvantage. The average viability index of 21 tested chromosomes was 45.91, the minimum 19.4 (weighted average of three tests—16.3, 21.6, 33.3).

The tests of heterozygotes for two third chromosomes from wild strains were made chiefly with lethal-bearing chromosomes. Twenty-five different combinations of lethal × lethal gave an average index of 53.0—i.e., more viable than the separate lethals heterozygous for or Sc pr. The lowest index among the 25 was 38.3.

The most probable conclusions are that

(1) lethals are wholly recessive;

(2) homozygotes for non-lethal chromosomes are slightly less viable, on the average, than heterozygotes;

(3) the viability of homozygotes is quite variable, some being definitely low, others apparently as high as the heterozygotes.

Comparison with Results from D. melanogaster

The frequency of lethals in wild populations of *D. melanogaster* has been studied by several authors. The most extensive work is that of Dubinin and co-workers (1934, 1936) on material from the Caucasus. These studies were carried out on a large scale. Altogether 4819 second chromosomes are recorded, with 470 (= 9.8 per cent) lethals. The frequencies range from 0 (in 92 tested, from Delizhan) to 16.1 per cent (161 tested, from Ordzhonikidze). From one locality, Gelendzhik, tests were made in three successive years, giving 7.98 per cent (877 tested), 12.86 per cent (616 tested), and 8.78 per cent (797 tested). Tests were also carried out for identity, but only within regions. The average value here was 2.2 per cent—i.e., an average of 2.2 per cent of the lethals found in one year at one locality may be expected to be allelomorphic to any one lethal found in that year at that locality. The average frequency for a given single lethal is 0.22 per cent (.022 × .098). There are no tests for identity of lethals from different localities, but those found at Gelendzhik in successive years were so tested, with the result that "among the 33 lethals of 1933 and the 55 lethals of 1934, nine lethals were common to both years." The conclusion is drawn that this represents survival of these lethals. The results reported in the present paper throw some doubt on this conclusion, for a comparable frequency of identity was found in *pseudoöbscura* for all years and all localities. One may surmise that, in *melanogaster* also, any two series of lethals would show a fairly large proportion of common members—i.e., that recurrent mutations are relatively frequent.

The absolute frequency of lethals found in the Caucasian *melanogaster* populations is only about half that in the American *pseudoöbscura,* though the chromosome concerned (II) is roughly twice as long as the III of *pseudoöbscura* and includes the same material plus that of the *pseudoöbscura* IV (Donald, 1936; Sturtevant and Tan, 1937). The preliminary experiment described above suggests a higher lethal mutation rate in *pseudoöbscura;* but until more data are available on this point it is scarcely possible to decide on the relative rates of elimination of lethals in the two species.

One difficulty in evaluating the data of Dubinin and his co-workers may be pointed out. The results are recorded by localities and years; but we are nowhere given more specific data. How large an area is included in a single "locality"? Over how long a period were specimens taken in one year? What kinds of places were collected—woods, grocery stores, garbage dumps, fruit orchards? These questions are of importance in judging the size of the populations sampled and the probable degree of relationship of the tested individuals.

Summary

1. Approximately 20 per cent of the third chromosomes found in wild populations carry lethals.

2. Less extensive data indicate a similar frequency for the second chromosome.

3. On the average, about 1.3 per cent of the lethals found may be expected to be allelomorphic to any given one.

4. The average frequency for any one lethal is about one-fourth of one per cent of the chromosomes of wild strains.

5. The lethals, so far as studied, are completely recessive.

6. Study of flies homozygous for non-lethal third chromosomes shows a considerable variation in their viability. Some are at a definite disadvantage, others apparently not. On the average they are not quite as viable as flies carrying two different third chromosomes.

LITERATURE CITED

DOBZHANSKY, TH., 1935: The Y chromosome of Drosophila pseudoöbscura. *Genetics,* **20:** 366.

DOBZHANSKY, TH., and R. D. BOCHE, 1933: Intersterile races of Drosophila pseudoöbscura Frol. *Biol. Zentralbl.,* **53:** 314.

DONALD, H. P., 1936: On the genetical constitution of Drosophila pseudoöbscura, Race A. *Jour. Genet.,* **33:** 103.

DUBININ, N. T., et al., 1934: Experimental study of the ecogenotypes of Drosophila melanogaster. *Biol. Zhurn.,* **3:** 166.

DUBININ, N. T., et al., 1936: Genetic constitution and gene-dynamics of wild populations of Drosophila melanogaster. *Biol. Zhurn.,* **5:** 939.

STURTEVANT, A. H., and TH. DOBZHANSKY, 1936: Inversions in the third chromosome of wild races of Drosophila pseudoöbscura, and their use in the study of the history of the species. *Proc. Nat. Acad. Sci.,* **22:** 448.

STURTEVANT, A. H., and C. C. TAN, 1937: The comparative genetics of Drosophila pseudoöbscura and D. melanogaster. *Jour. Genet.,* **34:** 415.

24

ESSAYS ON EVOLUTION

I. ON THE EFFECTS OF SELECTION ON MUTATION RATE †

New recessive lethal mutations arise constantly without any artificial treatment. The results of Muller (1928) and others summarized by Schultz (1936) indicate, for *Drosophila melanogaster,* a frequency of about 0.18 per 100 gametes in the X chromosome, and of about 0.5 per 100 gametes in the second chromosome. It is probable that the third chromosome gives about the same frequency as the second; we may conclude that the normal frequency, for all chromosomes taken together, is about 1 per 100 gametes. Timoféeff-Ressovsky (1934) shows that there is also a high frequency of new mutant genes that cause an appreciable reduction in viability. In the X-rayed X chromosomes that he studied, such viability mutants were from 2 to 2½ times as frequent as were lethals. There is also a class of mutations that are lethal or reduce viability even when heterozygous. It is difficult to judge the frequency of the dominant lethals, but there is no doubt that they are present. Perhaps the largest class of dominant genes that lower viability is that of the minute-bristled types, known to be rather frequent. Finally, there is another type of mutation, as yet little studied, that brings about sterility (Berg, 1937).

In general, therefore, one may conclude that each normal, untreated individual of *D. melanogaster* is subject to a lowering of its potential reproductive value through the occurrence of new mutations in its germ tract. Quantitative estimates of this reduction must be approximate, since the recessive autosomal lethals and viability and sterility mutants will have an unpredictable effect. This arises from the fact that they operate only when homozygous, and the frequency with which they become homozygous will depend on the degree of inbreeding—a factor not yet evaluated for wild populations of *Drosophila* or any comparable organism. It is also to be observed that these recessives will operate only after recombination has reduced the correlation of their distribution in the population with that of the other genes present in the individual in which the original mutation occurred.

It seems probable, however, that the loss in potential reproductive value

† Reprinted from The Quarterly Review of Biology, Vol. 12, 1937, pp. 464–467.

for the average individual is not far from one per cent, taking into account not only the immediate offspring but also later generations.

A body of data has been gradually accumulating that shows that different strains may deviate significantly from the usual values for mutation rate. This was indicated by Muller (1928), and has more recently been shown by Dubrolskij (1935) for material treated with X-rays and with heat, and by Demerec (1937) for untreated material. Suggestions of similar differences may be found in still other papers (see Schultz, 1936). It appears that, for the X chromosome at least, some strains give lethal mutation rates as much as five or six times as high as those shown by other strains. Demerec reports that in the strains studied by him the chief difference was due to something carried by the second chromosome, though the new lethals studied were themselves in the X.

There are, then, genes that affect general mutation rate, and stocks differ in their constitution with respect to such genes. It follows that, in wild populations, such genes must be subject to selection. If a population has the mutation rate arrived at above, which leads to an average of 1 per cent decrease in the potential reproductive value of each individual, then a gene causing a 10 per cent decrease in the general mutation rate will have a selective advantage of 1/1000. The data suggest that decreases even greater than 10 per cent may be found; but a selective advantage of 1/1000 is sufficient to be effective (Haldane, 1924).

Any external agents that increase the general mutation rate will evidently increase the average loss in potential reproductive value, and will therefore increase the intensity of the selection for genes that lower the rate. Such selection will always be present, but it must be more intense in regions with high natural ionization, and also in regions with high temperatures. (For a critical review of the effects of temperature on mutation rate see Timoféeff-Ressovsky (1934a).) It may, therefore, be expected that, as a general rule, strains found in regions of high ionization and high temperature will have more uniformly low mutation rates (under standard conditions) than will strains found in regions of low ionization and low temperature. These relations will probably be difficult to verify experimentally; both because the sign of the selective process is the same in both situations, and because ionization intensity commonly shows such great local differences that it will be difficult to judge the intensity to which a population has been subjected.

It is probable that different strains will have accumulated different genes lowering the mutation rate, and that most strains (in cross-fertilizing organisms) will be somewhat heterogeneous for genes affecting that rate. If this be so, then any interference with the operation of the usual selection for low rate will have the effect of increasing the rate. The situation may be compared with that which obtains with respect to genes affecting the general viability of the organism. Under natural conditions selection keeps the members of a population up to par, but on relaxation of selection a slow deterioration sets in. Such deterioration is, as is well known, greatly speeded up by inbreeding. If the analogy here is valid, it may be concluded that any artificial conditions, maintained over a number of generations, are likely to raise the general mutation rate of a strain. Are all mutation rates so far determined experimentally too high to be applicable to natural populations?

The method of action of genes affecting the general mutation rate is still unknown. Probably they act in various ways; but it remains possible that they have something in common, such for example as an influence on the relative duration in time of specific stages of mitosis, or some other general property. In this case it might well be that such effects would have greater selective value through other results than through their effects on mutation rate. While this possibility must be kept open until more is known about the way in which the genes in question affect mutation rate, it seems unlikely that it will invalidate the conclusions derived in this paper.

It seems at first glance that there should be a counter-selection, due to the occurrence of favorable mutations. It is true that favorable mutations furnish the only basis for improvement of the race, and must be credited with being the only raw material for evolution. It would evidently be fatal for a species, in the long run, if its mutation rate fell to zero, for adjustment to changing conditions would then not long remain possible. While this effect may occur, it is difficult to imagine its operation. It is clear that the vast majority of mutations are unfavorable, and Fisher (1930) has shown that the rare favorable ones must, in general, be genes with slight effects. In other words, for every favorable mutation, the preservation of which will tend to increase the number of genes in the population that raise the mutation rate, there are hundreds of unfavorable mutations that will tend to lower it. Further, the unfavorable mutations are mostly highly unfavorable, and will be more effective in influencing the rate than will the relatively slight improvements that can be attributed to the rare favorable mutations.

This raises the question—why does the mutation rate not become reduced to zero? No answer seems possible at present, other than the surmise that the nature of genes does not permit such a reduction. In short, mutations are accidents, and accidents will happen.

It may be noted that differences in general mutation rate may be characteristic of whole species. It has been the experience of all who have studied *Drosophila funebris* that it gives many fewer mutations than does *D. melanogaster* (see, for example, Timoféeff-Ressovsky, 1936). It has been suggested that this difference is due to failure of the same mutant gene to produce as distinct and easily seen a character in *funebris* as it does in *melanogaster*. My own experience leads me to disagree with this conclusion, since I found four types in *funebris* similar to known ones of *melanogaster,* and these four were at least as extreme and easily classifiable as the corresponding *melanogaster* types. How it has happened that *funebris* has come to have a lower mutation rate than *melanogaster* is not clear; but further analysis of this case may throw light on some of the questions raised in this paper. Possibly one might obtain more mutations by first crossing widely different races and then subjecting the hybrid lines to close inbreeding.

There are several characteristics of the organism that will influence the effectiveness of the selective process discussed in this paper. As in other types of selection, change will be more rapid in cases where selection can operate on haploid individuals, since here the large class of recessive mutations will be selected at once instead of only in the homozygous individuals. For this reason, the action of the suggested mechanism should be, in general, most marked in plants (where the gametophyte generation is haploid), and in animals, such as the Hymenoptera, with haploid males. In many animals and some plants there is one chromosome that is effectively (or actually) haploid in one sex. The number of genes that are carried by such a chromosome should show a positive correlation with the effectiveness of selection for low mutation rate.

Another general property that should influence this type of selection is the degree of inbreeding, since this will influence the rapidity with which unfavorable recessives are eliminated; and the sooner they are eliminated the greater will be the correlation between their elimination and the presence of the genes that stimulated the original mutation.

In polyploid organisms the elimination of unfavorable recessives will be prevented or greatly hindered, and in polyploid plants the gametophytic elimination of unfavorable types will be greatly reduced; this condition should therefore raise the mutation rate. However, polyploidy also makes the detection of mutations more difficult, so experimental determination of this relation can hardly be expected.

One may surmise, then, that minimum mutation rates may be expected in diploid self-fertilizing plants (excluding balanced lethal types, such as *Oenothera,* which constitute a special case), and in lower plants, where the haploid generation is well developed. Maximum rates (per individual, not per unit of time) may be expected in cross-fertilizing animals that have no sex-chromosome mechanism or in which the sex chromosomes are relatively small. By way of examples, the rate should be low in Bryophytes and in legumes, high in Lepidoptera and in mammals—to cite groups that furnish familiar objects for genetic study.

BIBLIOGRAPHY

BERG, R. L., 1937: The relative frequency of mutation in different chromosomes of *Drosophila melanogaster*. II. Sterility mutations. *Genetics* **22:** 241–248.

DEMEREC, M., 1937: A mutability stimulating factor in the Florida stock of *Drosophila melanogaster. Genetics* **22:** 190.

DUBROLSKIJ, N. V., 1935: On the question of the comparative mutability of stocks of *Drosophila melanogaster* of different origins. *C. R. Acad. Sci. U.S.S.R.* **4:** 95–97.

FISHER, R. A., 1930: The Genetical Theory of Natural Selection. 272 pp. *Oxford.*

HALDANE, J. B. S., 1924: A mathematical theory of natural and artificial selection. *Trans. Cambridge Phil. Soc.* 23, **2:** 19–41.

MULLER, H. J., 1928: The measurement of gene mutation rate in *Drosophila,* its high variability, and its dependence upon temperature. *Genetics* **13:** 279–357.

SCHULTZ, J., 1936: Radiation and the study of mutation in animals. In—Biological Effects of Radiation. *New York,* pp. 1209–1261.

TIMOFÉEFF-RESSOVSKY, N. W., 1934: Auslösung von Vitalitätsmutationen durch Röntgenbestrahlung bei *Drosophila melanogaster. Strahlentherapie* **51:** 658–663.

———, 1934a: The experimental production of mutations. *Biol. Reviews* **9:** 411–457.

———, 1936: Qualitätiver Vergleich der Mutabilität. von *Drosophila funebris* und *D. melanogaster. Zeits. ind. Abst. Vererb.* **71:** 276–280.

Author's Note, 1961

The postulated effect of ionizing radiation probably does not occur, since the evidence indicates that its effect on mutation frequency is additive, and is probably independent of the spontaneous frequency. It is not altogether clear whether or not this also applies to the effects of temperature.

25

ESSAYS ON EVOLUTION

II. ON THE EFFECTS OF SELECTION ON SOCIAL INSECTS †

There are two conflicting tendencies in recent developments of the mathematical theory of the behavior of wild populations. On the one hand there is the view that selection is ever-present and is the only effective agent in altering the constitution of populations. On this view selection operates only on individuals, and the members of a group are not thought of as occurring in semi-independent subgroups (Fisher, 1930). On the other hand there is the point of view expressed by Wright (1932), according to which the most favorable conditions for evolution are those in which the members of a large population do not interbreed at random, but exist in more or less independent subgroups, which undergo exchanges of genes at times but are also effectively in competition with each other. On this latter view there is an opportunity, lacking on the first view, for selection to bring about the establishment of characteristics favorable to the group but unfavorable to the individual.

The social insects, such as termites, ants, and many bees and wasps, furnish examples in which such characters unfavorable to the individual have in fact become established. In terms of natural selection, a "favorable" character is of course to be taken as meaning a character that leads to the production of more descendants. The sterile castes of the insects named have, therefore, developed a character that is unfavorable by definition. It is sometimes argued that in this special case the rule still holds, since here the colony, rather than the individual insect, is the unit in terms of natural selection, and it may be supposed that the colony produces more offspring as a result of the division of labor associated with the sterile castes. It is clear, however, that all the social insects have arisen from solitary forms in which the sterile caste was absent (Wheeler, 1928). It follows that evolution must have resulted in an increase of sterile individuals. At some point in the history of the race there must have been a change from the individual to the colony basis of selection. Unless this change be supposed to have been a sharp one and to have been associated from the first with the necessary genetic adjustments, there must

† Reprinted from The Quarterly Review of Biology, Vol. 13, 1938, pp. 74–76.

have been an intermediate stage in which some element other than the strict operation of reproductive selection was effective.

In fact, such intermediate stages still exist, and are familiar to all students of the social insects. In order that the colony be the unit in terms of natural selection, and that the existence of the sterile caste offer no difficulties to the advocates of pure selection, it is necessary that each colony have a single fertile queen and that the sterile individuals all be closely related to the queen (presumably her offspring). This is the standard account for the honey-bee, though even here it is clear that workers may at times produce offspring; in other social forms wide deviations from this condition are frequent.

In what follows I shall confine myself to the ants, though evidently somewhat similar relations occur in certain of the termites at least. For detailed summaries of the life-histories and for bibliographies the reader is referred to Wheeler (1910, 1928) and to Donisthorpe (1915). Fertile workers do not appear to be rare among the ants; usually they produce only male offspring, but there are some satisfactory records of female offspring—the latter presumably due to failure of chromosome reduction in parthenogenetic eggs, since it seems clear that such fertile workers do not mate. The production of males is, however, enough to put these workers into reproductive competition with the queens; and therefore to upset the view that the colony is the unit of selection, to the exclusion of individual selection within the colony. Even more to the point is that, in many ants, more than a single fertile queen is present in a given colony. In such dominant genera as *Myrmica, Cremato-gaster,* and *Tapinoma,* and in the *rufa* and *sanguinea* sections of the genus *Formica,* flourishing nests regularly contain many fertile queens. In some (probably in all) of these cases the colony is initiated by a single queen, but newly fertilized queens are adopted by established colonies after the mating flight occurs. In the case of *Formica rufa* at least, there is evidence that these adopted queens need not have been produced in the nest that adopts them—they may even belong to a distinct variety. Here and in *F. exsectoides* it seems probable that the colony long outlives the queen that originally established it, its fertility being dependent upon a whole series of queens that have no necessary genetic relationship to each other or to the founder.

Under these conditions there is evidently an opportunity for individual selection. A queen that produces unusually efficient worker offspring will have to share the benefits of their activities—i.e. her fertile offspring will have no advantage over those of the other queens living in the same nest. A queen that produces a relatively high proportion of fertile offspring will leave more descendants—at the expense of the nest economy, and in the long run to the detriment of the species. This possibility seems to constitute a serious danger to such ants, and may therefore be examined more closely.

There are ants in which the worker caste has disappeared entirely (*Anergates, Wheeleriella,* etc.), so that mutations of this sort are evidently possible. If such a one occurs, then a queen of the new type that is adopted by a normal colony will thrive at the expense of the worker offspring of her foster-sisters, but will produce more than her share of the sexual forms arising in the colony. Such a gene may be expected to spread through the population rapidly at first, but must soon be checked by the occurrence of too high a proportion of sexual forms to be supported by the reduced number of workers that will be available. It seems clear that the net result must be a decrease in the efficiency of the colonies and therefore of the total population of the species in the affected area. This will, evidently, leave such a region likely to be invaded by members of the species coming from places where such a process has not happened to occur.

Selection must, then, be thought of as operating, in these forms, on at least three different levels: on the individuals, on the colonies, and on the populations within an area. The interrelations between these three are obviously complex. Similar distinctions apply in the case of another social animal—namely man. Here, of course, the situation is even more complex, and I do not have the intimate knowledge of the data necessary for a profitable discussion.

BIBLIOGRAPHY

DONISTHORPE, H. ST. J. K., 1915: British Ants, Their Life History and Classification. 379 pp. *London.*

FISHER, R. A., 1930: The Genetical Theory of Natural Selection. 272 pp. *Oxford.*

WHEELER, W. M., 1910: Ants, Their Structure, Development, and Behavior. 663 pp. *New York.*

———, 1928: The Social Insects, Their Origin and Evolution. *New York.*

WRIGHT, S., 1932: The rôles of mutation, inbreeding, crossbreeding, and selection in evolution. *Proc. Sixth Internat. Congr. Genet.* **1:** 356–366.

26

ESSAYS ON EVOLUTION

III. ON THE ORIGIN OF INTERSPECIFIC STERILITY †

The problem of the origin of interspecific sterility has long presented the greatest difficulty in the way of constructing a complete theory of the evolution of species. This difficulty was felt by Darwin, and has, if anything, become more striking with the development of modern genetical theory. One solution, has, it is true, been developed. It now seems clear that, in certain cases, allopolyploidy may be effective in producing a new fertile form that is cross-sterile with its parents. This solution is, however, certainly a special one, applicable only to hermaphroditic organisms. There is no recognized general scheme, applicable to separate-sexed forms, that can be pictured in detail. It is the purpose of the present note to suggest such a general scheme.

It was pointed out by Fisher (1930) that, if a species is supposed to be broken up into two more or less isolated groups, and the hybrids between these two groups are relatively infertile, then any genes tending to decrease the cross-mating between the two groups will be selected. This is because cross-mating, since it leads to the production of less fertile offspring, will lead to a decrease in the potential ultimate reproductive value of the cross-mated individuals, as compared to those that mate within their own group. In other words, intra-group matings will, if we consider generations later than the first one, produce more descendants than will inter-group matings—and selection will therefore decrease the proportion of the latter.

On this basis, sterility of the F_1 comes first, and leads to an accumulation of genes that prevent cross-mating. The alternative assumption, more often made, is that cross-mating is prevented in some manner, and sterility of the F_1 follows as a more or less incidental consequence. This latter view has always remained rather vague, and presents serious logical difficulties when one attempts to make it more precise. There seems to be no good reason why the hybrid should become sterile, yet sterility of the hybrid is one of the most wide-spread characteristics of distinct species.

There is also a logical difficulty about the view that sterility of the F_1 is

† Reprinted from The Quarterly Review of Biology, Vol. 13, 1938, pp. 333–335.

primary. For, in this case, there must have been a fertile F_1 present in the beginning. If A and B give a sterile hybrid, how can B be derived from A, unless self-fertilization be assumed? There are two ways of escaping this difficulty: one may suppose the hybrid not to be completely sterile (in which case, as the sterility is decreased, it becomes easier for the condition to become established, but its effectiveness in leading to failure of cross-mating decreases), or one may assume that the sterility arose by two steps. In the latter case, one may assume two pairs of complementary genes, S,s and T,t, such that Ss Tt is sterile, but SS tt, SS Tt, ss TT, ss Tt, and ss tt are fertile. The derivation of the two cross-sterile homozygous types SS tt and ss TT then involves the production, at some stage, of a type (ss tt) that is fertile with both (see elaboration by Dobzhansky, 1937).

Fisher's analysis is based on the assumption that the two diverging groups have become adapted, through selection, to different conditions (usually geographical), and that the hybrids between them are less well adapted to either set of conditions.

There are, however, certain other systems which will lead to the same type of situation, without any difference in the selective agencies acting on the two populations—without any initial differences between them in gene-frequencies. Perhaps the simplest example is that of reciprocal translocations. Here the heterozygote has a reduced fertility, owing to the more or less frequent occurrence of "irregular" segregation. The smaller the pieces of chromosome exchanged, the greater is the frequency of irregular segregation—but also the greater is the chance that some of the products of such irregular segregation may have some degree of viability and fertility (since they will carry duplications or deficiencies for fewer genes). It is presumably for this reason that small, rather than large, translocations appear to have become established in the phylogeny of *Drosophila* species (Sturtevant and Tan, 1937). This system suffers from the disadvantage referred to above—the greater the sterility of the F_1, the greater the difficulty of establishing the translocation in the first instance. It has been shown by Dobzhansky and Tan (1936) that several small translocations have in fact occurred in the course of the differentiation between *Drosophila pseudoobscura* and *D. miranda;* accordingly this method must be considered as a possible one, though the inversion system now to be discussed seems more likely to be of general occurrence.

Another type of chromosome aberration that should have similar effects is furnished by inversions. If two chromosomes differ by a single inversion within an arm, there exists a mechanism such that there is only a negligible effect of crossing over on the fertility of the heterozygote (Sturtevant and Beadle, 1936). If, however, an inversion is followed by a second one not greatly different from it, the situation is otherwise. For example: suppose a chromosome ABCDEFGH is inverted to form AGFEDCBH, and the latter in turn gives rise to AGCDEFBH—it being assumed that the spindle-attachment is at A or H. A heterozygote carrying the first and the third of the above chromosomes may give crossing over in the CDEF region, and the chromatids resulting from such crossing over (one carrying two B's, no G, the other carrying two G's, no B) will pass to the functional egg cells at random (Gershenson, 1932; Sturtevant and Beadle, 1936). The result is that such a heterozygote has a definitely decreased fertility. In the limiting case the decrease will amount to 50 per cent.

Such a system is not imaginary. The properties outlined have been experimentally determined; and the existence of inversions in wild populations is now known to be frequent and widespread. It may be concluded that they have in fact often led to the production of relatively infertile combinations. Given these facts, there is present all that is needed for the initiation of the selective process outlined by Fisher. It is possible to postulate such an origin for the two races (A and B) of *Drosophila pseudoobscura* that give sterile F_1 hybrids. These two races do in fact show some degree of sexual isolation; and the most frequent sequence in the third chromosomes of race A (Arrowhead) is so related to the most frequent one of race B (Klamath) that crossovers of the type described should be relatively frequent (Dobzhansky and Sturtevant, 1938).

This case illustrates a difficulty encountered in such an analysis as is here attempted. The male hybrids between the two races are completely sterile, yet the postulated mechanism does not permit complete sterility at the initial stage. In fact, it is clear that in organisms that cannot self-fertilize, this must usually be so, as pointed out above. There thus arises a difficulty which was recognized by Darwin: the most that can be supposed in the early stages of differentiation is that the hybrid is relatively infertile. This infertility must be supposed to increase up to complete sterility—yet at first sight it seems that a decrease in fertility is precisely what selection cannot accomplish.

I have recently discussed (*Essay II* of this series) a case in which sterile individuals must be supposed to have increased in frequency because of selection. The general method invoked in that case may be so modified as to apply here also—namely, we may suppose that the area between the two incipient forms is occupied by a series of semi-independent small sub-populations, only occasionally interbreeding and ultimately competing with each other.

Within a mixed population, the hybrids themselves, when they do breed, will produce up to half their offspring like themselves. Therefore such fertility as they possess will be a source of new semi-sterile individuals. Selection will operate to increase the fertility of this class; but within a small sub-population, the chance establishment of any genes that decrease still further the fertility of the hybrids will increase the efficiency of the sub-population as a whole, and will give it an advantage over the competing sub-populations. Under these conditions, then, it may be supposed that the fertility of the hybrid may be decreased, while the frequency of crossing of the two races is still decreasing.

It is thus possible to formulate a scheme for the origin of interspecific sterility, starting with Fisher's demonstration that races giving a relatively infertile hybrid will tend to accumulate genes that tend to prevent crossing. Appropriate use of Wright's (1932) idea of competing sub-populations makes it possible to see how the infertility can be increased by selection. Inversions and small translocations furnish ever-present materials for initiating or for increasing such infertility of the hybrids—though it is clear that gene mutations must also be concerned, since not all interspecific sterility can be attributed to chromosome rearrangements (see Dobzhansky, 1937).

BIBLIOGRAPHY

DOBZHANSKY, T., 1937: Genetics and the Origin of Species. 364 pp. *New York:* Columbia University Press.

———, and A. H. STURTEVANT, 1938: Inversions in the chromosomes of *Drosophila pseudoobscura. Genetics,* **23:** 27–64.

———, and C. C. TAN, 1936: Studies on hybrid sterility. III. A comparison of the gene arrangement in two species, *Drosophila pseudoobscura* and *Drosophila miranda. Zeitschr. f. ind. Abst. Vererb.,* **72:** 88–114.

FISHER, R. A., 1930: The Genetical Theory of Natural Selection. 272 pp. *Oxford:* Clarendon Press.

GERSHENSON, S. M., 1932: Studies on the genetically inert region of the X chromosome of *Drosophila.* I. *Journ. Genet.,* **28:** 297–313.

STURTEVANT, A. H., and G. W. BEADLE, 1936: The relations of inversions in the X chromosome of *Drosophila melanogaster* to crossing over and disjunction. *Genetics,* **21:** 554–604.

———, and C. C. TAN, 1937: The comparative genetics of *Drosophila pseudoobscura* and *D. melanogaster. Journ. Genet.,* **34:** 415–432.

WRIGHT, S., 1932: The rôles of mutation, inbreeding, crossbreeding, and selection in evolution. *Proc. VI Internat. Congr. Genetics,* **1:** 356–366.

27

THE INTERRELATIONS OF INVERSIONS, HETEROSIS AND RECOMBINATION †

There is a rapidly growing body of evidence to show that wild strains of many organisms are heterogeneous for inverted sections of the chromosomes. The occurrence of specimens heterozygous for inversions, in one or more chromosomes, has now been recorded in so many species of plants and of insects that there can be little doubt that they are to be expected in almost any large sample, at least of organisms not normally propagating by self-fertilization.

Certain consequences of inversions, as concerned with the behavior of mixed populations, have not yet been adequately discussed. In particular, it was briefly mentioned by Dobzhansky and Sturtevant (1938—p. 61) that "heterozygosis for inversions decreases the amount of crossing over, and this may be of selective value in connection with heterosis effects." The present paper is a more detailed consideration of this question.

In individuals heterozygous for an inversion no effective crossing over occurs unless the inversion is, genetically speaking, a very long one. The single cross-over chromatids are unbalanced and in fact eliminated (Sturtevant and Beadle, 1936); effective recombination results only from two-strand or three-strand double crossing over. This is, of course, rare in any but long inversions. It has also been shown that, as a result of the elimination of the cross-over chromatids, the fertility of a female heterozygous for an inversion is not reduced, unless complementary, or four-strand, double crossing over occurs. Again this will be rare unless the inversion is long (Sturtevant and Beadle, *loc. cit.*). The loss of fertility from such double crossing over is obviously completely correlated with the effective recombination.

Thus, if a chromosome exists, in a population, in two sequences, differing by an inversion, it will in effect show two distinct lines of descent. There is free exchange of material within any line (*i.e.,* sequence), but none between the sequences. Therefore, fluctuations of the genic contents must occur almost independently in the two sequences. Under such conditions it is inevitable that in time the gene content of the two sequences will become different. It must be supposed that each sequence is susceptible to the same mutations, and with

† In collaboration with K. Mather. Reprinted from The American Naturalist, Vol. 72, 1938, pp. 447–452.

the same frequencies, but, as a result of the relative rarity in a population of a given mutant allelomorph at any one moment, certain genes will be present in one sequence and not in the other. Elimination of mutant genes, both by chance and by selection, will insure that the gene contents of the sequences are unlike. The gene content of any sequence must be always fluctuating, and over a great length of time must, on the average, be the same as that of any other sequence. The point which we wish to make is, however, that at any given moment, for any given population, the gene contents of the two (or more) sequences will be unlike, and each will carry mutant allelomorphs that are lacking in the other types.

It is thus clear, considering two sequences A and B, that the homozygotes AA and BB are more likely to be homozygous for deleterious recessive mutations than is the sequence heterozygote AB. It may be noted here that we need only consider deleterious recessives, as these are presumably by far the most common type of mutation in wild populations, as well as in laboratory stocks. Then in general the sequence heterozygote AB will be at a selective advantage with respect to either of the corresponding homozygotes.

It does not seem desirable to give a detailed algebraic analysis of this question, because, as pointed out above, there can be no stability in the exact relations of the sequences with each other. A few results are, however, of some importance.

In the first place a single gene difference can not in general cause such heterosis. The simplest effective condition is that in which each sequence contains a deleterious recessive not present in the other. Two recessives existing in the two sequences with unlike frequencies may cause heterosis, but this will lead to leveling of the frequencies of occurrence of the mutant genes in the two sequences, and so such heterosis will not be persistent. Secondly, if sequence A exists with frequency p, and B with frequency q($q = 1 - p$), in a population; and if, letting the fertility of type AB be 1, AA has fertility ($1 - a$) and BB ($1 - b$),

$$\frac{p}{q} = \frac{b}{a} \quad \text{or} \quad pa = qb = k$$

for the point of equilibrium. With more than two sequences the question is complicated by the inequality of the various heterozygotes in fertility. It is, however, clear that, at equilibrium, the sequence frequencies, p, q, r, etc., will be inversely proportional to a, b, c, etc., the loss in fertility of the corresponding homozygotes. In general, on the assumption that each sequence is subject to the same mutations with the same frequencies, a must tend to equal b, etc., over any length of time. In consequence there will be a tendency for the various sequences to exist with equal frequencies in the populations. This will, of course, be subject to the effect of large local and temporary fluctuations. Equality in frequency of the sequences is the mid-point of the fluctuation range.

Thus, when an inversion arises it will at first depend largely on chance for its opportunity to become established. After it has persisted for sufficient time to differentiate relative to the other sequences present, there will be a considerable selective action favoring individuals carrying it in the heterozygous state. In consequence, its chance of total loss will be much reduced. Equally, of course, its chance of completely replacing the other sequences already present will also be very small.

We have so far considered the fate of inversions when once established, but on turning attention to the origin of inversions it is immediately obvious that not all chromosomes in all wild populations are heterogeneous for sequence. In particular, whereas the third chromosome of *Drosophila pseudoobscura* shows a wealth of variation in this respect (Dobzhansky and Sturtevant, 1938), the X,[1] second and fourth only occasionally show heterogeneity in sequence. What is the explanation?

It would appear that this is but a specific case of a more general question, that of the advantage of crossing over, and also of recombination of non-homologous chromosomes. This has often been discussed (*e.g.*, by Muller, 1932, Darlington, 1932, p. 456, and Fisher, 1930, pp. 102–104). Two solutions have been proposed, viz.: (a) that recombination makes it possible to obtain favorable combinations of mutant types which have arisen independently, and (b) that recombination increases the "flexibility" of the species in its ability to adapt itself to ever-changing environmental conditions.

The former suggestion clearly indicates an advantageous property of recombination, but it would seem that the resulting selection in favor of crossing over would be small, since such recombination need occur but once. Indeed subsequent recombination would be disadvantageous in breaking up the new favorable combination. Further, the opposite effect of unfavorable mutations would appear to oppose and probably to outweigh this selective action favoring recombination (see discussion of a similar case where secondary effects of favorable mutations are opposite in sign to secondary effects of unfavorable ones—Sturtevant, 1937).

The second possibility, concerning the "flexibility" of the species, seems to be of greater moment, but it has never been analyzed in sufficient detail to bring out the essential prerequisites for its functioning.

The simplest system that we have been able to devise, having the required property of favoring recombination, is as follows: Two gene pairs, A,a and

[1] We are not aware of any case in which there is a heterogeneity of sequences in the sex-chromosomes of a wild population, except that associated with "sex-ratio" in certain species of *Drosophila* (Sturtevant and Dobzhansky, 1936). This constitutes a special case, where the heterogeneity for the sex-ratio gene is maintained by a mechanism quite different from the one discussed above. In fact, the present mechanism, being dependent on unfavorable recessive mutant genes, would be expected to be much less effective on the sex-chromosomes, since such recessives are regularly rapidly eliminated in the heterozygous sex.

B,b, exist in a population subjected frequently to three different sets of environmental conditions, D,E and F. Condition D favors A but acts unfavorably on B, whereas E favors B but lowers the frequency of A. These, if properly adjusted as to intensity of selection and number of generations over which they operate, will insure the perpetuation of both allelomorphs at each locus. Recombination will not, however, be of importance. For this it seems necessary to introduce the third condition, F, favoring the combination AB (with or without a similar effect on ab), but acting adversely on the single types A and B. Under such conditions it would appear that there will be a selective action favoring recombination, of the order of magnitude of the selection of the AB type, as opposed to Ab or aB, under condition F. Other combinations of two or more loci may be exerting similar action in similar or dissimilar environmental conditions, and so the net effect will probably be to favor recombination for the majority of loci under the majority of condition changes. The crucial point of the argument is, of course, that environmental changes which must be supposed to be each decisive for the fate of one or more generations, can not decrease recombination but can favor it. This is in sharp contradistinction to the earlier suggestion, where recombination, once having occurred, is acted against.

If we agree that recombination is favorable and selected in some such way as this, we can suggest a reason for the special case of the almost complete confinement of sequence heterogeneity to one chromosome in *Drosophila pseudoobscura.* The first inversion to arise in a chromosome will, on its becoming common in the population, severely reduce the effective recombination in that chromosome, and, of course, in the whole nucleus (its effect may be partially, although not completely, compensated by increases of recombination in other chromosomes). A second inversion in that chromosome will result in less reduction in recombination, a third in still less, and so on. On the other hand, a new inversion in a *different* chromosome must produce an effect more or less equal to that of the first sequence change, in its reduction of recombination. Consequently once one chromosome shows a sequence change, further similar changes are much more likely to become established in that same chromosome than they are in others. The first chromosome to show a change of sequence may achieve this distinction by pure chance, but subsequent events will then result in the reduction of the chance of inversion heterogeneity in other chromosomes, as these would severely reduce recombination. Any further heterogeneity that should arise in that same chromosome will, however, be maintained by the heterosis effect. The two effects, recombination reduction and heterosis, are in opposition, and their balance will determine the fate of inversions arising in a wild population. A similar diversification in the properties of chromosome sections, with respect to recombination frequencies, has been postulated by Darlington (1936) for Oenothera.

LITERATURE CITED

DARLINGTON, C. D., 1932: "Recent Advances in Cytology," (first edition). London. 1936. *Jour. Genet.,* **32:** 343–352.
DOBZHANSKY, T., and A. H. STURTEVANT, 1938: *Genetics,* **23:** 28–64.
FISHER, R. A., 1930: "The Genetical Theory of Natural Selection." Oxford.
MULLER, H. J., 1932: AM. NAT., **46:** 118–138.
STURTEVANT, A. H., 1937: *Quart. Rev. Biol.,* **12:** 464–467.
STURTEVANT, A. H., and G. W. BEADLE, 1936: *Genetics,* **21:** 554–604.
STURTEVANT, A. H., and T. DOBZHANSKY, 1936: *Genetics,* **21:** 473–490.

28

HIGH MUTATION FREQUENCY INDUCED BY HYBRIDIZATION †

The increase in variability that is a familiar result of the rearing of second generations from crosses between different races is certainly due chiefly to segregation and recombination of genes in which the parental races differed. There is, however, a persistent feeling that perhaps interracial crossing also induces the production of new mutations. The first attempt to test this view experimentally with *Drosophila* led to negative results (Duncan, 1915). Belgovsky (1937) reported an increased frequency of somatic mutations in the hybrids between *D. melanogaster* and *D. simulans,* in x-ray experiments. The differences reported were not very striking, and in any case have little bearing on questions concerning spontaneous frequencies.

Casual observations have led me to suspect an increased mutation frequency in the hybrids between the two semi-cross-sterile races (*A* and *B*) of *Drosophila pseudoöbscura.* The present note describes experiments planned to measure the rate quantitatively in such hybrids.

Two different interracial crosses were carried out, using appropriate "marker" genes. Females of the scutellar scarlet (race *B*) stock were mated to males of the beaded (race *A*) stock. The F_1 females were mated to yellow compressed short (race *A*) males, and the daughters (first back-cross —1 *BC*) of this mating were tested individually by yellow compressed short males, to give the 2 *BC* generation. In the second experiment, yellow compressed short (*A*) females were mated to pointed (*B*) males, and the F_1 females were mated to scutellar scarlet (*B*) males to give the tested 1 *BC* females. In both cases females of the 2 *BC* generation were also tested for lethals, by mating to males of the strain to which their respective fathers belonged. The table shows the sex-linked lethals recovered (referred to the generation to which the first detected heterozygous female belonged), and the number of "visible" mutant types observed in the males (referred to the generation to which the males themselves belonged—in most cases presumably one generation later than that in which the actual mutation occurred).

The totals (denominators) shown in the table represent, for lethals, twice

† Reprinted from the Proceedings of the National Academy of Sciences, Vol. 25, 1939, pp. 308–310.

the number of tested females. This is because 12 of the 31 recovered lethals were in paternal *X* chromosomes, which had evidently undergone mutation soon after fertilization had introduced them to the hybrid zygote. For the visibles the total given is that of the males counted in the generation concerned. Each separate mutation is counted once, though several of them appeared in several offspring of single mothers. These relations show that few, if any, of the mutations occurred at meiosis. They are, therefore, scarcely to be referred to crossing-over or recombination.

The "visible" types recovered were mostly previously known types. Thus, forked (or singed—probably both are present, but the tests are not yet completed) was found five times, beaded three times, scutellar twice and dusky, bobbed, ascute, slender and short each appeared once. In the 4 *BC*

Table 1

DETECTED MUTATIONS

	BACKCROSSES TO RACE *A*		BACKCROSSES TO RACE *B*		TOTAL (PERCENTAGES)	
GENERATION	LETHALS	VISIBLES	LETHALS	VISIBLES	LETHALS	VISIBLES
1 *BC*	0/38	1*/124	9/66	0/243	8.7	0
2 *BC*	6/106	0/911	16/152	10/1046	8.5	0.56
3 *BC*	. . .	5/2135	. . .	4/3796	. . .	0.15

* This was a somatic mutation—there was a large patch of forked (or singed) bristles on the thorax. The mutation is not included in the totals, since it is not comparable to the other "visibles," which presumably occurred in the mothers of the males recorded.

generation (not shown in the table because the pedigrees were varied in order to carry out tests of suspected mutant types) there were found white, bobbed, miniature and another forked (or singed).

The lethal frequencies given are to be taken as minimum values. As has been shown by Lancefield and others (see Sturtevant, 1937a), the sex-ratio from *A-B* hybrids is variable. For this reason all sex-linked lethals were tested, and none were entered in the table unless they resulted not only in a 2:1 sex-ratio but also in typical lethal ratios for the sex-linked genes heterozygous in the culture. This requirement was met in at least two successive generations. From several probable lethals no satisfactory test was obtained in the following generation, and these have not been recorded, though the chromosomes concerned do appear in the totals. Incidentally, this test also showed that the loci of the lethals were scattered throughout the *X* chromosome.

Owing to the complications just mentioned, and to the small number of cultures reared, it is not possible to judge whether there was an increased mutation frequency in the F_1 generation. The point is of theoretical interest, in connection with the hypothesis outlined below, but will have to be studied in experiments especially designed for the purpose.

In addition to the mutations shown in the table there were three dominants in the females of the 3 *BC* generation (Smoky, an extreme minute-bristle, and a sex-linked wing character). In the regular series no autosomal recessives could have been detected, but in testing one of the early sex-linked recessives, hybrids were mated together, and among their offspring was found the well-known second chromosome recessive, bithorax.

These experiments were carried out on a small scale, and no specific control is available. Nevertheless, the mutation frequencies—about 9 percent of lethals and one-half of 1 per cent sex-linked visibles—are so much higher than any other recorded spontaneous rates that their significance cannot be doubted. It may be added that I have detected only one sex-linked lethal in *pseudoöbscura* in experiments not concerned with *A-B* hybrids.

The theoretical implications of the results just described may be deduced from the effects that natural selection should have on mutation rate. I have argued (Sturtevant, 1937b) that natural selection must operate to keep the general mutation rate of a species at a minimum—a conclusion also reached by Shapiro (1938). I have suggested that different populations may achieve this result by different genetic means, and that crossing such different races should, through recombination, lead to an increase in mutation frequency. The results here recorded constitute a verification of this prediction. The increase here is unexpectedly great, but it may be pointed out that the cross made is about as extreme a one as it is possible to study in later generations.

These results indicate that a species may escape the impasse toward which the automatic decrease in mutation frequency is constantly pushing it, by the occasional interbreeding of semi-isolated populations within the species.

The general method of crossing may also be of use in the study of the nature of the mutation process, since there is no indication of an increase in major chromosome aberrations such as accompany x-ray treatment. Further study will be required to answer the question whether small rearrangements (deficiencies, small inversions) occur more frequently.

Summary

The offspring of the backcrosses from hybrids between the two races (*A* and *B*) of *Drosophila pseudoöbscura* show a large increase in mutation frequency.

LITERATURE CITED

Belgovsky, M. L., *Genetica,* **19:** 370–386 (1937).

Duncan, F. N., *Amer. Nat.,* **49:** 575–582 (1915).

Shapiro, N. J., *Zool. Jour.,* **17:** 592–601 (1938).

Sturtevant, A. H., these Proceedings, **23:** 360–362 (1937a); *Quart. Rev. Biol.,* **12:** 464–467 (1937b).

29

THE HOMOLOGIES OF THE CHROMOSOME ELEMENTS IN THE GENUS *DROSOPHILA* †

Introduction

It is becoming increasingly clear that certain elements in the chromosome complexes of the Drosophilas maintain their identity from species to species, though undergoing various changes in their relations to each other and in their internal structure (DONALD 1936; STURTEVANT and TAN 1937; STURTEVANT 1938a, 1940). The terminology for these various elements, however, is far from uniform, since several conflicting systems of lettering and numbering are in use. The most satisfactory general solution of the problem of nomenclature seems to be that suggested by MULLER (1940). According to this system the recognizable elements are lettered, in the sequence familiar in *melanogaster*. The X of that species becomes A; IIL, B; IIR, C; IIIL, D; IIIR, E; IV, F. It is not expected that this system will replace those established by long usage, as in *melanogaster,* but it is strongly recommended that it be applied in those cases where the change might more easily be made. Furthermore, arbitrary systems of numbering or lettering used to facilitate studies on the genetics and cytology of unexplored species should be selected with full appreciation of the ambiguities that may arise when the homologies of the various elements become known.

In those cases where species are so closely related that their salivary gland chromosomes show obvious similarities, or that their hybrids can be obtained, the task of homologizing the elements is greatly simplified; for all other species it is necessary to resort to a comparison of their mutant genes with those of species whose elements have been determined. (The term "element" seems the most appropriate one for this unit since chromosome, arm, and limb have all been used with other connotations.) Since only a fraction of the mutants can be regarded as "good" parallels, comparisons must be made with caution; on the other hand, the tendency to minimize otherwise good homologies which do not appear to be consistent with the scheme must be watched carefully.

It is our purpose in the present paper to extend these homologies as far as now seems possible, and to examine critically evidence contrary to the

† In collaboration with E. Novitski. Reprinted from *Genetics,* Vol. 26, 1941, pp. 517–541.

working hypothesis that the elements have remained essentially intact. The species to be so examined include, among others, *affinis, algonquin, ananassae, azteca, busckii, funebris, hydei, miranda, montium, pseudoobscura, simulans, virilis,* and *willistoni.*

Table 1

Compositions of the chromosomes in terms of the elements.

SPECIES	A	B	C	D	E	F	AUTHORITY
melanogaster	X	IIL	IIR	IIIL	IIIR	IV	
affinis	XL	IV	III	XR	II	V	STURTEVANT 1940
algonquin	XS	C	A	XL	B	D	MILLER 1939
ananassae	X part / IV part	IIIR	IIIL	IIR	IIL	IV	KIKKAWA 1938
azteca	XS	C	A	XL	B	D	DOBZHANSKY and SOCOLOV 1939
miranda	XL	IV	X2	XR	II	V	DOBZHANSKY and TAN 1936
pseudoöbscura	XL	IV	III	XR	II	V	LANCEFIELD 1922
simulans	X	IIL	IIR	IIIL	IIIR	IV	STURTEVANT 1922
virilis	I	VI	V	III	II	IV	METZ, MOSES and MASON 1923
virilis	V	II	IV	I	III		HEITZ 1934
virilis	X	IV	V	III	II	VI	CHINO 1936
virilis	X	D	E	C	B	M	HUGHES 1939

The Affinis Group

The metaphase chromosome configurations of females in this group (*affinis, algonquin, athabasca, azteca*) contain a large pair of V-shaped X chromosomes, two pairs of V-shaped autosomes, a pair of rods, and a pair of microchromosomes (METZ 1916). The original account (DOBZHANSKY in STURTEVANT and DOBZHANSKY 1936) of the salivary gland chromosomes in the members of the *affinis* group has been confirmed and extended by MILLER (1939). The observations of DOBZHANSKY and SOCOLOV (1939) on *azteca* failed to include the longer salivary gland chromosome arm of their "C" chromosome, a circumstance that invalidates their conclusion that "Some genes found in *D. azteca* in a single linkage group must belong to different ones in *D. pseudoobscura;* conversely, some genes linked in *D. pseudoobscura* may be expected to be independent in *D. azteca.*"

The salivary gland chromosomes of the different members of the *affinis* group are sufficiently similar so that one familiar with any one of these species can recognize the homologous elements in any other. [Since there is considerable variation in gene sequence within each species, the salivary gland chromosomes of interspecific hybrids from strains taken at random tend to possess quite remotely related chromosomes. These need not be interpreted as interspecific differences, as they have been (BAUER and

DOBZHANSKY 1937; DOBZHANSKY 1937), for it is possible that identical or simply related sequences exist within each species.] The homology of the *affinis* elements being known through a comparison of mutant types (STURTEVANT 1940), it was necessary only to correlate the genetic linkage groups with the salivary gland chromosomes of this species. XL of *pseudoobscura* agrees cytologically with the short arm of the X in *affinis* in having an interstitial heterochromatic region near the base and identical banding at the tip. It follows that the longer arms of the X chromosomes of the two species, containing the same genetic material, are homologous. For the identification of the genetic autosomes with the arbitrarily lettered salivary chromosomes of *azteca* (DOBZHANSKY and SOCOLOV 1939) and of *algonquin* (MILLER 1939), a study was made of nine X-ray induced translocations in *affinis,* with the result shown in table 1. From this correlation it is evident that both elements B and E have median centromeres, yielding the V-shape obvious in two of the metaphase and salivary gland chromosomes.

D. Ananassae

The metaphase chromosome configuration of *ananassae* indicates that, while the two large autosomes are not unlike those of *melanogaster,* the centromere of the X chromosome has been shifted from a terminal to a median position and the dot chromosome has been transformed into a small V-shaped chromosome (METZ 1916). The nucleolus-forming region and the bobbed gene, ordinarily related to the sex chromosomes in other species, are associated with the small V-shaped fourth chromosome (KAUFMANN 1937; KIKKAWA 1938). This may have been achieved by a simple exchange of centromeres of the dot and X chromosomes, along with adjacent heterochromatin. If this be the case, the exchange must have occurred prior to the median shift in the centromere of the X; that this sequence is correct is borne out by the cytological observations of KIKKAWA (1936) on *montium,* a relative of *ananassae,* in which the V-shaped fourth chromosome is present, but the X chromosome retains its rod shape. There are other possible explanations, involving an exchange of the Y and fourth chromosome centromeres.

A number of mutants in *ananassae* furnish means for comparing its chromosomes with those of *melanogaster* and with the elements. A more complete discussion of these mutants and their linkage data can, in general, be obtained from the papers of MORIWAKI (1935, 1938) and KIKKAWA (1938). *Melanogaster* mutants with which those of *ananassae* are compared are followed by a symbol showing the limb in which they occur.

The X Chromosome. Yellow, scute, white, Notch, cut, singed, miniature, dusky, forked, and Beadex seem good homologies with the same sex-linked mutants of *melanogaster*. The terminology for singed and forked may be reversed, as in *pseudoobscura,* but their value is not thereby impaired, since both occur in the same element. The close linkage of yellow with scute and of miniature with dusky (also found in *melanogaster, pseudoobscura,* and *virilis*) adds weight to already good comparisons. Vermilion, ordinarily a good parallel between the X chromosomes of two species, lacks the confirmation of a feeding or transplantation test.

Chromosome II. Cardinal in the left limb of this chromosome corresponds to cardinal (IIIR) and can be differentiated from its mimics, scarlet and cinnabar, by virtue of its striking change in color with age. Plexate probably represents Delta (IIIR), agreeing both in phenotype and in mutability. Off is much like Bare of *pseudoobscura* (element E), although its dominance, its effect of clipping and removing the bristles, and the occasional viability of the homozygote are not conclusive. Puffed has no homolog in *melanogaster,* but agrees with Puffed in *virilis,* which has small, reddish, and "inflamed" eyes (MORIWAKI 1938). This mutant is located on the second chromosome of *virilis,* which will be shown to correspond to element E. For this reason, we believe that the locus of this gene in *ananassae,* which has not been determined because of its association with an inversion, will prove to be in the left arm.

The right limb of this chromosome has no mutants suitable for homologies, but it may be concluded that it represents the element associated with E in *melanogaster*—that is, D—since elements B and C are both accounted for below.

Chromosome III. Plum, in the left limb of chromosome III, is like Plum (IIR) in *melanogaster* in having a dominant mottled eye color, a recessive lethal effect, and an association with an aberration involving heterochromatin (KIKKAWA 1938). Furthermore, it produces a mottled white eye color in combination with vermilion. This indicates its allelomorphism to brown.

Gap compares with gap (IIR) in weakening the fourth longitudinal vein and in giving the posterior crossvein an oblique direction. This is not conclusive in itself, but serves to confirm the identity of the left limb of the third chromosome of *ananassae* with the right limb of the second chromosome of *melanogaster.*

Only one mutant, plexus, has been found on the right limb of this chromosome. Since we would expect true plexus to occur in the same limb as Plum, we feel that this is in fact net (IIL), one of the most common mutant types in other species.

Chromosome IV. The fourth linkage group contains a dominant mutant, similar to Shaven. This dominant form is not known in *melanogaster,* but is in *simulans* (unpublished data). Haplo-IV individuals correspond quite closely to those of *melanogaster.* The linkage of bobbed with these two has already been discussed.

From the above homologies, it would seem that the nomenclature for the autosomal linkage groups has been reversed with respect to that found in *melanogaster* and, in each case, the significance of left and right as applied to the chromosome arms has also been reversed (table 1).

D. Busckii

The metaphase chromosome configuration here is essentially the same as in *melanogaster,* except for the dot chromosomes which are absent as such and may be attached to the X chromosomes as satellites (METZ 1916; KRIVSHENKO 1939). KRIVSHENKO has obtained a considerable number of mutants in this species after treatment with X-radiation. Forked, miniature, Notch, scute, singed, vermilion, white and yellow, all on the X chromosome, may be homologous to the sex-linked mutants in *melanogaster*. Autosomal dominants resulting from X-ray treatment, the most abundant type in KRIVSHENKO's studies, are usually unsuitable for drawing comparisons, partly because of their non-specific effects and partly because of their frequent association with chromosomal aberrations which make linkage studies difficult. One exception in this case is Delta, which occurred seven times, once associated with a IIL-IIIL translocation and once with a IIL-IIR-IIIR translocation, in each case one of the break points occurring in the same region of IIL. This would point to the correspondence of IIL of *busckii* with element E.

D. Funebris

The mutants in *funebris* available for drawing comparisons are generally unsatisfactory, both because of their indefinite nature and of their unknown linkages, with the exception of a few sex-linked types. Notch corresponds closely to Notch (X) (STURTEVANT 1918). Forked may be either forked (X) or singed (X) (MORGAN, BRIDGES, and STURTEVANT 1925). The similarity of bobbed to bobbed (X, Y) has been pointed out by LUERS (1937). STUBBE and VOGT (1940), by transplantation of eye disks, have shown vermilion to be homologous to vermilion (X). They have also confirmed the identity of autosomal cinnabar with cinnabar (IIR) by the same technique. Radius incompletus bears some resemblances to radius incompletus (IIIL) (TIMOFÉEF-RESSOVSKY 1927) and cubitus incompletus to cubitus interruptus (IV); but any attempt to elaborate these and other possible autosomal homologies seems unprofitable in the absence of linkage data.

D. Hydei

Notch, white, vermilion, and bobbed in the X chromosome of *hydei* (CLAUSEN 1923; SPENCER 1927) seem probable homologs of the same sex-linked mutants in *melanogaster*. For this species in particular and for many others there remains unavailable a considerable body of unpublished information, which, as far as we are aware, is not in disharmony with the scheme of permanent elements but which, in a great many cases, furnishes additional evidence in its favor.

D. Miranda

The homologies of the salivary chromosomes of *miranda* with those of *pseudoobscura* have been deduced from studies of their hybrids (DOBZHANSKY and TAN 1936). Besides a large number of differences between homologous elements arising, presumably, from the effects of cumulative inversions, DOBZHANSKY and TAN conclude that at least five translocations (involving all the chromosomes) have become established between the two species. In view of MACKNIGHT's criticism (1939) on cytological grounds, WRIGHT's mathematical calculations (1940) for the incorporation of one such translocation (see below) and the absence of conclusive evidence to the contrary, we are forced to conclude that these cases very probably do not represent exceptions to the rule of the indivisibility of the elements.

D. Montium

A number of mutants described by OSIMA (1940) provide means for drawing parallels between the linkage groups of *montium* and the elements. On the X chromosome, white, Notch, and vermilion compare favorably with the similar sex-linked mutants in other species.

Curled-b (left limb of II) simulates curled (IIIR) by curling the wings up and raising and crossing the posterior scutellar bristles. Confluent (left limb of II), obtained twice after X-ray treatment, would correspond to Delta (IIIR), one of the most frequent of mutants found after raying. Hairless, also on chromosome III, agrees with Hairless (IIIR) in its bristle and wing vein diminishing effect, in its dominance and in its lethality when homozygous. This mutant has not yet been allocated to a definite limb; it can be suggested here that it will prove to be on the left limb.

Curled (left limb of III) and jaunty (IIL) seem good parallels; plexus (right limb) would consequently suggest plexus (IIR) rather than net (IIL) since it falls on the anticipated element. On this basis, Plexate, located in the middle of chromosome III, would correspond to Plexate (IIR) and would be expected to occur on the right limb along with plexus. The centromere, therefore, might be expected to fall to the left of Plexate.

The metaphase chromosomes, reported by KIKKAWA and PENG (1938), suggest that this species may be like *ananassae,* but with a terminal instead of a median centromere in the X. The genetic data do not help in checking this supposition.

D. Pseudoobscura

The metaphase chromosomes of a *pseudoobscura* female consist of a pair of V-shaped X chromosomes, three pairs of autosomal rods, and a pair of dot chromosomes. LANCEFIELD (1922) suggested that a portion of the V-shaped X chromosome of *pseudoobscura* might exist in other species as autosomal material. CREW and LAMY (1935) pointed out, among other comparisons, that certain mutants in the right arm of the X chromosome could be homologized with similar mutants in IIIL of *melanogaster,* although this was not immediately obvious from this paper because of the authors' peculiar terminology. The argument that the elements have remained essentially intact in *pseudoöbscura* as compared with *melanogaster* was applied by DONALD (1936) who studied more mutants and implied the necessary corrections to the work of CREW and LAMY. STURTEVANT and TAN (1937) confirmed and extended the main conclusions, leaving no doubt of the identity of the elements in the two species (see table 1). Two additions (STURTEVANT and TAN, unpublished) may be made to the information published in this latter paper: The order of the loci on the X chromosome in the neighborhood of 72 has been shown to be forked, dusky, and bobbed, not dusky, forked, and bobbed as surmised before; also the distance between compressed and sepia is 30 units rather than the indicated 7, thereby increasing all loci from sepia to the right end inclusive by about 25 units.

TAN (1935) correlated the salivary gland chromosomes with the genetic linkage groups in this species.

D. Simulans

Since *simulans* hybridizes with *melanogaster,* suspected homologies can be tested directly by mating the corresponding mutants. Thus ordinarily insecure comparisons become unquestionable. The genetic evidence (STURTEVANT 1921a, 1921b, 1929) indicates a close agreement between the limbs of the *simulans* chromosomes and those of *melanogaster* (table 1). The hybrid salivary gland chromosome analyses by PATAU (1935) and by KERKIS (1936) confirmed the homologies of the chromosomes and verified the existence of an inverted sequence of genes in the third chromosome, the first inversion to be found in *Drosophila* (STURTEVANT 1921c). The more detailed examination by HORTON (1939) revealed that as many as 23 additional intra-chromosomal differences, all very small, may exist between the two species.

Six unpublished corresponding mutant types of *simulans* may be noted here: scute (X, 0), ocelliless (X, 24), javelin (III, 0), radius incompletus (III, 58), recessive hairless (III, 61), and dominant Shaven (IV). The sex-linked parallel vermilion (X, 31.8) was reported by STURTEVANT (1932), without its locus. All of these have been shown to be allelic to the *melanogaster* types whose names they bear and which they closely resemble.

D. *Virilis*

D. virilis remains the only species in the subgenus Drosophila (STURTEVANT 1939) with a sufficient array of mutants to make possible homologizing all the linkage groups with the elements. The six linkage groups agree with the metaphase configuration of five pairs of rods and one pair of microchromosomes. METZ, MOSES, and MASON (1923) and CHINO (1929, 1936) have drawn many comparisons of *virilis* mutants with those of *melanogaster*. *Melanogaster* mutants are followed by a notation of the chromosome limb in which they occur, all others being *virilis* types. We will attempt to evaluate these and, in some cases, make additional suggestions.

X Chromosome. Convincing parallels have been drawn in the cases of yellow, forked, singed, glazed (METZ, et al.), of crossveinless (WEINSTEIN 1920), of scute, white, Notch, miniature, dusky, rudimentary, Beadex, and bobbed (CHINO) with the identically-named mutants (except lozenge for glazed) in the X chromosome of *melanogaster*. HOWLAND, GLANCY, and SONNENBLICK (1937) have established the identity of vermilion of *virilis* with vermilion (X) by the transplantation of eye disks.

The comparisons of echinus with echinus (X), magenta with ruby (X), vesiculated with vesiculated (X), ragged with cut (X), decline with wavy (X), apricot with garnet (X), and small bristle with tiny (X) (CHINO) are open to doubt, but they do serve as corroborative evidence that the composition of the X chromosome of *virilis* and *melanogaster* is essentially the same.

Chromosome II. Confluent is very probably homologous to Delta (IIIR) (CHINO), although the weaker alleles described by METZ et al. show some discrepancies. Concave and crumpled (IIIR) constitute one of the most striking cases of homology (METZ et al.); the case is strengthened by the occurrence of similar types in element E of *affinis* and *pseudoobscura.* Our examination of varnished indicates that it is a good parallel to glass (IIIR).

It has been suggested that ebony corresponds to ebony (IIIR), brick to claret (IIIR), and broken to crossveinless-b (IIIR) or to crossveinless-c (IIIR) (CHINO). These are not diagnostic mutants and may be questioned, although they do fall in the proper limb.

Radius incompletus (IIIL) has been compared to both incomplete and detached of *virilis* (CHINO). From the published figures, it would seem that incomplete is quite like radius incompletus (IIIL), but this homology loses its force since it is not in the expected element. A similar situation is found in the case of lanceolate (see below).

Thus chromosome II of *virilis* corresponds to IIIR of *melanogaster,* or element E.

Chromosome III. Short veins suggest veinlet (IIIL) (CHINO) but the comparisons of rose with rose (IIIL), spread with dihedral (IIIL), and rolled with rolled (IIIL) (CHINO) are less convincing.

The striking resemblance of hunch to ascute (IIIL) has been pointed out (METZ et al.) but was discarded as a possible homology because a similar mutant was sex-linked in *D. pseudoobscura.* Since it is now known that the limb of the X chromosome of *pseudoobscura* in which this mutant is located corresponds to IIIL of *melanogaster* (DONALD 1936), this can be considered a convincing parallel. Telescoped has no homolog in *melanogaster,* but agrees closely with deformed and its allele serrate in *willistoni* (LANCEFIELD and METZ 1922) and to compressed in *pseudoobscura* where it is located in element D (STURTEVANT and TAN 1937). Garnet, like Henna (IIIL), is a dominant dark eye color; the only other possible homology would be Plum (IIR), which may be eliminated because of its mottling effect and allelism to brown (see eosinoid below).

The evidence favors the correspondence of *virilis* chromosome III with *melanogaster* IIIL or element D. On this basis, it may be safely predicted that cinnabar, after appropriate tests, will prove to be scarlet (IIIL) and not cinnabar (IIR).

Chromosome IV. Dachsous and dachsous (IIL), Star and Star (IIL) and reduced and reduced (IIL) seem good comparisons (CHINO). Clipped probably represents one of the truncate alleles of dumpy (IIL) (CHINO).

However, CHINO's identification of rough-4a with roughish (IIL), black with black (IIL), Squat with Squat (IIL) and flipper with pupal (IIL) are less convincing. Since the suggested homologs lie in the proper arm, they do tend to confirm the correspondence of these limbs. His figure of veinlet indicates that it is not veinlet (IIIL), as he suggests (see short vein, above).

Plexus might correspond to either net (IIL) or plexus (IIR). We would favor the first alternative since it falls in the proper limb. Lanceolate has been compared to lanceolate (IIR). This is one case where a mutant suitable for homologizing does not have an equally good or better comparison in the anticipated element. The force of this discrepancy is diminished by DONALD's observation (1936) that a similar mutant is found in element D in *pseudoobscura,* in addition to that found later in element C (STURTEVANT and TAN 1937). Also, KIKKAWA (1938) lists two such mutants, lance and lanceolate, one on the second, the other on the third chromosome of *ananassae.*

The bulk of the evidence favors the view that the third chromosome is element B.

Chromosome V. Eosinoid agrees with brown (IIR) in giving a white eye color in combination with scarlet or cinnabar (MORI 1937). Vestigial and vestigial (IIR) and straw and straw (IIR) seem excellent homologies (CHINO). Comparisons of ruffled with intwined (IIIL), fat with fat (IIL), dachsoid with four-jointed (IIL), mahogany with clot (IIL) and with sepia (IIIL), Beaded with Beaded (IIIR) and morula with morula (IIR) (CHINO) are questionable. Beaded may be compared with Jagged (C) of *pseudo-obscura.*

Branched might be either net (IIL) or plexus (IIR); the latter seems more likely on the basis of the other homologs. For the same reason, we believe that scarlet corresponds to cinnabar (IIR) and not to scarlet (IIIR) (see cinnabar above).

Chromosome VI. The comparison of abdomen rotatum with abdomen rotatum (IV) (CHINO) is convincing; Gap, which shortens the fifth longitudinal vein, may be questioned as a homolog to cubitus interruptus. To these may be added the parallel of stubby with shaven (IV). Thus the dot chromosomes of both species are apparently the same.

As far as the above data can show, the elements have remained essentially intact. We are therefore reluctant to accept in its entirety CHINO's conclusion that "it seems to be justified to assume that in the evolutional course of Drosophila there occurred many inversions, mutual translocations, and attachments or fragmentations of all chromosomes."

FUJII (1936), by means of four translocations and an inversion, has correlated the salivary gland chromosomes with the linkage groups, as indicated in table 1.

SPENCER (1940) has described a native American subspecies, *virilis americana,* in which elements D and E are united to form a V; elements A and B are also fused, the additional B necessary to the diploid complement of the male being present as a single rod (HUGHES 1939; STALKER 1940; PATTERSON, STONE, and GRIFFEN 1940). In another subspecies, *virilis texana,* elements B and D have a single centromere (PATTERSON, STONE, and GRIFFEN 1940). In each case the remaining elements are present as rods.

D. Willistoni

The X chromosome of this species is V-shaped, as in *pseudo-obscura,* while the autosomes are comprised of one V and one rod (LANCEFIELD and METZ 1921). From the metaphase chromosome length relationships, it is clear that one of the elements that is autosomal in most species is part of the X in *willistoni.* The homologies of the sex-linked mutants here are consequently of considerable interest.

Yellow has been compared to yellow (X), scute to scute (X), vermilion to vermilion (X), stubby to forked (X), triple to bifid (X), forked to singed (X) (LANCEFIELD and METZ 1922; FERRY, LANCEFIELD, and METZ 1923). In addition, square agrees with rudimentary (X); unpublished data show that a newly arisen vermilion (locus in X unknown) lacks the v^+ substance, as determined by the feeding technique. Deformed and its allele serrate correspond quite closely to telescope of *virilis* (element D) and to compressed of *pseudoobscura* (element D) (STURTEVANT and TAN 1937); stump compares favorably with radius incompletus (IIIL), and short and veinlet (IIIL) seem good parallels.

The evidence thus indicates that elements A and D have fused to form the X in *willistoni,* as in *pseudoobscura.* This conclusion is strengthened by the distribution of the mutants homologized: all those paralleling element A mutants are found on one half of the X (34.5 to 84) while all those paralleling element D mutants are on the other half (0 to 34). The genetic locus of the centromere on this basis would be between 34 and 34.5. Low interference values for this region, as deduced from the crossing over data of LANCEFIELD and METZ (1922) support this hypothesis.

The autosomal mutants do not yield a satisfactory account of the composition of the V and rod-shaped autosomes. Balloon (chromosome II) has been homologized with balloon (IIR) (FERRY, LANCEFIELD, and METZ 1923). Apterous (chromosome II) agrees well with apterous (IIR); both remove the wings, reduce the balancers, diminish the number of posterior scutellar bristles, deform the thorax, and sterilize both sexes. Clipped and Scalloped, both on chromosome II, probably represent the dominant vestigial deficiencies (IIR). Approximated and dachs (IIL) are close parallels. These homologies indicate that the second chromosome of *willistoni* includes elements B and C, as in *melanogaster.* The location of Knot, which bears some resemblances to Delta (IIIR), on the second chromosome casts doubt on this conclusion. In either case, however, chromosome II must be the autosomal V; this is supported by the slightly greater number of mutants in II than in III (11 in II as compared to 8 in III).

Neither the genetic nor the cytological evidence gives a clue as to the location of element F in *willistoni,* but it is to be noted that the salivary gland chromosomes have not yet been studied in this species. The most likely place to look for F would appear to be as a short arm on the apparently rod-shaped chromosome (III, element E?).

Homologous Genes in the Various Species

Table 6 lists the homologous genes in the species discussed, to which some previously unpublished cases have been added. Unpublished data of other workers, available particularly in Drosophila Information Service, tend to strengthen the suggested homologies given in the table.

The occurrence of "sex-ratio" in *melanica,* a member of the subgenus Drosophila, seems worthy of special mention. It has been previously noted only in the subgenus Sophophora (*obscura, pseudoobscura, affinis, athabasca,* and *azteca*) and in those cases analyzed, is located in element D, which is part of the X chromosome. It seems likely, then, that the X of *melanica* is a V, one arm of which is element D. The cytological observations of METZ (1916) indicate that a V-shaped chromosome of the required size is present, although no identification of the X has been made.

Lengths of the Elements in the Salivary Gland Nuclei

The percentage each element comprises of the total euchromatic length of the salivary gland chromosomes is listed in table 2. These values have been derived from the published lengths for *ananassae, melanogaster,* and *virilis,* from measurements of the published drawings for *algonquin* and *azteca,* and from a combination of two sets of drawings plus our estimate for element D for *pseudoobscura.* The figure given for element B of *azteca* has been obtained by assuming that it bears the same ratio to the other autosomal elements as do the homologous elements in *algonquin.*

Table 2

Relative lengths of the elements in the salivary gland nuclei.

	A	B	C	D	E	AUTHORITY
algonquin	16.4	18.5	19.7	20.0	25.4	MILLER 1939
ananassae	19.3	19.1	18.0	18.4	25.1	KIKKAWA 1938
azteca	15.1	18.8	19.0	21.4	25.8	DOBZHANSKY and SOCOLOV 1939
melanogaster	18.9	18.5	21.0	18.0	23.6	BRIDGES 1935
pseudoobscura	14.0	19.9	16.3	24.0	25.8	TAN 1936; DOBZHANSKY and TAN 1939
virilis	19.3	19.3	19.5	18.3	23.5	FUJII 1936
virilis	19.3	20.1	19.0	18.2	23.4	HUGHES 1939

Since these lengths are a function of both the elasticity of the chromosomes and the personal equation of the observer, the comparative lengths of the homologous elements are quite variable, with one striking exception: in every case element E is considerably longer than any of the others. This identification of the longest salivary gland element with element E may be of practical value in those instances where the correlation may be impracticable for other reasons. SLIZYNSKA and SLIZYNSKI (1941) report that in *funebris,* one of the elements is considerably longer than any of the others; it therefore probably corresponds to element E.

It is also to be noted that element D is about the same length as A when it is autosomal, but is distinctly longer in the three species in which it is part of the X. It seems clear from the table that the lengths of element A and D tend to show a negative correlation with each other. Numerous speculations are suggested by these relations, but they can scarcely be profitably discussed without more information than is now available.

Sequence of Corresponding Genes in Different Species

In general there is little similarity in the sequence of corresponding loci within each element, except when such closely related species as *melanogaster* and *simulans* are compared. In discussing *melanogaster* and *pseudoobscura,* STURTEVANT and TAN (1937) state: "The mathematical properties of series of letters subjected to the operation of successive inversions do not appear to have been worked out, so that we are so far unable to present a detailed analysis. It does appear, however, that the five arms (taken together) are definitely more alike in the two species than could result from chance alone." These statements now require some modification.

With the help of PROF. MORGAN WARD, a beginning has been made in the study of the mathematical consequences of successive inversions. Complete catalogs have been prepared, showing all the possible different arrangements of 2, 3, 4, 5, and 6 loci, respectively, together with the minimum number of successive inversions required to change each arrangement into a single arbitrarily chosen one. Actually, numbers were used, and the required arbitrary sequence was the ordinal one (1, 2, 3, 4, etc.). Table 3 shows the mean number of inversions required, together with the standard deviations for the respective populations.

Table 3

Mean number of inversions required to transform random arrangements of numbers into ordinal series.

NUMBER OF LOCI	MEAN	STANDARD DEVIATION
1	0	0
2	.500	.50
3	1.167	.69
4	1.750	.66
5	2.392	.70
6	3.036	.71
8	4.367 ± .092	.71 ± .06
9	4.975 ± .13	.82 ± .09

In the cases where more than six loci were involved, it became impracticable to make complete catalogs. Accordingly, for the two rows shown (eight and nine loci), random series of numbers (60 and 40 sequences, respectively) were chosen, and for each such sequence there was determined the minimum number of successive inversions required to reduce it to the ordinal sequence chosen as "standard." For numbers of loci above nine the determination of this minimum number proved too laborious, and too uncertain, to be carried out.

The table as it stands, however, gives a solution that seems safe to use for any numbers of corresponding loci likely to be encountered in such studies as these, since a plot of the values shows that there is an approximately linear relation between the number of loci considered and the average number of inversions required to reduce to ordinal sequence. For each additional locus considered, there is an increase of about .64 in the average number of inversions required. The curve is more irregular near its point of origin; and the values are exact up to six loci. Accordingly, for values above six it seems safest to start with the value 3.036 (for 6) and add to it $.64(n - 6)$, where n is the number of loci concerned.

The standard deviations shown in the table give a less regular curve when plotted, but it is clear that they are increasing only slowly. It may probably be assumed that σ will not be greater than 1 for values up to $n = 15$—beyond which it is unlikely that the present problem will require a solution for many years. This means that the spread is not great—in other words, that any random sequence is unlikely to require a number of inversions much different from the calculated one. If a number much lower is in fact encountered, it may be concluded that there is a significant degree of resemblance. This method has been applied to the sequence differences between *melanogaster* and *pseudoobscura,* with the result shown in table 4. Evidently the two species are not more alike than could easily result from chance alone.

Table 4

Comparison of the required and calculated numbers of inversions to change the melanogaster into the pseudoobscura sequences.

ELEMENT	A	B	C	D	E	TOTAL
Loci	13	6	6	6	7	
Inversions required	7	2	4	3	3	19
Inversions calculated	7.6	3.0	3.0	3.0	3.7	20.3

In any series of successive loci, each locus has two neighbors, and each of these may be either "right" or "wrong"—that is, may or may not be one that lies adjacent to the given locus in the arbitrarily chosen standard ordinal sequence. For a terminal locus there is only one adjacent locus, but the terminal position itself may be taken as constituting a "connection" and then may be treated in the same way, a "right connection" here meaning that the terminal locus is not only terminal in the chosen standard sequence but also lies at the same end (proximal or distal). There are then n + 1 "connections" in any sequence of n loci. It may be shown that in any series of random arrangements the average number of "right" connections is two, regardless of the value of n. It is also evident that any single inversion changes two and only two connections, since it has two ends, each of which must fall in a connection. These two relations are very helpful in working out the consequences of successive inversions. The conception of "right connections," however, is responsible for the incorrect conclusion drawn by STURTEVANT and TAN. One of the right connections appearing in their comparison has now disappeared, as a result of the revision of the *pseudoobscura* sequence recorded above (it may be noted that this revision does not change the number of inversions required to transform one sequence into the other). The result now is that there are 13 "right" connections (that is, identical in the two species) where ten are expected (two in each of the five elements) on chance alone. The difference is probably not a significant one.

This analysis does not take into account the distances concerned—a connection may be assumed to be alike in two elements regardless of the amount of crossing over shown. Thus, in element B, jaunty and hook are adjacent in the series of corresponding genes both in *melanogaster* and in *pseudoobscura;* but the crossover values are 5.2 and 23.1, respectively. In such a case, it is probable that further corresponding loci will show that the sequence resemblance is an accidental one. For certain pairs of loci that lie quite close together, the situation is different, as may now be shown.

Yellow and scute are quite near each other in *melanogaster, simulans, pseudoobscura, virilis, ananassae,* and *willistoni;* they are separated by other loci in *affinis.* (Unpublished data of DR. W. P. SPENCER indicate that they are also separated in *hydei.*) Notch and white are from one to four units apart in *melanogaster, simulans* (Notch is not known here, but facet serves to identify its locus), *ananassae, montium, hydei, virilis,* and *pseudoobscura.* Miniature and dusky give few if any crossovers in *melanogaster, ananassae, pseudoobscura,* and *virilis.* It may be noted that these are all short distances, not only in terms of crossing over, but also in terms of the salivary gland chromosome maps of *melanogaster,* which is the only species for which element A is adequately known in the salivary glands.

It may be concluded that inversions with end-points falling within these short sections have not become established during the differentiation of these species, except for those between yellow and scute in *affinis* and in *hydei.* It may be observed that all three sections appear to exist intact in both of the subgenera studied, since *virilis* and *hydei* belong to Drosophila while all the other forms named are members of the subgenus Sophophora. Evidently, then, these sections represent associations of loci that have been in their present conditions for a very long period of time.

Discussion

The essential argument showing the improbability of the incorporation of a translocation into a population has been published a number of times and has been prevalent in genetic thought since an earlier date; nevertheless, it seems opportune to recapitulate it briefly here, with specific reference to the Drosophilas.

Translocations have been observed to occur spontaneously under laboratory conditions (all the Drosophila translocations discovered prior to the advent of the X-ray technique fall in this category) and, in the few cases mentioned below, have been found in natural populations. It seems not unreasonable to assume therefore that they continually arise in nature with some very low frequency. Once having arisen, such a translocation would exist in the heterozygous state and thereafter chiefly in the heterozygous rather than homozygous state. The extensive studies on such rearrangements in Drosophila have amply demonstrated the selective disadvantage which their heterozygotes must suffer as a consequence of their production of unbalanced gametes. Thus a certain number of translocated chromosomes along with a number of normal chromosomes is lost in every generation; the percentage loss of the former may be considerable because of its low total frequency, whereas the percentage loss of the latter is usually insignificant. In this way selection discriminates against the less prevalent arrangement, progressively decreasing its frequency until it is wiped out.

It follows that the sole opportunity for a translocation to become established lies in its attaining a percentage frequency sufficiently great so that the discrimination of selection is not adequately expressed before chance changes the ratio of the frequencies in favor of the newer arrangement, whereupon the original is eliminated. A high percentage frequency can be reached only when the total number of chromosomes is low—that is, when the population size (N) is small. Furthermore, the population size must be small for chance to upset the ratio of the two arrangements in favor of the newer type.

For reciprocal translocations in which only the completely balanced types survive, WRIGHT (1940) has shown that fixation is difficult unless N is very low. Certain types of translocations might be expected not to encounter such drastic adverse selection, by virtue of the viability and fertility of individuals possessing an unbalanced (that is, heterozygous for a duplication or a deficiency or both) complement of chromosomes (STURTEVANT 1938b). The effect of this condition is to decrease the selection against the new arrangement and so to increase the maximum population size into which it might be incorporated. PROFESSOR WRIGHT has given us his kind permission to present here some more exact probabilities which he has calculated as an extension to his previous statements (WRIGHT 1940). For a translocation whose unbalanced products are completely inviable, the probabilities of fixation are of the order of 10^{-3} if $N = 10$, 2×10^{-6} if $N = 20$, and 3×10^{-14} if $N = 50$. In the most favorable case, when the heterozygous unbalanced products are both viable and fertile (only the homozygous deficiencies being eliminated), the probabilities are roughly 3×10^{-3} in populations of 20, 4×10^{-6} in populations of 50, and 3×10^{-10} in populations of 100. For the case of a small insertional translocation whose homozygous duplication product is not always accompanied by a homozygous deficiency and consequently may be viable and fertile, the probabilities must be somewhat greater than those given for the last case. Nevertheless, they are still very small and cannot greatly increase the maximum values indicated for N in the last case.

The various factors involved in the incorporation of a translocation into a population may be summarized as follows:

1. The translocation must occur in at least one of the individuals of a subpopulation. As N becomes smaller, the probability of this event becomes smaller, the relation being an approximately linear one.
2. The translocation must become incorporated into the whole subpopulation. N must be very small (see figures above); as N increases, the probability decreases rapidly.
3. The subpopulation must not become extinct. As the value of N increases, the probability that the subpopulation will survive increases.
4. The translocation must maintain itself in the descendants of the subpopulation, in competition with normal chromosomes that may be introduced by interbreeding with other subpopulations. As N increases, this probability also increases, since each introduced chromosome has a smaller effect on the composition of the subpopulation.

The third factor might be omitted, since it could be treated merely as decreasing the total number of subpopulations concerned. It is included here to emphasize the point that, if there is variability in the size of the subpopulations, the smaller ones are more likely to become extinct than the larger ones.

The final probability of the establishment of a translocation is the product of the four probabilities just enumerated. The first, third, and fourth components increase with an increase in N, the second decreases very rapidly with an increase in N. The result is that the product of the four must always be very small. The only situation in which the establishment of a translocation seems likely to occur with any considerable frequency—even if a long period of time be assumed—is that in which hermaphroditic individuals frequently self-fertilize—that is, when there are many subpopulations in which $N = 1$. It would thus seem desirable to have suspected cases of translocation in Drosophila supported by more conclusive evidence than has been considered necessary in the past.

For a translocation to be observed in a natural population, it must be found in the relatively short interval between its origin and its elimination. It seems significant that, despite the extensive genetic and cytological analyses of natural populations of Drosophila, not one translocation has been reported.

In the grasshopper Trimerotropis, CAROTHERS (1931) found a single individual heterozygous for a translocation among fifteen specimens from one locality. A more substantial case from the population standpoint is that reported by WHITE (1940). In 1934 he found a specimen of the grasshopper Metrioptera heterozygous for a translocation and in 1937 recognized the same translocation in some more individuals from the same locality. We are not familiar with the genetic properties of this translocation, with the statistical data for this case, or with the population mechanics of this genus; it therefore seems unwise to discuss the situation further.

Among hermaphroditic plants there are many instances of translocations that have become established. Here, however, the population size may frequently reach the theoretical minimum value of one and remain essentially that for several generations.

A class of "neutral" translocations which might be expected to give non-random segregation in the heterozygote includes those found in *virilis americana* and *virilis texana,* where two elements, ordinarily separate, have fused near the centromere. Such a fusion would seem likely to result, in the heterozygote, in almost completely regular disjunction of the individual rods from their attached homologs. If, in this particular case, there is some slight irregularity in disjunction, it would seem to be more than compensated for by the small sizes of the populations in which these two subspecies are found (PATTERSON, STONE, and GRIFFEN 1940).

RHOADES (1940) has demonstrated the instability inherent in a telocentric chromosome in maize; both the X and fourth chromosomes of *melanogaster,* once considered to have terminal centromeres, have been shown to have two arms (KAUFMANN 1934; GRIFFEN and STONE 1939, 1940). The view of NAWASCHIN (1916) and LEWITSKY (1931) that telocentric chromosomes do not exist is considerably strengthened. On this basis, apparent rods in Drosophila have a minute heterochromatic second arm. This arm would then serve as an anchor on which another element, less its centromere, could become attached by translocation, forming a V.

The reverse change, the formation of two rods from a V, demands an additional centromere. STURTEVANT and TAN (1937) suggest that "perhaps the spindle attachment of the Y chromosome is somehow utilized, since the properties of this chromosome allow it to be present as an extra or a fragment without damage to the organism." Means are thus provided for increasing the chromosome number. The one clear instance where such an increase has occurred, in *obscura* Gershenson, involves the breakage of a V-shaped element into two rods.

From the genetic data and metaphase chromosome configurations in *melanogaster, willistoni, immigrans, virilis,* and *pseudoobscura,* it is apparent that they all differ in the associations of their elements. Regardless of the metaphase chromosome configuration that may be assumed ancestral to that of all the Drosophilas, both fusion and fragmentation must have taken place. Although not in complete agreement with the systematic relationships, perhaps the simplest interpretation is that the ancestral type corresponded to the *virilis* configuration, five pairs of rods and a pair of dots. With slight modifications, mostly in the distribution of heterochromatin, this is the most common in the subgenus Drosophila. All other types can be derived by simple fusions of elements and changes in the positions of the centromeres within the elements by inversions, with the exception of *obscura* Gershenson and *ananassae,* as noted above. Within *virilis* itself, a single translocation of this kind has evidently given rise to the configuration found in subspecies *texana;* and two translocations from the *texana* arrangement have produced that of subspecies *americana.* The sequences found within the elements are in agreement with this conclusion (PATTERSON, STONE, and GRIFFEN 1940).

It seems clear that, in general, decrease in chromosome number is more easily brought about than is increase. Accordingly lower chromosome numbers may be regarded as probably (though not necessarily in every case) more recent than higher numbers. This conclusion will hold, however, only in forms—such as Drosophila—in which chromosomes frequently have nearly terminal centromeres, with one arm made up of heterochromatin.

A Note on the Relations Between the Species Groups of Sophophora

It was suggested by STURTEVANT (1940) that the *willistoni* (AD, BC, EF) and *melanogaster* (A, BC, DE, F) chromosome configurations have been derived from the type now found in *pseudoobscura* (AD, B, C, E, F). If this be so, the first event was probably the union of elements B and C, to give a hypothetical arrangement AD, BC, E, F. From this, a union of E and F presumably gave the *willistoni* type, while a translocation between AD and E gave the *melanogaster* one.

The hypothetical arrangement, which seems likely to have existed even if the sequence of events was different from that outlined above, has not been reported. In the absence of the identification of the X, it would be listed as the *melanogaster* type; accordingly we have been led to examine the descriptions of the phenotypic characters of the few Sophophoras for which such a report exists. One of them, *D. suzukii,* turns out in fact to be intermediate between the *melanogaster* and *obscura* species groups, as shown in table 5. We are inclined to surmise that this species (recorded by

Table 5

Comparison of some phenotypic characters of the melanogaster and obscura species groups and D. suzukii.

	COLOR	COSTAL INDEX	4TH VEIN INDEX	SECOND ORAL	MIDDLE ORBITAL
melanogaster group	yellow	1.2–2.2	2.2–2.7	long	short
suzukii	yellow	4.0*	2.2	long	long
obscura group	black	2.7–3.0	1.7–2.1	short	long

* The description given by KIKKAWA and PENG lists the costal index as 4.0; their illustration (Plate 32) suggests that this is perhaps a misprint for 3.0, which would make *suzukii* fall within the range of variation of the *obscura* species group.

KIKKAWA and PENG (1938) from Japan and China) will be found to have the hypothetical configuration.

The Effect of Pericentric Inversion on the Elements

None of the changes discussed above interferes with the integrity of the elements. There remains, however, one mechanism which may exchange genic material from one element to another. If two elements are united to form a V, an inversion across the centromere (pericentric) will shift genes from one to the other, and *vice versa*. Heterozygotes for this type of inversion are at a selective disadvantage because single crossovers within the inverted section produce inviable duplication-deficiency zygotes, thus leading to a situation similar to that of translocations. However, if crossing over be hampered, either by the nature of the inversion or by the presence of a crossover suppressor, the new arrangement may become established in the population. The products of a subsequent separation of these two limbs will simulate the effects of a reciprocal translocation.

MILLER (1939) has recorded such an inversion in *algonquin*, occurring with a high frequency in eight out of nine localities studied between Quebec, Canada, and Wooster, Ohio. In this case, only one element (E) was involved (this element being a V in this species), and another associated inversion acted as a crossover suppressor. The fact that this inversion has been found to have a high frequency makes more probable the occurrence of similar inversions in other species where a V is composed of two different elements.

It is self-evident that the scheme of the integrity of the elements must break down for the more distantly related species of the Diptera. Where this will first happen is not clear. The common metaphase chromosome configuration among the higher Diptera has six chromosomes; whether these are the Drosophila elements with a few alterations or whether they repre-

Table 6

Summary of the corresponding mutant types in the different species.

melanogaster	*simulans*	*pseudoobscura*	*affinis*	*virilis*	*ananassae*	other species
ELEMENT A						
yellow	yellow	yellow	yellow	yellow	yellow	*busckii, willistoni*
scute	scute	scutellar	scute	scute	scute	*busckii, takahashii, willistoni*
prune	prune					
white	white	white	white	white	white	*busckii, hydei, montium*
facet	facet					
Notch		Notch		Notch	Notch	*busckii, funebris, hydei,*
echinus		echinus		echinus?		[*montium*
bifid						*willistoni* (triple)
ruby	ruby			magenta?		
crossveinless	crossveinless			crossveinless		
vesiculated	vesiculated			vesiculated?		
cut	cut	beaded	cut	ragged?	cut	
singed	singed	forked		singed	singed?	*busckii, willistoni* (forked)
ocelliless	ocelliless					
lozenge	lozenge	glazed		glazed, rugose		[*montium? willistoni*
vermilion	vermilion	vermilion	vermilion	vermilion	vermilion?	*busckii? funebris, hydei?*
miniature		miniature	miniature	miniature	miniature	*busckii, takahashii*
dusky	dusky	dusky		dusky	dusky	*algonquin*
garnet	garnet			apricot?		
rudimentary	rudimentary			rudimentary		*willistoni* (square)
forked	forked	singed		forked	forked?	*busckii, funebris? willistoni*
Beadex		Pointed		Beadex	Beadex	[(stubby)
fused	fused					
bobbed	bobbed	bobbed	bobbed	bobbed	on chrom. IV	*hydei, funebris*
ELEMENT B						
melanogaster	*simulans*	*pseudoobscura*	*affinis*	*virilis*	*ananassae*	other species
dachsous				dachsous		
net		tangled	net	plexus	plexus	
Star		Rough		Star		
Curly		Curly				
dumpy	Truncate		truncate	Clipped		
dachs						*willistoni* (approximated)
abrupt		incomplete				
black	black			black?		
jaunty		jaunty	jaunty			*montium* (curled)
reduced				reduced		
hook		hook				
ELEMENT C						
melanogaster	*simulans*	*pseudoobscura*	*affinis*	*virilis*	*ananassae*	other species
straw	straw		straw	straw		
apterous						*willistoni*
cinnabar		orange	cinnabar	scarlet?		*athabasca, algonquin, funebris*
vestigial	vestigial-nick	Jagged		vestigial		*willistoni* (Clipped, Scalloped)
gap					gap	
curved		curved				
arc	arc					
plexus		plexus		branched?		*montium*
brown		purple		eosinoid		
Plum					Plum	
lanceolate		narrow	narrow			
	polychaete	polychaete, Scute				

ELEMENT D

melanogaster	*simulans*	*pseudoobscura*	*affinis*	*virilis*	*ananassae*	other species
veinlet		short	veinlet	short veins		*willistoni* (short)
javelin	javelin	slender				
Henna				Garnet		
sepia	sepia	sepia				
hairy			hairy			
tilt		tilt	tilt			
scarlet	scarlet	scarlet	scarlet	cinnabar?		
ascute		ascute	ascute	hunch		
r. incompletus	r. incompletus	snapt				*willistoni* (mstup)
		compressed		telescoped		*willistoni* (deformed, serrate)
		sex-ratio	sex-ratio			*athabasca, azteca, melanica, obscura*

ELEMENT E

melanogaster	*simulans*	*pseudoobscura*	*affinis*	*virilis*	*ananassae*	other species
pink	peach	pink, claret?	pinkish, claret?			
curled		upturned				*montium* (curled-b)
Stubble		Stubble				
bithorax		bithorax	bithorax			*athabasca*
aristapedia	aristapedia	aristapedia				
glass		glass		varnished		
Delta	Delta	Smoky	Delta	Confluent, Delta	Plexate	*busckii, montium* (Confluent
Hairless	hairless					*funebris, montium*
cardinal		cinnabar			cardinal	*azteca*
crumpled		crumpled	crumpled	concave		
claret	claret	claret, pink?	pinkish, claret?	brick?		
		Bare			Off?	
		pauciseta	pauciseta			
				Puffed	Puffed	

ELEMENT F

melanogaster	*simulans*	*pseudoobscura*	*affinis*	*virilis*	*ananassae*	other species
abdomen rotatum			abdomen rotatum	abdomen rotatum		
grooveless		grooveless				
shaven	Shaven		reduced?	stubby	Shaven	
Minute-4	Minute-4					

sent quite different gene combinations can be answered only after a more thorough analysis of their genetics is made.

Summary

The six chromosome arms of *D. melanogaster* (X, IIL, IIR, IIIL, IIIR, IV) retain their essential identity among the species of Drosophila so far studied. They are here called "elements" and are designated by the letters A to F in the order just stated.

The corresponding mutant genes on which this conclusion is based are summarized in table 6 of this paper; and the relation of the letter designations to the various chromosome terminologies previously used is summarized in table 1.

Element E is consistently the longest of the six, when studied in the salivary gland chromosomes; element F is always very short. The others are so nearly the same in size that their lengths are not diagnostic.

Analysis of the mathematical properties of sequences subjected to repeated inversions indicates that, in the specific case of *melanogaster* and *pseudoobscura,* the sequences are not significantly more alike than might result from chance.

Three pairs of loci—yellow and scute, Notch and white, miniature and dusky—apparently represent sections which have remained intact over a long period, since the two members of each pair are associated in numerous species and in representatives of both subgenera studied. Yellow and scute are separated, however, in *affinis* and in *hydei.*

The conclusion that the elements remain intact within the group means that few or no translocations have become established in the history of these two subgenera. Analysis of the conditions necessary for such establishment leads to the conclusion that the supposed instances of its occurrence as between species are not adequately established—with the exception of the attachment of part of element A to element F in *ananassae.*

The various ways in which the elements are attached to each other may be supposed to result from a special type of translocation in which both breaks occur in heterochromatin near the centromeres. Under the conditions found in most Drosophila species, this type of translocation more easily decreases chromosome number than increases it. It may be surmised that the configuration found in *virilis* (where all six elements are separate) is the most primitive one among those yet analyzed in Drosophila.

LITERATURE CITED

BAUER, H., and T. DOBZHANSKY, 1937: A comparison of gene arrangements in *Drosophila azteca* and *D. athabasca.* Genetics **22:** 185.

BRIDGES, C. B., 1935: Salivary chromosome maps. J. Hered. **26:** 60–64.

CAROTHERS, E. E., 1931: The maturation and segregation of heteromorphic homologous chromosomes in Acrididae (Orthoptera). Biol. Bull. **61:** 324–349.

CHINO, M., 1929: Genetic studies on the Japanese stock of *Drosophila virilis.* Report 1. Japan. J. Genet. **4:** 117–131.

1936–1937: The genetics of *Drosophila virilis.* Japan J. Genet. **12:** 187–210, 257–277; **13:** 100–120.

CLAUSEN, R. E., 1923: Inheritance in *Drosophila hydei.* Amer. Nat. **57:** 52–58.

CREW, F. A. E., and R. LAMY, 1935: Linkage groups in *Drosophila pseudoobscura.* J. Genet. **30:** 15–29.

DOBZHANSKY, T., 1937: Genetics and the origin of species. 364 pp. New York: Columbia Univ. Press.

———, and D. SOCOLOV, 1939: Structure and variation of the chromosomes in *Drosophila azteca.* J. Hered. **30:** 3–19.

———, and C. C. TAN, 1936: Studies on hybrid sterility. III. A comparison of the gene arrangements in two species, *Drosophila pseudoobscura* and *Drosophila miranda*. Z. i. A. V. **72:** 88–114.

DONALD, H. P., 1936: On the genetical constitution of *Drosophila pseudoobscura,* race A. J. Genet. **33:** 103–122.

FERRY, R. M., R. C. LANCEFIELD, and C. W. METZ, 1923: Additional mutant characters in *Drosophila willistoni*. J. Hered. **14:** 372–384.

FUJII, S., 1936: Salivary gland chromosomes of *Drosophila virilis*. Cytologia **7:** 272–275.

GRIFFEN, A. B., and W. S. STONE, 1939: The demonstration of a new chromosome arm in *Drosophila melanogaster*. Genetics **24:** 73.

1940: Studies in the genetics of Drosophila. IX. The second arm of chromosome 4 in *Drosophila melanogaster*. Univ. Texas Pub. No. **4032:** 201–207.

HEITZ, E., 1934: Über α- und β-Heterochromatin sowie Konstanz und Bau der Chromomeren bei Drosophila. Biol. Zbl. **54:** 588–609.

HORTON, I. H., 1939: A comparison of the salivary gland chromosomes of *Drosophila melanogaster* and *Drosophila simulans*. Genetics **24:** 234–243.

HOWLAND, R. B., E. A. GLANCY, and B. P. SONNENBLICK, 1937: Transplantation of wild type and vermilion eye disks among four species of Drosophila. Genetics **22:** 434–442.

HUGHES, R. D., 1939: An analysis of the chromosomes of the two subspecies *Drosophila virilis virilis* and *Drosophila virilis americana*. Genetics **24:** 811–834.

KAUFMANN, B. P., 1934: Somatic mitoses of *Drosophila melanogaster*. J. Morph. **56:** 125–155.

1937: Morphology of the chromosomes of *Drosophila ananassae*. Cytologia (Fujii Jubilee Volume) 1043–1055.

KERKIS, J., 1936: Chromosome configuration in hybrids between *Drosophila melanogaster* and *Drosophila simulans*. Amer. Nat. **70:** 81–86.

KIKKAWA, H., 1936: Two races of *Drosophila montium*. Japan. J. Genet. **12:** 137–142.

1938: Studies in the genetics and cytology of *Drosophila ananassae*. Genetica **20:** 458–516.

———, and R. T. PENG, 1938: Drosophila species of Japan and adjacent localities. Japan. J. Zool. **7:** 507–552.

KRIVSHENKO, Y. D., 1939: A comparative genetic study on *Drosophila busckii* and *Drosophila melanogaster*. Acad. Sci. Ukrainian SSR, Inst. Zool., Memoirs on Genetics **3:** 15–89.

LANCEFIELD, D. E., 1922: Linkage relations of the sex-linked characters in *Drosophila obscura*. Genetics **7:** 335–384.

LANCEFIELD, R. C., and C. W. METZ, 1921: Non-disjunction and the chromosome relationships of *Drosophila willistoni*. Proc. Nat. Acad. Sci. **7:** 225–229.

1922: The sex-linked group of mutant characters in *Drosophila willistoni*. Amer. Nat. **56:** 211–241.

LEWITSKY, G. A., 1931: The morphology of the chromosomes. Bull. Appl. Bot. **27:** 19–173.

LUERS, H., 1937: Ein Beitrag zur vergleichenden Genetik an Hand des dominanten Bobbed im Y-Chromosome von *Drosophila funebris*. Z. i. A. V. **74:** 70–90.

MACKNIGHT, R. H., 1939: The sex-determining mechanism of *Drosophila miranda*. Genetics **24:** 180–201.

METZ, C. W., 1916: Chromosome studies on the Diptera. II. The paired association of chromosomes in the Diptera, and its significance. J. Exp. Zool. **21:** 213–279.

———, M. S. Moses, and E. D. Mason, 1923: Genetic studies on *Drosophila virilis* with considerations on the genetics of other species of Drosophila. Carnegie Inst. Wash. Pub. **328,** 94 pp.

Miller, D. D., 1939: Structure and variation of the chromosomes in *Drosophila algonquin*. Genetics **24:** 699–708.

Morgan, T. H., C. B. Bridges, and A. H. Sturtevant, 1925: The genetics of Drosophila. Bibliogr. Genet. **2:** 1–262.

Mori, K., 1937: A study on the development of pigments of various eye color mutants of Drosophila. Japan. J. Genet. **13:** 81–99.

Moriwaki, D., 1935: Some mutant characters in *Drosophila ananassae*. Genetica **17:** 32–46.

1938: The genetics of some mutant characters in *Drosophila ananassae*. Japan. J. Genet. **14:** 1–22.

Muller, H. J., 1940: Bearings of the Drosophila work on systematics. The New Systematics, 185–268. Oxford: Clarendon Press.

Nawaschin, S., 1916: Sur quelques charactères de l'organisation interne des chromosomes. Rec. d'Art. Sci. dédié à C. Timiriaseff. Moscow. 185–214.

Osima, T., 1940: Cyto-genetic studies of *Drosophila montium*. (Report 1.) Cytologia **10:** 450–457.

Pätau, K., 1935: Chromosomenmorphologie bei *Drosophila melanogaster* und *Drosophila simulans* und ihre genetische Bedeutung. Naturwissenschaften. **23:** 537–543.

Patterson, J. T., W. S. Stone, and A. B. Griffen, 1940: Studies in the genetics of Drosophila XI. Evolution of the *virilis* group in Drosophila. Univ. Texas Pub. No. **4032:** 218–250.

Rhoades, M. M., 1940: Studies of a telocentric chromosome in maize with reference to the stability of its centromere. Genetics **25:** 483–520.

Slizynska, H., and B. M. Slizynski, 1941: A salivary gland chromosome map of *Drosophila funebris*. Fabr. Proc. Seventh Int. Genetical Cong. 226.

Spencer, W. P., 1927: The X chromosome of *Drosophila hydei*. J. Exp. Zool. **47:** 441–466.

1940: Subspecies, hybrids and speciation in *Drosophila hydei* and *Drosophila virilis*. Amer. Nat. **74:** 157–180.

Stalker, H. D., 1940: Chromosome homologies in two subspecies of *Drosophila virilis*. Proc. Nat. Acad. Sci. **26:** 575–578.

Stubbe, A., and M. Vogt, 1940: Über die Homologie einiger Augenfarbgene bei *Drosophila funebris* und *D. melanogaster* an Hand von Augentransplantationen. Z. i. A. V. **78:** 251–254.

Sturtevant, A. H., 1918: A parallel mutation in *Drosophila funebris*. Science **48:** 72–73.

1921 a: Genetic studies on *Drosophila simulans*. II. Sex-linked group of genes. Genetics **6:** 43–64.

1921 b: Genetic studies on *Drosophila simulans*. III. Autosomal genes. General discussion. Genetics **6:** 179–207.

1921 c: A case of rearrangement of genes in Drosophila. Proc. Nat. Acad. Sci. **7:** 235–237.

1929: The genetics of *Drosophila simulans*. Carnegie Inst. Washington Publ. **399:** 1–62.

1932: The use of mosaics in the study of the developmental effects of genes. Proc. Sixth Int. Cong. Genetics **1:** 304–307.

1938 a: The homologies of the chromosome arms of the different species of Drosophila. Genetics **23:** 173–174.

1938 b: Essays on evolution III. On the origin of interspecific sterility. Quart. Rev. Biol. **13:** 333–335.

1939: On the subdivision of the genus Drosophila. Proc. Nat. Acad. Sci. **25:** 137–141.

1940: Genetic data on *Drosophila affinis,* with a discussion of the relationships in the subgenus Sophophora. Genetics **25:** 337–353.

———, and T. DOBZHANSKY, 1936: Observations on the species related to *Drosophila affinis,* with descriptions of seven new forms. Amer. Nat. **70:** 574–584.

———, and C. C. TAN, 1937: The comparative genetics of *Drosophila pseudoobscura* and *D. melanogaster.* J. Genet. **34:** 415–432.

TAN, C. C., 1935: Identification of the salivary gland chromosomes in *Drosophila pseudoobscura.* Proc. Nat. Acad. Sci. **21:** 200–202.

1937: The cytological maps of the autosomes in *Drosophila pseudoobscura.* Z. Zellf. mik. Anat. **26:** 439–462.

TIMOFÉEFF-RESSOVSKY, N. W., 1927: Studies on the phenotypic manifestation of hereditary factors. I. On the phenotypic manifestation of the genovariation radius incompletus in *Drosophila funebris.* Genetics **12:** 128–198.

WEINSTEIN, A., 1920: Homologous genes and linear linkage in *Drosophila virilis.* Proc. Nat. Acad. Sci. **6:** 623–639.

WHITE, M. J. D., 1940: A translocation in a wild population of grasshoppers. J. Hered. **31:** 137–140.

WRIGHT, S., 1940: Breeding structure of populations in relation to speciation. Amer. Nat. **74:** 232–248.

30

A GENE IN *DROSOPHILA MELANOGASTER* THAT TRANSFORMS FEMALES INTO MALES†

A culture, counted in April, 1944, gave 20 males of an unexpected phenotype (wild type for a series of sex-linked alleles) that could not be accounted for by any of the usual interpretations current in Drosophila genetics. These males were sterile, but studies of descendants of their sibs have shown their nature.

It is now clear that these were XX diploids—that is, had the usual composition of normal females, but were homozygous for a third chromosome recessive gene that may be called "transformer" (symbol *tra*). Heterozygotes (tra^{+}/tra) do not differ from wild type in any way yet detected, nor do XY *tra/tra*—that is, the gene is recessive and also has no effect on males. XX *tra/tra* individuals appear like normal males. They have fully developed sex-combs, male-colored abdomen, normal male abdominal tergites, anal plates, external genitalia, genital ducts, sperm pump, and paragonia, and show the usual rotation of the genital and anal segments through 360°. They mate with females readily and normally. The testes, however, while normal in color, elongate, curved, and attached to the ducts, are much reduced in size. The testis size is variable, but in no case does it approach that found in the normal brothers. In two respects these specimens resemble females: they are often larger than their XY brothers, and they more often appear in first counts from cultures, suggesting that their development rate approaches that of females.

These XX *tra/tra* individuals are all wholly sterile. It seemed possible that this was due to their not carrying a Y chromosome. Since the gene is an autosomal recessive, it was easy to transfer it to an attached-X line, and to obtain $\widehat{XX}$ Y *tra/tra* specimens. These differ in no way from those without a Y, and are equally sterile. These same cultures should give $\widehat{XX}$ X *tra/tra*—that is, superfemales homozygous for the new gene. In a large series of cultures only one such specimen has been obtained, indicating that the gene does not increase the usual very low viability of superfemales. This one specimen had the beaded wings characteristic of superfemales, but had the male genitalia of XX *tra/tra;* it did, however, have sex-combs with only

† Reprinted from Genetics, Vol. 30, 1945, pp. 297–299.

about half the normal number of teeth, and accordingly may be taken as indicating that superfemales are not as strongly modified toward maleness as are ordinary females—perhaps the best support yet produced for the view that superfemales are more "female" than are females.

The gene has also been introduced into a triploid line. 3X3A *tra/tra/tra* triploids resemble the corresponding diploids—that is, they are males with large wing-cells and ommatidia, with small testes, and they are sterile. Unfortunately, as the experiment was carried out, the 2X3A (triploid intersex) classes could not all be identified with certainty; however, large numbers of possible 2X3A *tra/tra/tra* were tested, and no fertile males were encountered. It seems certain that such specimens are sterile, and it is probable that they still have the abnormal wings and eyes usual among triploid intersexes. It is not known whether one or two doses of *tra* have any influence on the sex grade of triploid intersexes.

A few preliminary tests having indicated that *tra* lies in chromosome III, it was balanced against *"Dcx,"* a third chromosome carrying the dominant Dichaete (lethal when homozygous) and a complex rearrangement that usually prevents effective crossing over except near the ends of the chromosome. A few sample results follow:

♀ w^a/w^a *Dcx/tra* × ♂ w^+ *Dcx/tra* → 108 ♀ *D*, 78 ♂ +, 120 ♂ *D* w^a, 63 ♂ w^a (plus 1 ♂ *D*, due either to non-disjunction of the X's of the mother, or to crossing over between *D* and *tra*).

♀ w^a/w^a *Dcx/tra* × ♂ w^+ *tra/tra* → 64 ♀ *D*, 50 ♂ +, 47 ♂ *D* w^a, 60 ♂ w^a.

♀ *yv Dcx/tra* × ♂ w^a *tra/tra* → 137 ♀ *y v D*, 1 super-♀ + (described in text), 126 ♂ *y v*, 194 ♂ w^a *D*, 190 ♂ w^a (plus 1 ♂ *y v D* evidently due either to detachment of the X's of the mother or to crossing over between *D* and *tra*).

Using this strain it is easy to get males known to be homozygous for *tra*, and accordingly tests for its locus in chromosome III can readily be made. The methods used are not new in principle, so it need only be stated that they show the gene to lie between *st* (locus 44.0) and *cp* (locus 45.3). This places it in the left limb of III (= element D).

The effects of *tra* on the phenotypes of mutant characters showing sexual dimorphism have been studied in a few cases. Scute-3 is lethal in females but fully viable in males (DUBININ 1933). It appears that this lethal effect is not overcome by homozygous *tra*—that is, sc^3/sc^3 *tra/tra* did not survive.

Eosin (w^e) is paler and more yellowish in males than in females (BRIDGES and BREHME 1944), and facet (*fa*) is more extreme in males than in females (BRIDGES and BREHME 1944). In both these cases it is known that the normal sexual dimorphism is not due to the presence of a Y in the males and its absence in the females. In both these mutant types, the XX *tra/tra* individuals resembled their XX tra^+/*tra* sisters, and differed distinctly from their XY brothers.

In all three tested instances, therefore, the XX *tra/tra* individuals were like the females rather than the males with respect to sexual dimorphism dependent on mutant genotypes—though they were like males rather than females in the sexual characters (other than size and development rate) in which normal males and females differ.

One other similar test was carried out, with the semi-dominant sex-linked gene Bar (*B*). *B/B* and *B/+* in XX *tra/tra* closely resemble the same two types in their sisters.

It may be argued that the transformed individuals are better described as intersexes, rather than as males. I have preferred to call them males simply because they are indistinguishable from males in ordinary experiments—unless one dissects out their testes, and even the testes show no female characteristics. Perhaps the best reason for considering them intersexes is the short sex-combs found on the single transformed superfemale that was seen. In any case, the gene is more extreme in its shifting of the sex of the individual than is any other yet described in any Drosophila. Only that recorded by LEBEDEFF (1934) for *D. virilis* was shown to be due to a gene in element D. The specimens here described resemble the most extreme types recorded by LEBEDEFF, but the extreme variability that he observed is here absent. It remains doubtful whether or not the genes in the two species represent corresponding loci. The present gene is certainly different from the one previously described intersex gene in *D. melanogaster,* since that lies in the second chromosome (L. V. MORGAN, in MORGAN, REDFIELD, and MORGAN 1943).

A mutant type of this sort, if the gene is autosomal, may easily be overlooked. It will give a sex ratio of 3 ♀ : 5 ♂ if both parents are heterozygous, but this will be difficult to distinguish from 1:1. The mating of heterozygous female × homozygous male gives 1 ♀ : 3 ♂, but this mating will be less likely to occur. The most likely situation for its detection will be that which actually led to the discovery of this one—namely, the presence of males of classes, with respect to sex-linked genes, that appear impossible from the known parentage. In those species where element D is part of X, detection through the distortion of the sex ratio should be easier.

Summary

A recessive gene, *tra,* lying between 44.0 and 45.3 in the third chromosome of *Drosophila melanogaster,* transforms diploid females into sterile males.

The gene is without effect on males (that is, on XY diploids).

A single specimen indicates that *tra/tra* superfemales are intersexual but strongly male-like.

No fertile triploids homozygous for *tra* were obtained.

The gene is without effect on the sexual dimorphism of the mutant types scute-3, eosin, and facet—in all three of these cases the XX *tra/tra* specimens resemble normal females, rather than normal males, in the manifestation of the mutant phenotypes.

LITERATURE CITED

BRIDGES, C. B., and K. S. BREHME, 1944: The mutants of *Drosophila melanogaster*. Carnegie Instn. Washington, Publ. **552.** 257 pp.

DUBININ, N. P., 1933: Step-allelomorphism in *Drosophila melanogaster*. J. Genet. **27:** 443–464.

LEBEDEFF, G. A., 1934: Genetics of hermaphroditism in *Drosophila virilis*. Proc. Nat. Acad. Sci. **20:** 613–616.

MORGAN, T. H., H. REDFIELD, and L. V. MORGAN, 1943: Yearb. Carnegie Instn. **42:** 171–174.

31

INTERSEXES DEPENDENT ON A MATERNAL EFFECT IN HYBRIDS BETWEEN *DROSOPHILA REPLETA* AND *D. NEOREPLETA*†

Drosophila repleta Wollaston and *D. neorepleta* Patterson and Wheeler are closely similar species, the former widely distributed and the latter known from Guatemala. It was found by Dr. E. Novitski that these species occasionally cross, and that the F_1 females sometimes give a few offspring when mated to repleta males (see Wharton 1942 and Sturtevant 1946).

I have found a sex-linked recessive white-eyed mutant type in *D. repleta* (actually not quite white, but retaining only a slight tinge of color). At least 5000 neorepleta females have been crossed to white repleta males (in a few cases the repleta males carried singed, another sex-linked recessive, rather than white). These matings included at least 500 mass cultures, of which 74 produced hybrid offspring—a total of 532 females and 635 males, all wild type for the sex-linked mutant characters used. The males had very narrow testes, and were wholly sterile. The females were variable; most of them had bristles somewhat reduced in size ("minute"), and many of them had three anal plates instead of the usual two—this last character suggesting intersexuality.

About 400 of these F_1 females were mated to white repleta males, and offspring were obtained from 34 of them—the total output being 70 wild-type females, 9 white females, 42 wild-type males and 58 white males. The expectation is for these classes to be equal; there is evidently a great deficiency of white females. Of the 9 recorded in this class, notes made on two make it possible to be certain (in the light of later results, to be described below) that they were intersexes, and it is possible that some of the other 7 were also of this nature, as were probably a few of the white males.

The wild-type males from this first backcross resembled the F_1 males in having long narrow testes and in being sterile. The wild-type females, however, included some moderately fertile individuals, which were again crossed to white repleta males. Such successive backcrosses of wild-type females to white repleta males have now been continued through many generations, and

† Reprinted from the Proceedings of the National Academy of Sciences, Vol. 32, 1946, pp. 84–87.

have resulted in strains presumably pure repleta in composition except for a section of the *X* chromosome near the locus of white. In some of these strains the double recessive, white singed, has been used instead of white; and in these strains all the wild-type males are sterile and have narrow testes, while all the white-singed males have normal testes and are as often fertile as are white-singed males with no neorepleta chromosomes in their immediate ancestry. Tests of the few crossovers between white and singed indicate that the locus of the narrow-testis gene (or genes) lies between white and singed, and very near white.

The wild-type females from the backcrosses, when mated to white repleta males, appeared to fall into two classes; some of them gave the four types of offspring in approximately equal numbers, whereas others continued to give a marked deficiency of white daughters. The daughters of the first type of female always repeated the first type of result; but the females of the second type commonly yielded few offspring, and their very existence as a distinct type was at first uncertain because of the confusing effects of sampling errors in the small families obtained from them. Finally, however, after four successive backcrosses, a single female of this type was obtained which was more fertile, and her descendants have retained this fertility. Presumably the neorepleta gene responsible for the unusual ratios was at first linked to another gene that decreases fertility in females largely repleta in constitution, and in this case the infertility gene was lost by crossing over. The later studies have all been carried out with descendants of this more fertile female. There are, however, enough fragmentary data to make it clear that the behavior is essentially the same in lines derived from other F_1 hybrids—i.e., the results are really due to genes derived from neorepleta, rather than to a mutation that occurred in this line.

From females of this more fertile line, backcrossed to white repleta males, a total of 33 wild-type daughters (of females giving a deficiency of white daughters) have been tested by white males. Of these, 16 gave approximately 1 + ♀ :1 w ♀ :1 + ♂ :1 w ♂; the remaining 17 all gave a deficiency of white females. The total counts from these 17 were: 472 + ♀, 5 w ♀, 63 w intersexes, 482 + ♂, 339 w ♂. The 5 white females presumably represent crossing over between the loci of white and of the critical gene in the *X* derived from neorepleta. Their nature can be more accurately determined when other sex-linked genes are introduced in the crosses; it is, however, already clear that the gene concerned is not the same as that responsible for the narrow testes of the hybrid males.

The intersexes are of an extreme type, with gonads very small (rudimentary ovaries in those cases where they were found at all); external genitalia missing or of abnormal male type; wings usually not expanded, and, when they are, usually with thickened veins; one or more (sometimes all four) scutellar bristles often absent. They are weak individuals—and evidently usually die before emergence. It is to be supposed that such preimaginal mortality is responsible for the difference in number of wild-type females (472) and white intersexes (63). While no systematic study has been made, white-eyed flies have been found dead in their puparia in such cultures.

These results suggest that there is an autosomal dominant gene, derived from neorepleta, that so conditions the eggs (before meiosis) that two repleta X chromosomes result in the development of intersexes rather than females. Evidently the action comes before meiosis, and the autosomal gene in question may be absent in the intersexes themselves. This interpretation has been confirmed by tests of the white brothers of intersexes. Such males, when crossed to pure repleta females, gave normal offspring of both sexes; but some of their daughters (presumably half of the daughters from half of these males, though the data are not extensive enough to establish this), when mated to repleta males, gave only intersexes and males.

It will be seen that this last experiment shows that the intersexes are not dependent on the presence of neorepleta cytoplasm, since their mothers were offspring of pure repleta females. The experiment also shows that the autosomal gene from neorepleta has no phenotypic effect on females that have received it from their father. The result likewise agrees with the earlier ones in indicating that an individual with two repleta X's is intersexual whether or not it carries this gene, provided the gene was present in its mother.

Conclusions

D. neorepleta carries an autosomal gene which, when present in single dose in a hybrid female, makes her eggs male in potentiality. This predisposition to maleness is only partially overcome by two repleta X's, and male-like intersexes result. One repleta X and one neorepleta X are sufficient to produce normal females. This autosomal gene is present in two doses in pure neorepleta eggs. Two neorepleta X's are sufficient to cause such eggs to develop into normal females; but in the F_1 hybrids, having one neorepleta X and one from repleta, some of these eggs develop into females with male-like anal plates.

Presumably the essential sex-determining mechanism is the same in the two species, but both the male-determining action of the autosomal gene and the female-determining action of the X are stronger in neorepleta.

LITERATURE CITED

STURTEVANT, A. H., "On the Dot-Chromosomes of *Drosophila repleta* and *D. hydei*," *Genetics*, **31** pp. 259-268 (1946).

WHARTON, L. T., "Analysis of the Repleta Group of *Drosophila*," *Univ. Texas Publication* **4228**, pp. 23–52 (1942).

32

A MAP OF THE FOURTH CHROMOSOME OF *DROSOPHILA MELANOGASTER*, BASED ON CROSSING OVER IN TRIPLOID FEMALES †

Under ordinary conditions there is so little crossing over in the fourth chromosome of *Drosophila melanogaster* that the usual method of constructing a map is not practicable. Deduction from the behavior of translocations has been utilized, but as will be shown here, has led to an incorrect result. Bridges and Brehme (1944) give the seriation *bt* (bent), *sv* (shaven), *ci* (cubitus interruptus), *gvl* (grooveless), *ey* (eyeless), with 0.2 per cent crossing over for the whole series. This crossover value is certainly too high; it may be doubted if as many as five crossovers have ever been detected from diploid females. The results presented below show also that the above sequence is altogether incorrect, the true order being *ci, gvl, bt, ey, sv,* (with a possibility that the positions of *ci* and *gvl* should be reversed).

Table 1

CROSSING OVER IN DIPLO-IV TRIPLOID FEMALES

MOTHER	0	1	2	3	1, 2	1, 3	2, 3	TOTAL
+/*ci gvl ey sv*	2017	3	34	20	1	1	1	2077
gvl sv/ci ey	305	0	4	2	0	0	0	311
Total	2322	3	38	22	1	1	1	2388
			TESTED CROSSOVERS					
+/*ci gvl ey sv*		2	23	12	0	0	0	
gvl sv/ci ey		0	3	2	0	0	0	
Total		2	26	14	0	0	0	

I have reported (Morgan, Sturtevant and Morgan, 1945) that there is an appreciable amount of crossing over in the fourth chromosomes of diplo-IV triploid females. This has made it possible to build up multiple recessive stocks, which have now been used in four-point test-cross experiments

† Reprinted from the Proceedings of the National Academy of Sciences, Vol. 37, 1951, pp. 405–407.

(table 1). (The mutant alleles used were *ci, gvl*, ey^{R}, and sv^{na}.) The crossover classes are designated by regional numbers based on the assumption that the sequence is as given, and are clearly inconsistent with any other sequence, except that for region 1 (between *ci* and *gvl*) there are three single crossovers and two doubles, which means that the seriation *gvl, ci, ey, sv* would give nearly as good a fit.

It happens that *ci*—and to a lesser extent *gvl* and *ey*—may at times be somewhat difficult to classify with certainty. I have, therefore, carried out genetic tests on many of these crossovers (usually by mating them to *ci gvl ey sv*, but in some cases by getting individuals homozygous for the crossover chromosome). In these tests the determination was, in each case, based on numerous individuals, thereby eliminating the uncertainty of classification. As the table shows, no such test was obtained on any of the three individuals recorded as double crossovers; it may be that all of them were wrongly classified—though the evidence now to be presented makes this unlikely.

Before the *ci gvl ey sv* stock was available several crosses were carried out in which females heterozygous for all four loci were test-crossed to males with two of the four recessives, and the crossovers between these two were then tested both for confirmation of the classification with respect to these two, and also for the presence of the other two recessives. These data do not permit any simple direct calculation of crossover values, but as table 2 shows, they do include four tested double crossovers. The occurrence of true doubles thus seems clear, though it is of course wholly unexpected when the total amount of crossing over is so low. No satisfactory explanation of this high coincidence has yet been produced.

Table 2

Tested Crossovers from 4-Point Test-Crosses

MOTHER	FATHER	1	2	3	1, 2	1, 3	2, 3
gvl sv/ci ey	*ci ey*	0	2	0	0	0	0
gvl sv/ci ey	*gvl ey*	0	6	0	0	0	1
gvl sv/ci ey	*gvl sv*	0	9	1	0	0	0
gvl ey/ci sv	*ci sv*	0	2	1	0	0	0
gvl ey/ci sv	*gvl ey*	0	6	0	1	0	2
Total		0	25	2	1	0	3
Table 1		2	26	14	0	0	0
Grand total		2	51	16	1	0	3

Table 2 shows one 1.2 double, which suggests that the true sequence is *gvl ci ey sv*. It is to be noted, however, that none of the experiments of table 2 involved the use of *ci gvl* males for test-crossing, with the result that single crossovers in region 1 (if the sequence is *ci gvl ey sv*) could have been detected only in the tests to *ci* males. These included only five of the 31 tested crossovers.

In some of the cultures included in the first row of table 1, the "+" chromosome actually carried the gene *bt*. In these cases most of the crossovers were tested not only for *ci, gvl, ey* and *sv*, but also for *bt*, with the results shown in table 3.

These tests (which are all included in the lower half of table 1, disregarding *bt*) show that *bt* lies between *gvl* and *ey*, and is probably nearer *ey*. This conclusion is not in agreement with the locus usually given (Dubinin, Sokolov and Tiniakov 1935, Bridges and Brehme 1944). The incorrect locus seems to have arisen from acceptance of the report of Bolen (1931) that the break in translocation white-mottled-5 lies between the loci of *ey* and of *bt*. I have recently reinvestigated this translocation, which involves breaks near the base of IV and the tip of X. There is a viable hyperploid type, which carries a duplication for the tip of X and the base of IV. Bolen reported that this duplication did not suppress the eyeless phenotype when added to flies of the composition *ey/ey*, but did suppress bent in flies of the composition *bt/bt*. I can confirm the result for *ey*, and can extend it to *gvl, sv*—and *bt*. The duplication males have spread wings, which makes classification for *bt* rather uncertain in them; but I have seen a duplication-carrying female with all the various aspects of the bent phenotype fully expressed.

Dubinin, Sokolov and Tiniakov (1935) concluded that the "heterochromatic effect" of a translocation could be spread through the *bt* locus without affecting the bent phenotype; the present results, showing that *bt* was wrongly

Table 3

FROM $\frac{bt}{ci\ gvl\ ey\ sv}$ ♀ × *ci gvl ey sv* ♂. CROSSOVERS TESTED FOR ALL FIVE LOCI

REGIONS	CROSSOVERS	
1	1 *gvl ey sv*	0 *ci bt*
2	5 *ey sv*	6 *ci gvl bt*
3	3 *bt ey sv*	2 *ci gvl*
4	9 *bt sv*	0 *ci gvl ey*

located, indicate that their data have no bearing on this question. It should be added that the statement by Bridges and Brehme (1944) that white-mottled-5 "when tested over *ci*, shows no position effect on dominance of ci^{+}" is incorrect, as had already been reported by Dubinin, Sokolov and Tiniakov (1935)—a result confirmed in the present work. The position effect is present in rather extreme form.

Summary. A map of the fourth chromosome of *Drosophila melanogaster*, based on crossing over in diplo-IV triploid females, shows the following relations (calculated from the upper half of table 1, with *bt* inserted on the basis of the data of table 3): *ci* (0); *gvl* (0.2); *bt* (1.4); *ey* (2.0); *sv* (3.0).

The sequence shown is definitely established except that it is still possible (though unlikely) that *ci* and *gvl* should be reversed. The uncertainty arises from the occurrence of unexpected double crossover classes.

The sequence given is in agreement with that reached by Fung and Stern in the accompanying paper.

LITERATURE CITED

BOLEN, H. R., *Am. Nat.,* **65:** 417–422 (1931).

BRIDGES, C. B., and BREHME, K. S., *Carnegie Inst. Wash. Pub.* **552:** 1944, 253 pp.

DUBININ, N. P., SOKOLOV, N. N., and TINIAKOV, G. G., *Biol. Zhurn.,* **4:** 716–720 (1935).

MORGAN, T. H., STURTEVANT, A. H., and MORGAN, L. V., *Carnegie Inst. Wash. Yearbook,* **44:** 157–160 (1945).

33

A HIGHLY SPECIFIC COMPLEMENTARY LETHAL SYSTEM IN *DROSOPHILA MELANOGASTER* †

In January, 1954, two prune-3[1] females (*Drosophila melanogaster*) were mated to three males from the *S/E-S* stock. There resulted 339 females (all, as expected, pn^+), and one bobbed male that was also pn^+ (it was later found that the *S/E-S* stock carried *bb,* so this male was clearly a nondisjunctional exception). The absence of the class of pn^3 males was wholly unexpected, since the usual result would be that such males would be as numerous as their pn^+ sisters. Several possible interpretations were tested, but the only one that turned out to be adequate was that there is a dominant autosomal gene (now called "Prune-killer," symbol *K-pn*) in the *S/E-S* stock, that kills all prune flies but is without effect on pn^+. Later experiments have shown that the *S/E-S* stock is homozygous for *K-pn*; rather extensive tests have failed to show its presence in any other stock—both mutant stocks and wild strains from widely scattered localities having been tested. It has also been found that *K-pn* is equally effective in killing pn^1, pn^2, or pn^3—the only independently arisen alleles at present available for study—and that *pn* females are killed as effectively as are *pn* males. No other effects of *K-pn* have been found. It has no detected effect on the phenotype in any combination tried; in particular, it has no modifying effect on the color of the eyes in any of the numerous mutant eye-colors with which it has been tried.

The experiments leading to these and other conclusions may now be outlined.

Genetic Tests

Seventeen females from the original culture were tested. One of them was shown, by further testing, to carry an ordinary sex-linked lethal. Of the others, 8 were mated to *pn* males: they produced 745 + ♀♀, 393 *pn* ♀♀, 723 + ♂♂, 351 *pn* ♂♂—i.e., about 2 + :1 *pn* among both sons and daughters. Seven females were mated to males from the *S/E-S* stock, and

† Reprinted from Genetics, Vol. 41, 1956, pp. 118–123.

[1] The names of mutants and the symbols for the genes concerned are the standard ones used in Drosophila. For descriptions, loci, etc. see BRIDGES-BREHME 1944.

produced (disregarding *S*): 756 + ♀♀, 414 + ♂♂—i.e., the expected *pn* sons were absent. The remaining female was mated to an unrelated + male, and produced: 125 + ♀♀, 60 + ♂♂, 34 *pn* ♂♂—i.e., half her *pn* sons were missing.

These results were sufficient to indicate that a dominant autosomal gene was involved; it was also apparent that this gene could not be closely linked to *S*, since in the above tests some of the females used were *S*/+, and there was no obvious deviation from the expected proportions for *S* and S^+ among their offspring. However, *S* is near one end of the second chromosome, and is therefore not an efficient marker for that chromosome in heterozygous females. Accordingly a test was made that depended on the absence of crossing over in male Drosophila.

The uniformity of the *S/E-S* stock was first tested. Twelve matings (including the original one that led to the present study) were made of the type pn^3 ♀ × *S/E-S* ♂. There resulted 1629 + ♀♀, 0 *pn* ♂♂, 4 *bb* ♂♂ (nondisjunctional and XO). Two matings of pn^1 ♀ × *S/E-S* ♂ gave 198 + ♀♀, 0 ♂♂; one pn^2 ♀ × *S/E-S* ♂ gave 121 + ♀♀, 0 ♂♂; one pn^1/pn^3 ♀ × *S/E-S* ♂ gave 198 + ♀♀, 0 ♂♂. These tests, plus those listed above from matings to F_1 females, show that at least 23 males (probably more, since more than one male was used in most cultures) from the *S/E-S* stock were unable to produce *pn* offspring. It has therefore been assumed that this stock was homozygous for the gene concerned—and all experiments designed on this assumption have given results consistent with it.

The *S/E-S* stock was mated to males carrying the dominants *Pin* (chromosome II) and *Ubx* (chromosome III—formerly called bx^D), and F_1 *Pin*; *Ubx* males were mated to pn^3 females. There resulted 227 pn^+ ♀♀, 92 *pn* ♂♂—i.e., half of the *pn* offspring were absent. Further, the 92 *pn* males were classified for the dominant markers: there were 50 *Pin Ubx*, 42 *Ubx*. *K-pn* showed no recombination with *Ubx*, and is therefore in the third chromosome.

The determination of the locus of *K-pn* in the third chromosome has been somewhat more troublesome. The first test was with the two dominants Lyra (*Ly*, locus 40.5) and Stubble (*Sb*, locus 58.2). Males pn^3; *Ly Sb*/+ were mated to females from the *S/E-S* stock. The resulting females of the constitution $+/pn^3$; *S*/+; *Ly Sb/K-pn* were mated to pn^3 males. There resulted (disregarding *S*) 525 pn^+, and the following 252 *pn*: *Ly Sb* 117, *Ly* 18, *Sb* 24, +93. These data show only that *K-pn* is not between *Ly* and *Sb*, nor is it near either one—i.e., it is near one end or the other of the chromosome.

This result showed that it would be necessary to use multiple recessive stocks to locate *K-pn*, and for this reason it was desirable to establish a strain of it free of *S* (which interferes with the classification of some third chromosome markers) and of *E-S* (which carries an inversion in II that might affect crossing over in III). Accordingly, from the test just described, pn^+; *Sb* males were selected and tested individually by pn^+/pn females. One such male gave no *pn*; *Sb* sons. His pn^+; *Sb* sons were mated to females from the *S/E-S* stock, and the *S*; *Sb* offspring were mated together. These flies were of the composition *S*/+; *Sb K-pn*/*K-pn*. From their offspring S^+ were selected, and a stock maintained that has in fact been shown to be homozygous for *K-pn*. *Sb* has been maintained in it by selection, since for some experiments it is useful to have the *K-pn* chromosome marked by a dominant.

Two methods have been used to locate *K-pn*. One is to mate *Sb K-pn*/*K-pn* to a multiple recessive, mate the F_1 *Sb* females to the same multiple recessive, and test various types of *Sb*-carrying crossover sons individually by *pn* females. In case the tested crossover carries *K-pn*, no *pn*; *Sb* sons are produced, whereas if the crossover does not carry *K-pn* the *pn* sons are *Sb* and Sb^+ in equal numbers. Tests of this sort soon showed that *K-pn* lies near the right end of the chromosome, and accordingly the second method was used to get a more precise location.

This method is like that used in the *Ly Sb* test described above, and gives a direct testcross for *K-pn*; but it requires the production of somewhat more complex stocks. Several such tests were made; two of them may be recorded here, since the others gave no additional essential information.

Females of the composition $\frac{+}{pn}$; $\frac{Sb\ \ ca\ \ K\text{-}pn}{e^s}$ were mated to e^s *ca* males. The *pn* offspring were of the following classes (numbers in parentheses indicate the intervals in which crossing over occurred): (0) e^s, 88; (1) *Sb* e^s, 14: (2) *Sb*, 38; (3) *Sb ca*, 4; (1, 2) +, 2; total, 146. It follows that *K-pn* lies to the right of *ca*. There is only one marker gene now available that lies further to the right than *ca*—namely, brevis (*bv*). Accordingly, two tests were carried out: (A) $\frac{+}{pn}$; $\frac{ca\ \ K\text{-}pn}{bv}$ and (B) $\frac{+}{pn}$; $\frac{K\text{-}pn}{ca\ \ bv}$, both kinds of females being mated to *ca bv* males—which in most cases were also *pn*. The following results were obtained:

	pn^+ offspring		*pn* offspring			
	0	1	0	1	2	1, 2
A	971	25	204 (*bv*)	3 (*ca bv*)	0 (*ca*)	0 (+)
B	620	8	443 (*ca bv*)	9 (*bv*)	1 (+)	0 (*ca*)

(The relative numbers of pn^+ and *pn* here are not significant, since in some cultures the fathers were pn^+ and in others they were *pn*—and also the pn^+ were not classified for *ca* and *bv* in all cultures.)

These data give the locus of *K-pn* as 0.2 units to the right of *bv*. The latter is listed by BRIDGES-BREHME (1944) at 104.3, with only one locus—*M*(3)*g*—to the right of it, at 106.2. This latter mutant has been lost, and in view of its slight phenotypic effect, dependence on a linked enhancer, and phenotypic resemblance to brevis, one may question the accuracy of the determination of its locus. Brevis is listed as 3.6 units from *ca*; the present data give $100 \times 45/2284 = 2.0$ for the recombination percentage. In view of BRIDGES' custom of using maximum observed recombination values for map distances, this is a reasonable agreement with expectation, and the locus of *K-pn* may be put at 104.5, making it the rightmost locus now available for study in the third chromosome.

Dr. E. B. Lewis has kindly examined the salivary gland chromosomes of *K-pn* larvae; he finds the right end of chromosome III normal in appearance.

Various kinds of tests other than those already described have been carried out, and have confirmed the conclusions there drawn. Attached-X *pn* females × *K-pn/K-pn* males gave pn^+ sons and a few pn^+ superfemales; attached-X pn^+/pn^+; *K-pn/K-pn* females × *pn* males gave pn^+ daughters and no sons. The first of these two results shows that $pn/pn/pn^+$; *K-pn/+* survives—a conclusion confirmed by the use of two different short duplicating fragments containing pn^+. In this way it is possible to obtain pn/pn^+; *K-pn/+* males, and $pn/pn/pn^+$; *K-pn/+* females, that are viable and fertile, since the pn^+ is contained in a small duplicating fragment that does not seriously affect viability or fertility. (The duplications used were *Dp (1;f)* X^{c2} and *Dp (1;4)*^{m5} —see Bridges-Brehme 1944).

Properties of K-pn. From the mating of *pn/pn* ♀♀ × *K-pn/K-pn* ♂♂, all the regular daughters are viable and all the regular sons die. Examination of such cultures, when the parents are transferred daily, shows that the eggs hatch and the larvae appear normal. However, the male larvae (*pn*; *K-pn*) do not develop beyond the second instar. They appear superficially normal until the time when their sisters molt at the end of the second instar; at this time they usually crawl out of the food, and all of them die without undergoing the molt. No detailed study of them has been made, but no structural abnormality has been observed.

It is clear from some of the tests described above that *pn* females carrying *K-pn*, and *pn* males that are homozygous for *K-pn*, also die; no attempt has been made to determine the time of death for these classes.

It was hoped that mosaics might be obtained that would contain parts that were *pn*; *K-pn* in composition, in order to determine the phenotype of specific regions of that composition. Two different tests were carried out, using *y pn* so that yellow bristles and hairs would serve as an index of the presence of *pn*, and using ca^{nd} (Lewis and Gencarella 1952) as a source of mosaicism. It does not seem necessary to present the details of these experiments, since they were not wholly conclusive. There was a deficiency of mosaics carrying *K-pn* and with male parts *pn* in composition—but it is not certain that this deficiency was statistically significant. All such mosaics obtained had the yellow (i.e., male and genetically *pn*) parts confined to the thorax and abdomen—but again it is not certain (though it is probable) that the absence of *pn* eye-color was significant. In any case, the experiments did not answer the question of the possible effect of *K-pn* on eye color.

There is one reason for suspecting that *K-pn* may be related to pigment formation: the *pn*; *K-pn* larvae, that die at the end of the second instar, have been observed to have slightly paler Malpighian tubes than control *pn* larvae at the same age. However, it may well be that this is a symptom of approaching death, rather than a specific effect of *K-pn* on pigment synthesis.

K-pn/+ has been combined with a wide variety of eye-colors other than *pn,* and in no case has it been found to have any modifying effect on them. The list tried includes, among others: *v*, *cn*, *st* (which remove the brown pigment); *bw* (which removes the red pigment); w^a, w^e, zeste (which are known to be very susceptible to modifiers); *ca* (which lies near *K-pn*); and chocolate, *se*, and *pd* (which somewhat resemble *pn*). In the latter case, BRIDGES-BREHME (1944) described a special modifier of *pd* (purpleoid), called Purpleoider (*Pdr*) which has effects on *pd* somewhat similar to the effects of *K-pn* on *pn*. Tests have shown that *K-pn* does not affect *pd,* and that *Pdr* does not affect *pn*.

If *K-pn* kills *pn* through an action on some substance present in *pn* but not in other flies, it seemed possible that this hypothetical substance might be involved in the production of eye-pigment, and that it might be absent in white-eyed flies. Accordingly *pn w,* and *pn; cn bw* (both with eyes without pigment) were tried; *K-pn* was fully effective in killing both types. Such a result was perhaps to have been expected, since the *pn*; *K-pn* larvae die at a stage where the synthesis of eye-pigment has presumably not yet begun—although the pigment of the Malpighian tubes (also absent in *w* and in *cn bw*) is already present in *pn* larvae at that stage.

Discussion

The case of *pd* and *Pdr*, briefly referred to above, is somewhat like the present one. It differs in that each mutant gene has an effect on the color of the eyes in the absence of the other one, and in that the lethal effect is found only when both genes are homozygous (BRIDGES-BREHME 1944). No detailed account has ever been published.

Another analogous situation is that found with crossveinless (*cv*) and fused (*fu*) in *Drosophila simulans* (STURTEVANT 1929). Fused has reduced viability, and this is greatly exaggerated in *cv fu*, which has a phenotype that appears to be the simple sum of *cv* and *fu*, but has such a low viability that it rarely survives to the adult condition.

Perhaps a closer analogy is provided by the "synthetic lethals" described for *D. pseudoobscura* by DOBZHANSKY (1946). In this case it was found that two second chromosomes, neither of which had a lethal effect, might produce a lethal chromosome by crossing over—presumably through the bringing together of complementary recessive genes.

The situation is also similar to the well-known examples of genes that are lethal or cause tumors in species hybrids but not in the species in which they are normally found. (See, for example, HOLLINGSHEAD 1930 and GORDON 1931.)

There are other more or less similar examples scattered through the literature—cases where genes, that themselves may or may not have observed phenotypic effects, produce unexpectedly great effects on phenotypes conditioned by other genes. One has only to refer to the extensive literature on "enhancers" or "specific intensifiers" on the one hand, or "suppressors" on the other hand, to locate numerous examples.

K-pn may be thought of as an extreme "enhancer"—though it is not known whether it has any effect on the most obvious phenotypic effect of *pn* (the eye color); and if *pn* has any effect on viability (as it presumably does), it is not one that is easily detected in ordinary experiments.

It may be that specific "killers" are more frequent than would be inferred from the previous failure to recover any as effective and as specific as *K-pn*. This will become evident if it is recognized that the present case owes its discovery to the facts that *pn* is a sex-linked recessive and that *K-pn* is an autosomal dominant. The analysis would probably have failed if it had not happened that the *S/E-S* stock was homozygous for *K-pn*. It seems probable that these rather special requirements for discovery and analysis will not be met with in many cases—i.e., it is probable that recessive "killers" exist, that autosomal mutants are subject to such killing, and that most "killers" exist in heterozygous form in the stocks where they occur.

There is one use to which the present system may be put, in the design of experiments. In the two matings

pn ♀ × *K-pn*/*K-pn* ♂
attached-X *K-pn*/*K-pn* ♀ × *pn* ♂

no males are produced except by nondisjunction (in the first case) or detachment of attached-X's (in the second case). These two matings therefore give convenient and efficient methods of producing large numbers of all-female cultures—and therefore of necessarily virgin females. For some types of experiments the procedure may be found useful.

One possible hypothesis is that *pn* leads to the production (or accumulation) of some substance that is absent, or present only in small amounts, in pn^+ flies, and that *K-pn* converts this substance to a different and toxic substance. Such a hypothesis at least suggests experiments to be carried out, and offers some hope of relating the phenomena to the reactions concerned in the synthesis of pigments.

Summary

1. The *S/E-S* stock was found to be homozygous for a gene called "Prune-killer" (symbol, *K-pn*), that is located at 104.5 in chromosome III.

2. *K-pn* has only one known effect: it causes the death of all prune flies that carry it, this effect being dominant.

3. This lethal effect of *pn*; *K-pn* occurs with pn^1, pn^2, or pn^3—the only separately arisen *pn* mutants available.

4. In the only case studied—males that are *pn*; *K-pn*—death occurs at the end of the second larval instar, and no gross abnormality was detected in the dying larvae.

LITERATURE CITED

Bridges, C. B. (edited by K. S. Brehme), 1944: The mutants of *Drosophila melanogaster*. Carnegie Inst. Wash. Publ. **552,** 257 pp.

Dobzhansky, T., 1946: Genetics of natural populations. XIII. Recombination and variability in populations of *Drosophila pseudoobscura*. Genetics **31:** 269–290.

Gordon, M., 1931: The hereditary basis for melanosis in hybrids of Mexican killifishes. Proc. Nat. Acad. Sci. U. S. **17:** 276–280.

Hollingshead, L., 1930: A lethal factor in Crepis effective only in an interspecific hybrid. Genetics **15:** 114–140.

Lewis, E. B., and W. Gencarella, 1952: Claret and nondisjunction in *Drosophila melanogaster*. Genetics **37:** 600–601.

Sturtevant, A. H., 1929: The genetics of *Drosophila simulans*. Carnegie Inst. Wash. Publ. **399:** 1–62.

Publications of Alfred H. Sturtevant

1910 On the inheritance of color in the American harness horse. Biol. Bull. **19:** 204–216.

1911 Another sex-limited character in fowls. Sci. **33:** 337–338.

1912 A critical examination of recent studies on colour inheritance in horses. Jour. Gen. **2:** 41–51.

Is there association between the yellow and agouti factors in mice? Amer. Nat. **46:** 368–371.

Federley's breeding experiments with the moth Pygaera. Amer. Nat. **46:** 565–568.

An experiment dealing with sex-linkage in fowls. Jour. Exp. Zool. **12:** 499–518.

1913 The linear arrangement of six sex-linked factors in Drosophila, as shown by their mode of association. Jour. Exp. Zool. **14:** 43–59.

The Himalayan rabbit case, with some considerations on multiple allelomorphs. Amer. Nat. **47:** 234–238.

A third group of linked genes in Drosophila ampelophila. Sci. **37:** 990–992.

1914 The reduplication hypothesis as applied to Drosophila. Amer. Nat. **48:** 535–549.

Linkage in the silkworm moth. Amer. Nat. **48:** 315–317.

A new gene in the second chromosome of Drosophila and some considerations on differential viability. (with C. B. Bridges). Biol. Bull. **26:** 205–212.

1915 The behavior of the chromosomes as studied through linkage. Zeits. ind. Abst.-Vererb. **13:** 234–287.

The mechanism of Mendelian heredity. (with Morgan, T. H.; H. J. Muller and C. B. Bridges) H. Holt Co. 258 p.

No crossing over in the female of the silkworm moth. Amer. Nat. **49:** 42–44.

A sex-linked character in Drosophila repleta. Amer. Nat. **49:** 189–192.

Experiments on sex recognition and the problem of sexual selection in Drosophila. Jour. Animal Behav. **5:** 351–366.

Castle and Wright on crossing over in rats. Sci. **42:** 342.

1916 Notes on North American Drosophilidae with descriptions of twenty-three new species. Ann. Ent. Soc. Amer. **9:** 323–343.

1917 Genetic factors affecting the strength of linkage in Drosophila. Proc. Nat. Acad. Sci. **3:** 555–558.

Crossing over without chiasmatype? Genetics **2:** 301–304.

1918 Acalypterae (Diptera) collected in Mobile County, Alabama. Jour. N.Y. Ent. Soc. **26:** 34–40.

A synopsis of the Nearctic species of the genus Drosophila (Sensu Lato). Bull. Amer. Mus. Nat. Hist. **38:** 441–446.

Flies of the genus Drosophila as possible disease carriers. Jour. Parasit. **5:** 84–85.

An analysis of the effects of selection. Carn. Inst. Wash. Publ. No. 264, 68 p.

A parallel mutation in Drosophila funebris. Sci. **48:** 72–73.

1919 A new species closely resembling Drosophila melanogaster. Psyche **26:** 153–155.

Inherited linkage variations in the second chromosome. Carn. Inst. Wash. Publ. No. 278, pp. 307–341.

The spatial relations of genes. (with Bridges, C. B. and T. H. Morgan). Proc. Nat. Acad. Sci. **5:** 168–173.

A semi-lethal in Drosophila funebris that causes an excess of males. (with O. L. Mohr). Proc. Soc. Exper. Biol. Med. **16:** 95–96.

1920 Intersexes in Drosophila simulans. Sci. **51:** 325–327.

The vermilion gene and gynandromorphism. Proc. Soc. Exp. Biol. & Med. **17:** 70–71.

The dipterous genus Zygothrica of Wiedemann. Proc. U. S. Natl. Mus. **58:** 155–158.

Genetic studies on Drosophila simulans. I. Introduction. Hybrids with Drosophila melanogaster. Genetics **5:** 488–500.

1921 Genetic studies on Drosophila simulans. II. Sex-linked group of genes. Genetics **6:** 43–64.

Genetic studies on Drosophila simulans. III. Autosomal genes. General discussion. Genetics **6:** 179–207.

The North American species of Drosophila. Carn. Inst. Wash. Publ. No. 301, 148 p.

A case of rearrangement of genes in Drosophila. Proc. Nat. Acad. Sci. **7:** 235–237.

Linkage variation and chromosome maps. Proc. Nat. Acad. Sci. **7:** 181–183.

Study of the constitution of the germ-plasm in relation to heredity. (with Morgan, T. H. and C. B. Bridges). Carn. Inst. Year Book No. 20: 375.

1922 Study of the constitution of the germ-plasm in relation to heredity. (with Morgan, T. H. and C. B. Bridges). Carn. Inst. Year Book No. 21 pp. 325–328.

1923 New species and notes on synonymy and distribution of Muscidae acalypteratae (Diptera). Amer. Mus. Nat. Hist. Novitates No. 76 12 p.

Reverse mutation of the bar gene correlated with crossing over. (with Morgan, T. H.) Sci. **57:** 746–747.

The probable occurrence of parthenogenesis in Ochthiphila polystigma (Diptera). Psyche **30:** 22–27.

Inheritance of direction of coiling in Limnaea. Sci. **58:** 269–270.

The mechanism of heredity. (with Morgan, T. H., H. J. Muller and C. B. Bridges). Revised ed., H. Holt Co. 357 p.

Laboratory directions for an elementary course in genetics. (with Morgan, T. H., H. J. Muller and C. B. Bridges) H. Holt Co. 16 p.

The constitution of the germ-material in relation to heredity. (with Morgan, T. H. and C. B. Bridges). Carn. Inst. Year Book No. 22 pp. 283–287.

1924 An interpretation of orthogenesis. Sci. **59:** 579–580.

1925 Notes on the ant fauna of oak galls in the Woods Hole region. Psyche **32:** 313–314.

The effects of unequal crossing over at the bar locus in Drosophila. Genetics **10:** 117–147.

The genetics of Drosophila. (with Morgan, T. H. and C. B. Bridges). Martinus Nijhoff 262 p.

The constitution of the germ material in relation to heredity. (with Morgan, T. H. and C. B. Bridges). Carn. Inst. Year Book No. 24 pp. 286–288.

1925–1926 The seminal receptacles and accessory glands of the Diptera, with special reference to the Acalypterae. Jour. N.Y. Ent. Soc. **33:** 195–215 and **34:** 1–21

1926 The constitution of the germ material in relation to heredity. (with Morgan, T. H. and C. B. Bridges). Carn. Inst. Year Book No. 25 pp. 308–312.

Fauna of Penikese Island, 1923. Ed. by Coker, R. E. Biol. Bull. **50:** 17–37.

Renner's studies on the genetics of Oenothera. Quart. Rev. Biol. **1:** 283–288.

Sequence of corresponding third-chromosome genes in Drosophila melanogaster and D. simulans. (with Plunkett, C. R.). Biol. Bull **50:** 56–60.

A crossover reducer in Drosophila melanogaster due to inversion of a section of the third chromosome. Biol. Zentralbl. **46:** 697–702.

1927 Philippine and other Oriental Drosophilidae. Philip. Jour. Sci. **32:** 361–374.

The effects of the bar gene of Drosophila in mosaic eyes. Jour. Exp. Zool. **46:** 493–498.

The social parasitism of the ant Harpagoxenus americanus. Psyche **34:** 1–9.

The constitution of the germ material in relation to heredity. (with

Morgan, T. H. and C. B. Bridges). Carn. Inst. Year Book No. 26 pp. 284–288.

1928 A further study of the so-called mutation at the bar locus of Drosophila. Genetics **13:** 401–409.

The constitution of the germinal material in relation to heredity. (with Morgan, T. H. and C. B. Bridges). Carn. Inst. Year Book. No. 27 pp. 330–335.

1929 The genetics of Drosophila simulans. Carn. Inst. Wash. Publ. No. 339 pp. 1–62.

The claret mutant type of Drosophila simulans: a study of chromosome elimination and cell-lineage. Zeits. wissensch. Zool. **135:** 323–356.

1930 A peculiar sex-ratio in Drosophila obscura. The Collecting Net. (Woods Hole) **5:** 275.

Reciprocal translocations in Drosophila and their bearing on Oenothera cytology and genetics. (with Dobzhansky, T.) Proc. Nat. Acad. Sci. **16:** 533-536.

1931 Genetic and cytological studies on Oenothera. I. Nobska, Oakesiana, Ostreae, Shulliana, and the inheritance of old-gold flower-color. Zeits. ind. Abst.-Vererb. **59:** 365–380.

Genetic and cytological studies on Oenothera. III. The translocation interpretation. (with Emerson, S.) Zeits. ind. Abst.-Vererb. **59:** 395–419.

Ants collected on Cape Cod, Massachusetts. Psyche **38:** 73–79.

Contributions to the genetics of certain chromosome anomalies in Drosophila melanogaster. (with Dobzhansky, T.) Carn. Inst. Wash. Publ. No. 421, 81 p.

The inadequacy of the sub-gene hypothesis of the nature of the scute allelomorphs of Drosophila. (with Schultz, J.) Proc. Nat. Acad. Sci. **17:** 265–270.

1932 The use of mosaics in the study of the developmental effect of genes. Proc. Sixth Internat. Cong. Genetics **1:** 304–307.

The linkage relations of certain genes in Oenothera. (with Emerson, S.). Genetics **17:** 393–412.

1933 Chromosome mechanics. Nature Jan. 7, pp. 5–6 (a review).

1934 Preferential segregation of the fourth chromosomes in Drosophila melanogaster. Proc. Nat. Acad. Sci. **20:** 515–518.

1935 X-chromosome inversions and meiosis in Drosophila melanogaster. (with Beadle, G. W.). Proc. Nat. Acad. Sci. **21:** 384–390.

Further data on maternal effects in Drosophila pseudoobscura hybrids. (with Dobzhansky, T.). Proc. Nat. Acad. Sci. **21:** 566–570.

1936 Inversions in the third chromosome of wild races of Drosophila pseudoobscura, and their use in the study of the history of the species. (with Dobzhansky, T.). Proc. Nat. Acad. Sci. **22:** 448–450.

Preferential segregation in triplo-IV females of Drosophila melanogaster. Genetics **21:** 444–466.

Geographical distribution and cytology of "sex-ratio" in Drosophila pseudoobscura and related species. (with Dobzhansky, T.). Genetics **21:** 473–490.

The relations of inversions in the X chromosome of Drosophila melanogaster to crossing over and disjunction. (with Beadle, G. W.). Genetics **21:** 554–604.

Problèmes Génétiques. (trans. by J. Monod). Exposés de Génétique No. 408 20 p.

Observations on the species related to Drosophila affinis, with descriptions of seven new forms. (with Dobzhansky, T.). Amer. Nat. **70:** 574–584.

1937 Culture methods for Drosophila. In Galtsoff et al., Culture Methods for Invertebrate Animals, pp. 437–446.

Autosomal lethals in wild populations of Drosophila pseudoobscura. Biol. Bull. **73:** 542–551.

The comparative genetics of Drosophila pseudoobscura and D. melanogaster. (with Tan, C. C.) Jour. Gen. **34:** 415–432.

An effect of the Y-chromosome on the sex-ratio of interracial hybrids of Drosophila pseudoobscura. Proc. Nat. Acad. Sci. **23:** 360–362.

Essays on evolution I. On the effects of selection on mutation rate. Quart. Rev. Biol. **12:** 464–467.

1938 Essays on evolution II. On the effects of selection on social insects. Quart. Rev. Biol. **13:** 74–76.

Essays on evolution III. On the origin of interspecific sterility. Quart. Rev. Biol. **13:** 333–335.

Inversions in the chromosomes of Drosophila pseudoobscura. (with Dobzhansky, T.). Genetics **23:** 28–64.

The interrelations of inversions, heterosis and recombination. (with Mather, K.). Ameri. Nat. **72:** 447–452.

1939 An introduction to genetics. (with Beadle, G. W.), W. B. Saunders Co. 391 p.

On the subdivision of the genus Drosophila. Proc. Nat. Acad. Sci. **25:** 137–141.

High mutation frequency induced by hybridization. Proc. Nat. Acad. Sci. **25:** 308–310.

1940 Genetic data on Drosophila affinis, with a discussion of the relationships in the subgenus Sophophora. Genetics **25:** 337–353.

A new inherited character in man. Proc. Nat. Acad. Sci. **26:** 100–102.

1941 Physiological aspects of genetics. Ann. Rev. Physiol. **3:** 41–56.

Sterility in crosses of geographical races of Drosophila micromelanica. (with Novitski, E.). Proc. Nat. Acad. Sci. **27:** 392–394.

The homologies of the chromosome elements in the genus Drosophila.

(with Novitski, E.). Genetics **26:** 517–541.

1942 The classification of the genus Drosophila, with descriptions of nine new species. Univ. Texas. Pub. No. 4213 5–51.

1943 Genetics and the systematist. (A review.) Journ. Hered. **34:** 307–308.

1944 Drosophila pseudoobscura. (A review). Ecol. **25:** 476–477.

Can specific mutations be induced by serological methods? Proc. Nat. Acad. Sci. **30:** 176–178.

Maintenance of a Drosophila stock center in connection with investigations on the constitution of the germinal material in relation to heredity. (with Morgan, T. H.). Carn. Inst. Wash. Year Book No. 43 pp. 164–165.

1945 A gene in Drosophila melanogaster that transforms females into males. Genetics **30:** 297–299.

1946 Thomas Hunt Morgan. Amer. Nat. **80:** 22–23.

On the dot-chromosomes of Drosophila repleta and D. hydei. Genetics **31:** 259–268.

Thomas Hunt Morgan (1866–1945). Year Book Amer. Philosoph. Soc. **1945:** 387–389.

Iris genetics. (with Randolph, L. F.). Bull. Amer. Iris Soc. **99:** 52–66.

Intersexes dependent on a maternal effect in hybrids between Drosophila repleta and D. neorepleta. Proc. Nat. Acad. Sci. **32:** 84–87.

Maintenance of a Drosophila stock center, in connection with investigations on the constitution of the germinal material in relation to heredity. (with T. H. Morgan and L. V. Morgan). Carn. Inst. Wash. Year Book **44:** 157–160.

T. H. Morgan. Records Genetics Soc. Amer. **14:** 14–17.

Thomas Hunt Morgan. Anat. Record **95:** 480–481.

Thomas Hunt Morgan. Genetics **32:** 1–2.

White-eyed mutants of Diptera. Nature **160:** 754–755.

An outline of the genetics of tall bearded Iris. In "The Iris, an Ideal Hardy Perennial," G. Douglas, editor. Pp. 96–105.

Caltech's Division of Biology is nineteen years old. Eng. & Sci. Monthly **10:** 6.

Drosophila. Eng & Sci. Monthly **10:** 7, 26.

1948 The evolution and function of genes. Amer. Scientist, Spring Issue, Vol. **36:** 225–236. (Based upon Sigma Xi National Lecture, 1947.) Reprinted in "Science in Progress," vol. 6, 250–265 (1949) and in Smithsonian Report for 1948. Pp. 293–304.

1949 The Beaded Minute combination and sex-determination in Drosophila. Proc. Nat. Acad. Sci. **35:** 311–313.

1951 The relation of genes and chromosomes. In "Genetics in the twentieth century" (L. C. Dunn, editor, The Macmillan Co., New York). Pp. 101–110.

A map of the fourth chromosome of Drosophila melanogaster, based on crossing over in triploid females. Proc. Nat. Acad. Sci. **37:** 405–407.

Three kinds of white bearded Iris. Bull. Amer. Iris Soc. **123:** 99–100.

Notes on the tangerine beard. Bull. Amer. Iris Soc. **123:** 101–102.

1953 Lilian V. Morgan. Records Genetics Soc. Amer. **22:** 16.

1954 Synopses of Nearctic Ephydridae (Diptera). Trans. Amer. Ent. Soc. **79:** 151–257. (with M. R. Wheeler).

Nearctic flies of the family Periscelidae (Diptera) and certain Anthomyzidae referred to the family. Proc. U.S. Nat. Mus. **103:** 551–561.

Social implications of the genetics of man. Science **120:** 405–407.

1955 The genetic effects of high energy irradiation of human populations. Engineering and Science **18:** 9–12.

Evaluation of recombination theory. J. Cell. Comp. Physiol. **45:** 237–242.

1956 A highly specific complementary lethal system in Drosophila melanogaster. Genetics **41:** 118–123.

1959 Thomas Hunt Morgan, 1866–1945. Biographical Memoirs, Nat. Acad. Sci. **33:** 280–325.

1960 Effects of high energy radiation on the human body. In "Hubris, man and education," pp. 65–72. Western Wash. College of Education.

Index